Multilevel Selection

"Multi-Level Selection is the only logically coherent and empirically supported theory that can explain human *ultrasociality*—the capacity of humans to cooperate in huge groups of genetically unrelated individuals. Yet influential critics, including Richard Dawkins and Steven Pinker, continue to reject it. This timely and important book is a welcome entrant to this intense scientific debate. The stakes are high, because understanding how cooperation evolved and can be maintained is key to solving the Tragedy of the Commons problems at both local and global levels."

—Peter Turchin, *author of* Ultrasociety (2015) *and Professor at University of Connecticut, USA*

Steven C. Hertler
Aurelio José Figueredo
Mateo Peñaherrera-Aguirre

Multilevel Selection

Theoretical Foundations, Historical Examples, and Empirical Evidence

palgrave
macmillan

Steven C. Hertler
College of Saint Elizabeth
Morristown, NJ, USA

Aurelio José Figueredo
University of Arizona
Tucson, AZ, USA

Mateo Peñaherrera-Aguirre
University of Arizona
Tucson, AZ, USA

ISBN 978-3-030-49519-0 ISBN 978-3-030-49520-6 (eBook)
https://doi.org/10.1007/978-3-030-49520-6

Cover illustration: Niday Picture Library / Alamy Stock Photo

This Palgrave Macmillan imprint is published by the registered company Springer Nature Switzerland AG.
The registered company address is: Gewerbestrasse 11, 6330 Cham, Switzerland

Preface: A Pan-Selectionist Model of Evolution and Development

Introduction

Open an atlas, turn the page to a map of a city, and find, in some far corner, a smaller map of the nation containing that city. This provides necessary perspective. While viewing a map of London, for instance, the page will be dominated by the Thames meandering north and south as it pushes east; also prominent will be labyrinthine streets, designated municipal buildings, parks, landmarks, regions, and districts. Again, however, somewhere high or low, in this corner or that, there may well be a miniature map of the United Kingdom, on which London appears as a dot in the Southeast of England. This introductory chapter is much like that small map. It provides necessary perspective for the 12 content chapters that follow. While this book is about multilevel selection, this introductory chapter is about evolution more generally. However, multilevel selection progresses from possible to exceedingly plausible when contextualized within the overarching pan-selectionist model of evolution presented in this preface. This pan-selectionist view of evolution permeates boundaries between evolutionary and developmental processes within individuals, while yoking cultural to biological evolutionary change. Evolution is taking place, not only across generations of organisms but within organisms and among groups of organisms, hence the need to preface *selection* with *pan*, which is of course a Greek prefix meaning *all*. By this preface's end, one will see how this pan-selectionist theory of evolution and development is, in effect, an integrative meta-theoretical framework in which groups are classed as one among several hierarchically nested levels of biological organization. For all this, pan-selectionism has precedents in the past, to which we will first turn.

Historical Precedence

Several decades ago, B. F. Skinner and E. O. Wilson noted that the object of study in the fields which they founded, *Radical Behaviorism* and *Sociobiology*, respectively, consists exclusively of organismal characteristics, developed through the lifespan in response to genetic processes governing development coupled with one's experiential history with an environment (Naour, 2009). In other words, the environment acts on the organism through the medium of genes and genetic change. Thus, ultimately, genetic processes influence all characteristics selected through development. Less commonly discussed, but also recognized by Skinner and Wilson, is that phenotypes selected during development can have their genetic regulation modified over evolutionary time, ultimately producing changes in phenotypic plasticity. Thus, processes of *selection by consequences* are pivotal to understanding the evolution and development of behavior. Indeed, the common currency of selection unites an integrated conception of behavioral evolution and development (e.g., Naour, 2009). Based on that conceptualization, the objective of the present prefatory essay is to unify our theories of evolution and development under the overarching framework of a pan-selectionist logic.

It is important here to understand that Darwin (1859) used the word "selection" as an anthropomorphic metaphor personifying "Nature" as an intentional agent. In reality, all that happens is the differential survival and reproduction of alternative variants. Darwin was quite clear that he used this expression as just a figure of speech:

> Every one knows what is meant and is implied by such metaphorical expressions; and they are almost necessary for brevity. So again it is difficult

> to avoid personifying the word Nature; but I mean by Nature, only the aggregate action and product of many natural laws, and by laws the sequence of events as ascertained by us. (p. 81)

In bridging the gap between such ultimate and metaphorical concepts with proximate and mechanistic ones, we therefore need to construct more rigorous definitions of what exactly is being selected and on what basis. When doing so, we note that selective processes have been subdivided various ways according to different organizational schemata, making reducing all the competing categories into one superordinate schema quite challenging.

We begin by adopting Dawkins' (1976) admittedly imperfect but useful distinction between *replicators* and *vehicles*. As discussed in Chap. 1, replicators, such as genes, are ultimately the only things that might truly survive across multiple generations. This realization led to the redefinition of evolution by population biologists in terms of changes in gene frequencies over time within a given population of living things. Subsequently, candidates for replicators other than genes have been proposed, including by Dawkins himself, but we start with the case of the gene as the replicator *par excellence*.

Genes, however, do not exist in a material vacuum, naked and exposed to the environment. They are instead ensconced within vehicles of varying complexity, from cells to tissues, to organs, to organ systems, to organisms, and beyond, all the way to complex ecosystems and biospheres. These vehicles are analogized to organic machines or instruments constructed by the replicators for their survival and reproduction. It is safe to say that genes are selected based on the *direct* effects that they exert upon the structure and function of their vehicles, as well as the *indirect* effects that they exert upon the environment as mediated by the former. Thus, as genes are *selected*, the vehicles *evolve*.

Selection by *Sources* of Consequences Versus Selection by *Vehicles* of Consequences

The most venerable scheme for categorizing evolutionary forces has been to organize them by the *sources* of the selective pressures that are operative. The oldest such taxonomy is thus: (1) *Natural Selection*, (2) *Social Selection*, and (3) *Sexual Selection*. *Natural Selection* names the outcome of a population of causal processes that produce changes in gene frequencies occasioned by the average differential survival and reproduction of variant individuals, attributable to what Darwin (1859) called their differential abilities to combat the "hostile forces of nature." To survive and reproduce, one typically must produce behavior that overcomes threats of morbidity and mortality from the physical elements, such as cold or drought, as well as from other species, such as predators, parasites, and prey. In contrast, *Social Selection* names the outcome of a population of causal processes that produce changes in gene frequencies occasioned by the differential survival and reproduction of variant individuals that are attributable to their differential abilities to produce behavior that competes successfully or cooperates socially, as required by the circumstances, with members of one's own species, regardless of their sex. Although the term was coined much later (Nesse, 2009; West-Eberhard, 1979), Darwin (1871) described *social selection* in the process of articulating his theory of human descent with modification (Nesse, 2009; West-Eberhard, 1979). *Sexual Selection*, on the other hand, names the outcome of a population of causal processes that produce changes in gene frequencies occasioned by the average differential survival and reproduction of variant individuals to what Darwin (1871) androcentrically called their differential abilities to "charm the females" and "to conquer other males in battle." In our more modern sexually neutral terms, the phrase refers to the differential abilities of variant individuals to behave in ways that obtain and retain sexual partners, a prerequisite for propagation in sexually reproducing species. Sexual selection includes *intrasexual* selection, the average differential success produced by the behavior of variant individuals in direct competition with same-sex members of one's own species for access to sexual partners, and *intersexual* selection, the average differential success produced by the behavior of variant individuals in acceptance or rejection as sexual partners by opposite-sex members of one's own species.

We now turn from selection by the *sources* of consequences to selection by the *vehicles* of consequences: It is increasingly evident that selection by consequences also drives development, including development of the brain and behavior, by means of analogous processes. This gives rise to an alternative organizational scheme for evolutionary processes, classifying them by the level of the instrumental vehicles employed to transact with the environment, such as: (1) *Clonal Selection*; (2) *Neuronal Selection*; (3) *Respondent* and *Operant/Instrumental Selection*; (4) *Cultural Selection*; and (5) *Biocultural Group Selection.*

Clonal Selection, which shapes immunological function, consists of changes in cellular variability with differential mortality, retention, and proliferation of specific types of white blood cells (lymphocytes) attributable to their relative encounter rates with various specific antigens for which each type of white blood cell possesses a different specific antibody. This selective process involves vast numbers of *T* and *B* lymphocyte types, each with a unique surface antibody, which serve as templates for all antibodies the immune system can ever make. The organism develops its antibody immunological repertoire prenatally. Upon encountering an antigen to which it can bind, each antigen-specific lymphocyte begins repeated rounds of mitosis (clonal expansion) producing more identically antigen-specific *B* and *T* lymphocytes. Clonal expansion eventually produces a large pool of *effector cells*—plasma cells (antibody-secreting *B* cells) that synthesize and secrete massive amounts of antigen-specific antibody—and *memory cells* (*B* lymphocytes) with the same specific receptors as those on the original (activated) *B* cell. Memory cells serve as a form of long-term immunity to the specific antigen. Put metaphorically, an antigen encounter *selects*, from a large number of cells with unique antigen receptors, a restricted number of specific lymphocyte types that eventually produce a massive amount of antigen-specific antibodies (e.g., Jerne, 1955).

Neuronal Selection, which shapes brain development, consists of changes in cellular and connective variability with differential mortality, retention, and proliferation of neurons and their synapses, attributable to their relative degree of use or disuse during development. This process consists of early rapid proliferation of brain cells, followed by a massive die-off of neurons over developmental time; these two stages recapitulate those of variability and then selection. For example, the majority of neurons present in the brain of a two-year-old human, with their redundant and diffuse synaptic connections, are no longer present by the age of sixteen because of a subsequent die-off of cells and synapses, which apparently follows a "use it or lose it" principle (Changeux, 1985):

> The 10,000 or so synapses per cortical neuron are not established immediately. On the contrary, they proliferate in successive waves from birth to puberty in man… One has the impression that the system becomes more and more ordered as it receives 'instructions' from the environment. If the theory proposed here is correct, spontaneous or evoked activity is effective only if neurons and their connections already exist before interaction with the outside world takes place. Epigenetic selection acts on preformed synaptic substrates. To learn is to stabilize preestablished synaptic combinations, and to eliminate the surplus. (p. 248)

The differential mortality, retention, and proliferation of neurons and their synapses lead to a finely tuned system that makes both perceptions and adaptively complex behaviors possible. For example, it appears that visual system axons *compete* for space in the visual cortex, with the result of the competition determining the amount and type of sensory stimulation carried by the axons.

Respondent and *Instrumental* conditioning can also be understood as forms of selection, in that they create differential retention and elimination of *respondent* or *instrumental* behaviors that produce either adaptive reactions to or adaptive effects upon the individual's environment. A seminal conversation between B. F. Skinner and E. O. Wilson (Naour, 2009) began a productive dialectic between two superficially incompatible fields wherein meaningful correspondences between *nurture* and *nature* were found. For instance, as Wilson and Skinner concluded, it is no coincidence that some reinforcers are considered "primary" in that they became reinforcing through evolutionary time because such reinforcers are wedded to resources necessary for survival and, ultimately, genetic replication.

Reinforcement mechanisms, shaped by the genes that govern epigenetic rules of development, permit new behavioral phenotypes to be generated during ontogeny; thus a variety of behavioral phenotypes can develop through operant conditioning, herein recast as *operant selection*, supplementing the survival value of more biologically canalized behaviors that guide one to, and are directly reinforced by, primary reinforcers. Through a process called *chaining*, these *new* behavioral phenotypes typically consist of various combinations of *old* phenotypes, rather than springing into existence out of nothing. Behavioral phenotypes elected by secondary reinforcers also facilitate access to primary reinforcers. In other words, if behavioral phenotypes selected by secondary reinforcers become a part of the behavioral repertoire of an individual, it is because the new phenotypes increase both the availability and probability of primary reinforcers. Such behavioral modifications imply epigenetic modifications within developmental time and, as Wilson and Skinner point out, may eventually involve changes in gene frequencies over evolutionary time. This is because those genes more likely to be transmitted across generations predispose individuals to acquire behavior reinforced by secondary reinforcers, which in turn increases the probability of accessing primary reinforcers.

Cultural Selection, which describes cognitive and behavioral changes in populations both within and between generations, represents the differential retention and elimination of *socially transmitted* cognitions and behaviors. These cognitions and behaviors may be initially acquired by social modeling and social learning, including the acquisition of rule governance, but are selected at the phenotypic level based on the adaptive or maladaptive consequences that they produce within the context of the individual's environment.

MacDonald (2008) revisited the characteristics of the cognitive systems produced, respectively, by *nature* and *nurture*. On the *nature* side, there are the *implicit* systems, which are characterized as not reflectively conscious, automatic, faster, evolved earlier, higher capacity, and effortless. These implicit cognitions are represented as consisting primarily of evolutionary (presumably meaning genetic) adaptations. On the *nurture* side, there are the explicit systems, which are characterized as conscious, controllable, slower, evolved later, resource-limited, and effortful. These explicit cognitions are represented as acquired primarily through learning and culture. MacDonald acknowledges, however, that behaviors may acquire some of the automaticity of implicit systems by practice, such as in overlearned *habits* (Andrews, 1903), which are not purely genetic in origin. In the view of the present authors, this generates a category of behavior that is intermediate in characteristics between those of the implicit and the explicit cognitive systems and is of great relevance to cultural selection.

This pivotal category of behavior is that of *social norms*, and adherence thereto, and specifically *injunctive norms* (Jacobs, Sisco, Hill, Malter, & Figueredo, 2012), which are

> normative rules or beliefs regarding what constitutes morally approved or disapproved conduct... Such norm giving and following keeps us civil, permits civilized cooperation, and lets us reap the benefits of the experience of others without the costs of trial-and-error learning. We also assume that social norms, and the fact that humans in general follow them, not only guide personal interactions, they form the fabric of our societies. Although what is normative varies among cultures, cooperation within human groups (group cohesion) depends upon following appropriate–often poorly stated but reasonably well understood–social norms and moral principles. (pp. 356–357)

One such simple social norm might be stated as follows: "Stop on red, go on green, exert caution on yellow," in reference to traffic signals. Within that rule is embedded an implicit verbal statement in the form of a syllogism: "if ***A***–and–if ***B***–then ***C***," pointing to contingent relations among ***A***ntecedent conditions, target ***B***ehaviors, and their probabilistic ***C***onsequences. For example, *if* the light is red, *and if* you do not stop, *then* the police will issue you a traffic ticket, associated with a monetary fine or penalty. The monetary fine or penalty, however, is merely the proximal or *social* consequence of nonadherence to the social norm. Behind it is a more distal or *natural* consequence of nonadherence, one of possibly being involved in a collision with another moving vehicle—one that presumably *is* adhering to the norm.

Through overlearning, adherence to social norms may become a habit, as described above, in that the injunctive norms are followed blindly rather than with explicit reference to consequences, whether social or natural. Driving home after a hard day's work, one often simply stops at every red light, without consciously considering the various contingencies of noncompliance. One may even wind up in one's driveway without even considering or even realizing that one has dutifully stopped at every red light along the way.

Thus, social norms might even evolve under conditions wherein the distal or natural consequences of a behavior are not even explicitly represented in awareness. Norms may be culturally transmitted simply through the power of the proximal or social consequences, albeit ultimately as a result of the selection pressures exerted by the distal or natural consequences of noncompliance. This condition might lead to cases of "moral dumbfounding" (Haidt, 2012) when norm-adherents are challenged to explicitly justify the rule.

Biocultural Group Selection. Culturally transmitted cognitions and behaviors are also indirectly selected at the genotypic level by producing longer-term fitness consequences that interact systematically over evolutionary time with the gene pool of the population to generate the dynamics of *gene-culture coevolution* (Lumsden & Wilson, 1980).

Spencer (1855) had previously posited that ingrained behavioral habits, through repeated practice, could eventually work their way into the heritable structure of organisms by means of the Lamarckian mechanism known as the *inheritance of acquired characteristics* through the purported dynamics of *use* and *disuse*. We know today that this is simply not possible via that mechanism. However, we must revisit the possibility that a habit, such as adherence to a social norm, could be transformed into a biologically prepared response via the perfectly respectable Neo-Darwinian process of *genetic assimilation* (Waddington, 1953), also known as the *Baldwin* (1896) *effect.* By converting a normative habit into a fully implicit and genetically guided process, it would afterwards give rise more reliably to *moral dumbfounding* as the distal or natural consequences of nonadherence would no longer even be encoded anywhere within the mechanism as newly implemented at the proximate level.

The genetic assimilation of social norms could thereby give rise to the phenomenon of biocultural group selection, wherein different biocultural groups might be discriminable both culturally and genetically based on their divergently evolved normative structures and might thus be *selected among* as higher-order vehicles for their inner replicators (see Sarraf, Woodley of Menie, & Feltham, 2018). Thus, the epigenetic rules of development controlling gene-culture coevolutionary processes reflect systematic and lawful patterns of gene-environment interaction, ultimately controlled by the molecular biology of environmentally contingent gene expression. It becomes increasingly apparent that these various processes of selection by consequences at the phenotypic level are more than merely *analogous*, but turn out to be *homologous*, in the full evolutionary sense of the word.

The Pervasiveness of Epigenetic Selection

Inspired by the emerging *Neo-Darwinian Synthesis* of the time, primarily Waddington's genetic assimilation (1957; see de Beaugrande, 1994), Jean Piaget developed an approach to development he termed *genetic epistemology*, which represents behavioral development as continuing adaptation of an organism to its environment (e.g., Piaget, 1936, 1945, 1970, 1975). Through the complementary processes of *assimilation* and *accommodation*, cognitive schemata interact systematically and predictably with evolved species-typical mechanisms to produce a match between a developing organism and the environment in which that organism must survive and ultimately reproduce. Following Piaget (1970), the theoretical work in the present preface simultaneously considers the evolutionary and developmental origins of behavior and remains anchored in his prescription: "We shall do as biologists do and turn to ontogenesis" (p. 13).

We therefore review our current understanding of the evolutionary processes by which epigenetic selection by consequences occurs. According to the West-Eberhard (2003) model of developmental plasticity, phenotypic changes triggered by environmental changes typically precede genetic changes in evolutionary processes, although, as described above, the influences of "other genes" and of "the environment" may be functionally and logically interchangeable. The subsequent genetic selection that ensues entails genetic *assimilation* and *accommodation*, which represent selection among genetic *regulators* of epigenetic processes that direct the future *norms of reaction* to environmental changes in the population by means of heritable biological preparedness or what the Lumsden and Wilson's (1980) model of gene-culture coevolution referred to as *epigenetic rules of development.*

We note once again that clonal selection, the selection among cells identical in nucleotide sequence (DNA code), but nonetheless represent epigenetically modified variants of each other, governs many developmental processes. We are therefore introducing a general term for this process, referring to this type of selection among epigenetic variants of clonal lines as *epigenetic selection*, and postulate that epigenetic differentiation must necessarily precede selective processes among the variants so generated. In this view, epigenetic diversification of clonal lines necessarily precedes all systematic clonal selection, and all adaptive clonal selection is thus *epigenetic* selection. This is logically necessary because any selection among clonal lines that are *not* epigenetically variant from each other would be expected to produce no systematic phenotypic effect, as any one such clonal line would be functionally identical to another. We thereby infer that this fundamental process underlies most developmental changes we observe in living organisms and serves as the shared proximate mechanism for all forms of *selection by consequences.*

Selection by consequences is always necessarily selection *among* the variants available, speaking purely functionally. There are, however, different mechanisms by which such selective processes may operate. Selection pressures *for* names a category of causal mechanisms that entail the proliferation of one type of variant, without necessarily entailing the decimation of the other alternative types; selection pressures *against* names a category of causal mechanisms that entail an absolute reduction in the population of one type of variant, without necessarily entailing any absolute increase in the population of the other alternative types.

In the mature immune system, for example, the selection *among* processes can be best characterized as selection *for* certain types of variants, in that the epigenetically modified clonal lines of B lymphocyte that encounter any antigen binding with their epigenetically variant receptors (that they are thus able to "recognize"), and to which they produce epigenetically variant specific antibodies, are selectively proliferated. This occurs via two different causal pathways: (1) the B-lymphocyte cells encountering the antigens are themselves triggered to proliferate by increased rates of mitosis; and (2) these same cells send paracrine chemical signals to the pluripotent hematopoietic progenitor cells in bone marrow, prompting them to produce more of the same specific variety of epigenetically variant lymphocyte. There appear to be no mechanisms selecting *against* other types of variants, which may continue to persist in their previous numbers, albeit in reduced proportions with respect to the former as they are not proliferated.

In the nervous system, on the other hand, the selection *among* processes can be best characterized as selection *against* certain types of variants, in that any epigenetically modified neurons that do not fire at a rate above their base rates are selectively eliminated by the selective withdrawal of a *transcription factor* called nerve growth factor (NGF) that is needed for their continued survival. A transcription factor is a protein that binds to the target DNA molecule and either upregulates or downregulates the synthesis of its specific gene product. Withdrawal of NGF thereby activates the mitochondrial pathway of *apoptosis* (among others) and results in the selective and programmed cell death of the underperforming neurons. In this case of selection *against*, there appears to be no selection *for* other epigenetically variant neurons, as they may continue to persist in their previous numbers, albeit in increased proportions with respect to the ones eliminated, but they are not in any way proliferated by this process.

These considerations all converge upon the logical implication that *selection by consequences* at the phenotypic level is ultimately regulated biologically by means of genetic and epigenetic mechanisms of development and thus subject to adaptive evolution. For example, biological *preparedness* is the evolved genetic programming of developmental switching mechanisms that bias development disproportionally toward alternative developmental outcomes that, in the case of reinforcers, influence which stimuli in the environment do and do not have reinforcing or punishing properties relative to individual behavior and to the relative degrees they each possess such values.

Obviously, *primary reinforcers* or *punishers* are biologically prepared and are ultimately subject to natural, social, and sexual selection. Biological preparedness also figures into the acquisition of *secondary reinforcers*, which acquire consequential value in behavioral development through associations of either: (1) cues with fitness-relevant environmental events, in the case of respondent selection (e.g., where they can serve as *unconditional stimuli*, which either increase or decrease the utilization of those cues); or (2) means with fitness-relevant environmental consequences of those tactics, in the case of instrumental selection (e.g., where they serve as primary reinforcers that increase the functional utilization of those means or punishers that decrease the functional utilization of those means). At the proximate level, primary reinforcers and punishers constitute an indirect but final functional objective of all instrumental behavioral tactics even if intermediate functional goals (secondary reinforcers or punishers) are also acquired along the way as an indirect mechanism for achieving them (see e.g., Bolles, 1970; Chung, Barot, Kim, & Bernstein, 2011; Garcia & Koelling, 1966; Jacobs & LoLordo, 1977, 1980; LoLordo & Jacobs, 1983; Rozin & Kalat, 1971; Seligman, 1970; Shapiro, Jacobs, & LoLordo, 1980).

Biological preparedness and developmental plasticity are thus complementary, where the interplay between *nature* and *nurture* is essentially a process of feedback from the environment in the service of continuing adaptation throughout behavioral development (Figueredo, Hammond, & McKiernan, 2006; West-Eberhard, 2003). Thus, all "payoffs" (proximate consequences), be they losses or gains (equivalent to punishers or reinforcers), associated with instrumental behaviors are directly or indirectly reducible to *fitness* payoffs (ultimate consequences) because behavioral evolution systematically biases learning processes (through natural, sexual, and social selection). That is, punishment and reinforcement serve as proximate mechanisms at the psychological level, signaling the fitness consequences (losses and gains with respect to survival and reproduction) that govern the ultimate adaptive function of behavior.

In Brunswikian terms (see Figueredo et al., 2006), these differential psychological payoffs signal fitness losses or gains produced by instrumental behaviors proportionally to their differential ecological validities (effectiveness within a given context) of those vicarious and intersubstitutable means (alternative tactical elements) in producing fitness-enhancing distal achievements (the ultimate functional objectives of behavior). Thus, at every level of behavioral evolution and development, what we have are mechanisms serving *selection by consequences*. Adaptation to environmental variations consists of processes analogous to those underpinning epigenetic cellular differentiation and those that construct multicellular organisms. As with the processes of cellular differentiation during organismic development, learning processes follow an evolved and preprogrammed design governed by *epigenetic rules* of development (the biologically prepared biasing of learning processes toward more fitness-enhancing outcomes; Alexander, 1990; Lumsden & Wilson, 1980, 1981).

With recent advances in scientific knowledge and the introduction of novel concepts (such as biologically prepared phenotypic plasticity and other epigenetic phenomena), it has been suggested that the current paradigm in biology, the Neo-Darwinian Modern Synthesis, may need to be extended (Pigliucci, 2007) by incorporating the ideas regarding the role of developmental plasticity in the evolutionary process that were laid out by West-Eberhard (2003). At its core, the theory of evolution by natural selection is a simple and elegant theory. When you have variation in a heritable trait that influences differential reproductive success, evolution will inevitably happen. In contrast, West-Eberhard's theory of phenotypic plasticity and evolution is more nuanced in that it incorporates development, responsiveness, selection, genetics, and evolution into one comprehensive model, without really contradicting the Modern Synthesis in any particular point of fact. This theory nevertheless shifts the focus in importance from the genotype to the phenotype at a fundamental level. Under this model, a certain proportion of the variability that natural selection works on is not exclusively at the molecular genetic level. Instead, the developmental history of the organism is both a major source of phenotypic variability and an organizing force within the organism that moderates gene expression through evolved epigenetic rules of development.

Phenotypic accommodation precedes genetic accommodation: West-Eberhard argues that development starts with a formed phenotype (e.g., fertilized egg or *zygote*). This phenotype changes during ontogenetic time due to both environmental and genomic influences. At every stage of the organism's life cycle, phenotypic development depends on the preexisting structure of the phenotype that is developmentally sensitive to environmental inputs as well as genotypic variants found on other sections of the organism's genome. Any novel gene-environment (*GxE*) or gene-gene (*GxG*) interaction can cause developmentally plastic phenotypes to appear, and, as a consequence, these phenotypic variants are themselves exposed to selective pressures. Depending on the fitness cost or benefits of each particular epigenetic response, the genes involved in the epigenetic mechanism that produces any given phenotypic variant will either decrease (due to a fitness cost) or increase (due to a fitness benefit) in the population. This process of natural selection on these heritable and hence evolvable epigenetic mechanisms, which West-Eberhard terms *genetic accommodation*, can lead to evolutionary adaptation.

According to this model, genetic accommodation consists primarily of selection on the *threshold* of environmental input at which each of these epigenetic responses will be triggered. Threshold selection will thus generate a continuum of such epigenetic rules of development, lying between what might otherwise appear to be *genetically fixed* phenotypes (that are either always absent or always present in the population) and a state of affairs where there are varying degrees of polyphenism among *developmentally plastic* phenotypes in the population based on the environmental circumstances of each individual. The degree to which each such phenotype may vary may thus be partially autonomous from the frequency of any particular allele or a set of alleles but instead dependent on the epigenetically controlled developmental threshold that is under selection.

The application of West-Eberhard's theory to the functional taxonomy of selectionist processes that we have outlined thus serves to somewhat blur the distinction between *genetic* selection (whether natural, social, or sexual) and *epigenetic* selection (whether clonal, neuronal, respondent, operant/instrumental, or cultural). Functionally stated, in terms of the consequences for population biology, this distinction boils down to what types of variant are being driven *up* in absolute frequency (via *selection for*) or *down* in absolute frequency (via *selection against*). In the case of genetic selection, the variants being selected among are genotypes (or portions thereof), whereas in the case of epigenetic selection, the variants being selected among are alternative phenotypes (or portions thereof) that are all based upon identical genotypes. Mechanistically stated, however, in terms of exactly *how* the environment is acting upon such variants (of either type), selection pressures *for* and *against* represent two distinct causal mechanisms, which are not mutually exclusive as the possibility exists that both might be operating simultaneously.[1]

Given these fundamental definitional distinctions, the line between genetic and epigenetic selection is thus in no way blurred by *what* is being selected for or against, functionally speaking, but in *how* those variants are being selected, mechanistically speaking. If we take West-Eberhard's theory seriously, the immediate target of any form of selection is always the phenotypic variant. Where phenotypic variants differ in the nucleotide sequences, and represent genetically induced differences in phenotypes within identical environmental contexts, *genetic* selection among them will be operative; where phenotypic variants do not differ in the nucleotide sequences, but represent environmentally induced differences in the phenotypic expressions of identical genetic substrates, only *epigenetic* selection among them will be operative. In both cases, the immediate loci of direct causal *transaction* with the selecting environments are the *phenotypic variants* that have been generated by either source of variation that might be operative or both of them in combination. Thus, all *selection by consequences* operates directly upon the phenotypic products of epigenesis, and the only differences between the two major types (*genetic* and *epigenetic*) are entirely attributable to a specification of the *proximate causal mechanisms* that generated the variation. It is thus not complete or comprehensive to render a purely functional description of selectionist processes, in terms of the relative frequencies of variants within a given biological population. An account of the *proximate mediation* of the phenotypic variation is absolutely required for a satisfactory understanding of the evolutionary processes involved and constitutes a pivotal element in this or any other substantive taxonomy of selectionist processes. A purely functional account of selection only offers a description of the *outcomes*; a mechanistic account of selection also provides a specification of the causal pathways *mediating* these functional outcomes.

The Molecular Biology of Epigenesis

Genes and environments are more correctly characterized as existing in a perpetual state of mutual *transaction*, meaning mutual causal influences upon each other, and only some of these transactions conform to the technical definition of an interaction, as classically defined by Johnson and colleagues (Johnson, Deary, & Iacono, 2009; Johnson, Turkheimer, Gottesman, & Bouchard, 2009). To more fully explain how these gene-environment transactions operate, however, we must explore recent advances in molecular biology that reveal what appears to be a handful of biochemical mechanisms responsible for developmental changes in gene expression, meaning either the upregulation or downregulation of the synthesis of specific gene products. These *epigenetic* modifications are proximally triggered by changes in the intracellular environment. As the functioning of other genes modifies the intracellular environment, all such triggers (whether ultimately genetic or environmental in origin) may thus be said to originate from the environment (*sensu lato*), whether intracellular or extracellular, of the gene being regulated. We therefore briefly describe epigenetic mechanisms known to exist in metazoan organisms, although this does not constitute an exhaustive list of all mechanisms present in other taxa.

Three main classes of epigenetic mechanisms have been identified in multicellular animals (Craig & Wong, 2011): (1) covalent bonding that modifies primary or higher-order structure of chromatin, (2) downstream modifications involving RNA translation and protein transcription factors, and (3) physical repositioning of nuclear chromatin. The first class (*covalent chromatin bonding*) includes: (1a) DNA methylation and (1b) histone modification. The second class (*translational and transcriptional modifications*) includes (2a) self-regulation of DNA via transcription factors, (2b) self-regulation of RNA via micro RNA (miRNA), and (2c) regulation of RNA via methylation. The third class (*nucleosome repositioning*) entails relocation, within the nucleus, of the most fundamental units of chromatin and is involved in cellular differentiation. We now briefly review what each of these epigenetic mechanisms entails.

The most common class of epigenetic mechanisms involves the first of these three classes, *covalent chromatin bonding*. These modifications consist of either direct bonding of molecules to nucleotides or changes to higher-order DNA structures, such as the modification of histone proteins. Methylation is the best-studied mechanism for covalent modification of chromatin and applies to both DNA and histone modification, as DNA and histone proteins are two basic components of chromatin. When DNA itself is methylated, the methyl groups may be attached to either cytosine or adenine nucleotides, although adenine methylation is widespread only in prokaryotes and possibly plants (Ratel, Ravanat, Berger, & Wion, 2006). When cytosine is methylated, it ceases to be active; in other words, it cannot be transcribed, and the site is thereby "turned off." This serves to downregulate, as opposed to completely shut down, the expression of any phenotypic trait for which that gene is partially responsible (as most traits are polygenic).[2] The second general class of epigenetic mechanisms in animals, *translational and transcriptional modifications*, involves the behavior of RNA and its products. Transcription factors are proteins that maintain or modify the rate at which a gene is read and transcribed into mRNA, by modifying the aforementioned covalent mechanisms. The ways in which transcription factors operate is of interest because this process potentially allows genes to regulate themselves, generally leading to self-enhancement by upregulation. RNA can also self-regulate, as certain noncoding sections of mRNA called micro RNA (miRNA) cause mRNA to decay, thereby preventing translation from those mRNAs and hence the synthesis of the coded proteins.[3] The third main class of epigenetic mechanism, *nucleosome repositioning*, refers to *where* the chromatin is during stem cell differentiation. Cellular differentiation (the specialization of stem cells) involves nucleosome repositioning as well as many of the other epigenetic mechanisms listed above and multiple intercellular signaling pathways (Teif et al., 2014).

An important point to be made regarding the proximate behavior of epigenetic control mechanisms is that these processes operate at all levels of biological organization. Just as epigenetic mechanisms are hierarchical, going from DNA to RNA to protein, and yet interact among all levels, so too are their effects, both in terms of those upon normal development and pernicious epigenetic pathologies. From intracellular organelles and metabolic functions, through tissues, biological systems, and the overall structure of the organism, epigenetic regulation is crucial to development and function.

Implications of Epigenetic Mechanisms for Gene-Environment Transactions

Based upon all the foregoing insights, we suggest a radically parsimonious reconceptualization of the nature of gene-environment transactions. Although our knowledge of epigenetic mechanisms is incomplete, enough of a pattern has emerged that it presents the possibility of a shift in perspective regarding gene-environment relations. Our increasing understanding of the molecular mechanisms governing differential gene expression opens a window into the inner workings of gene-environment interactions and helps demystify processes that genes and environments use to causally transact with each other.

We consider it likely that changes in gene expression regulate *all* the biochemical processes within living cells and that these changes in gene expression are *all* mediated by epigenetic modifications of one kind or another. This implies that epigenetic effects are the rule, not the exception. Epigenetic effects are the way that *all* genes work, *all* of the time. Pondering the meaning of this realization, it follows that *all* so-called environmental influences represent changes in gene expression by means of one epigenetic mechanism or another, which implies there can be no purely "environmental" effects upon the phenotype without genetic involvement. It also follows that *all* so-called genetic influences are subject to environmental context dependency, which implies there can be no purely "genetic" effects upon the phenotype without entailing the possibility of environmental influence.

This understanding goes beyond mere facile sophistry claiming that genes cannot perform their function without existing in some environmental setting, as any DNA-containing organism would doubtlessly suffer explosive decompression in a total vacuum, and that environments can have no effects without some genetic material existing within the

Table 1 Schematic diagram for a generalizability analysis of genes and environments

	Environment 1	Environment 2	Environment 3	Row Means
Gene A	A.1	A.2	A.3	$\overline{\mathbf{G.A}}$
Gene B	B.1	B.2	B.3	$\overline{\mathbf{G.B}}$
Gene C	C.1	C.2	C.3	$\overline{\mathbf{G.C}}$
Column Means	$\overline{\mathbf{E.1}}$	$\overline{\mathbf{E.2}}$	$\overline{\mathbf{E.3}}$	**Grand Mean**

organism (Wahlsten, 1990). In short, genes moderate *all* causal processes attributable to environmental effects, and environments moderate *all* causal processes attributable to genetic effects. This can be better expressed in terms of the schematic diagram for a generalizability theory (GT) analysis shown in Table 1.[4]

Furthermore, a biochemical understanding of gene-environment transactions also implies these are deterministic processes, jointly specified by the genetic and environmental inputs involved. These lawful gene-environment transactions, called *epigenetic rules of development* (Lumsden et al., 1981), are subject to selective pressures, based on the fitness consequences of the phenotypic effects to which they give rise. This is because the nucleotide sequences in genes code for the substrate that permits epigenetic selection and modification of observable characteristics through development. The performance of these developmental mechanisms, as reflected in their relative success at shaping characteristics that facilitate survival and reproduction, in turn influences frequencies of genes in subsequent generations. This necessarily implies there are recursive effects of epigenetic selection on genetic selection (Waddington, 1953).

As changes in gene frequencies occur across generations, genetic selection does not take place during the developmental trajectory of an individual organism. As environmental conditions are rarely invariant, however, selection operates on genetically influenced mechanisms to optimize local adaptation by fine-tuning biobehavioral mechanisms of development to systematic variations in environmental conditions (Figueredo et al., 2006). Thus, evolution of sensitivity to environmental cues depends upon: (a) generation of genetic variants for epigenetic rules of development for which variability exists; and (b) subsequent selection among those variants.

Conclusions

We have ranged far and wide across theories and theorists, disciplines, and time in succession to construct this pan-selectionist model. In doing so, we reviewed: (1) selection by sources of consequences, including natural, social, and sexual selection; (2) selection by vehicles of consequences, including clonal, neuronal, respondent, behavioral, cultural, and biocultural group selection; and (3) the pervasiveness of epigenetic selection including phenotypic and genetic accommodation and the molecular biology of epigenesis in all its forms. We then attempted to review the implications of epigenetic mechanisms for gene-environment transactions. While all those chapters in this book's first section will provide a more traditional introduction as it directly relates to multilevel selection, we have presently mapped out a meta-theory in which the processes of multilevel selection can be properly situated among the hierarchically nested levels of biological organization. It is this meta-theory of pan-selectionist evolution that justifies our subsequent treatment of historical, social, and cultural change as convergent with, if not actually homologous to, biological evolution at the higher levels of organization. When we discuss the aggregation, decline, and collapse of complex societies in the context of modern Europe or Ancient Rome, it will be within the presumed meta-theoretical matrix of pan-selectionism, such that synergistic relations replace static divisions.

University of Arizona
Tucson, AZ, USA

Aurelio José Figueredo
W. Jake Jacobs
Pedro S. A. Wolf
Zachary J. Hohman
Heitor B. F. Fernandes

Notes

1. It is theoretically important to note that we specify *absolute* not *relative* frequency here. *Selection for* one variant necessarily drives down the *relative* frequency of the other, and *selection against* one variant necessarily drives up the *relative* frequency of the other, so the distinction becomes meaningless when considering *relative* frequencies.
2. These changes are maintained due to the action of DNA methyltransferase (DNMT), which ensures that daughter DNA strands have the same methylation pattern as does the parent strand. When DNMT reaches a segment of DNA where only one of the two strands is methylated, the DNMT enzyme attaches synonymous methyl groups to the other half. Although the cause of *variation* in the fidelity of methylation maintenance is not well understood, unregulated de novo methylation may have serious consequences such as carcinogenesis. If genes responsible for tumor suppression via selective apoptosis are hypermethylated, aberrant cells can accumulate into tumors, which may be malignant. The converse has also been demonstrated; hypomethylation of oncogenes is linked to the occurrence of cancer as well, the existence of which is often the result of de novo mutations in genes responsible for programmed cell death (Gonzalo, 2010). In that case, the development and survival of malignant cells are the products of both DNA replication errors and the failure to epigenetically silence these errors.

   Any type of covalent modification can have similar effects on phenotypes because the biochemical mechanisms involved are closely analogous. For example, methylation can also modify chromatin by altering the structure of the histone "backbone" that dictates secondary and tertiary DNA structures. Methylation of histone proteins effectively hides certain regions (called *compartments*) of DNA from being read by DNA transcriptase. These modifications also have the effect of hiding those same compartments from themselves being methylated, which means that de novo histone modification will affect the fidelity of DNA methylation. Histone modification may also occur via lysine acetylation, in which a change to chromatin structure may either upregulate or downregulate transcription or both simultaneously, by exposing certain DNA compartments and hiding others (Bernstein et al., 2005).

3. As with DNA, mRNA itself is also sometimes methylated and interrupts the process at the translational stage and can lead to profound downstream consequences. For example, dysfunction in mRNA regulation has been linked to obesity and diabetes in humans due to a failure to maintain energy homeostasis (Bennett et al., 1997).
4. To estimate an overall genetic effect, in this conception of the problem, the gene's influence upon the phenotype must generalize across alternative developmental environments. The main effect of a gene's influence upon the phenotype is the mean observed score for that genetic effect, averaged across a representative sample of alternative environments, as represented in the diagram by the "Row Means" (G.1, G.2, G.3); the main effect of an environmental influence upon the phenotype is the mean observed score for that environmental effect, averaged across a representative sample of alternative genes (alleles), as represented in the diagram by the "Column Means" (E.1, E.2, E.3). A gene by environment interaction represents how each individual gene (allele) responds in each different environment, as represented by each individual cell in the cross-tabulation (A.1, A.2, A.3, B.1, B.2, B.3, C.1, C.2, C.3). This latter term serves as our measure of the variability in the effect of each gene across different environments. Thus, there is no contradiction between speaking of additive variance components of "heritability" and "environmentality" in addition to the unique nonadditive variance components contributed by the gene-environment interactions. To make pronouncements such as the oft-repeated claim that the heritability coefficient has been rendered meaningless or obsolete by the discovery of gene-environment interactions is therefore mathematically illiterate.

References

Alexander, R. D. (1990). Epigenetic rules and Darwinian algorithms: The adaptive study of learning and development. *Ethology and Sociobiology, 11,* 1–63.

Andrews, B. R. (1903). Habit. *The American Journal of Psychology, 14*(2), 121–49.

Baldwin, J. M. (1896). A new factor in evolution. *The American Naturalist, 30*(354), 441–451.

Bennett, S. T., Wilson, A. J., Esposito, L., Bouzekri, N., Undlien, D. E., Cucca, F., Nisticò, L., Buzzetti, R., Bosi, E., Pociot, F., Nerup, J., Cambon-Thomsen,

A., Pugliese, A., Shield, J. P., McKinney, P. A., Bain, S. C., Polychronakos, C., & Todd, J.A. (1997). Insulin VNTR allele-specific effect in type 1 diabetes depends on identity of untransmitted paternal allele. The IMDIAB Group. *Nature Genetics, 17*(3), 350–352.

Bernstein, B. E., Kamal, M., Lindblad-Toh, K., Bekiranov, S., Bailey, D. K., Huebert, D. J., McMahon, S., Karlsson, E. K., Kulbokas, E. J., Gingeras, T. R., Schreiber, S. L., & Lander, E. S. (2005). Genomic maps and comparative analysis of histone modifications in human and mouse. *Cell, 120*(2), 169–181.

Bolles, R. C. (1970). Species-specific defense reactions and avoidance learning. *Psychological Review, 71,* 32–48.

Changeux, J. P. (1985). *Neuronal man: The biology of mind.* New York: Pantheon Books.

Chung, A., Barot, S. K., Kim, J. J., & Bernstein, I. L. (2011). Biologically predisposed learning and selective associations in amygdalar neurons. *Learning & Memory, 18,* 371–374.

Craig, J., & Wong, N. C. (Eds.). (2011). *Epigenetics: A reference manual.* Norfolk, UK: Caister Academic Press.

Darwin, C. (1859). *On the origins of species by means of natural selection.* London: Murray.

Darwin, C. (1871). *The descent of man, and selection in relation to sex.* London: Murray.

Dawkins, R. (1976). *The selfish gene.* London: Oxford University Press.

de Beaugrande, R. (1994). Function and form in language theory and research: The tide is turning. *Functions of Language, 1/2,* 163–200.

Figueredo, A. J., Hammond, K. R., & McKiernan, E. C. (2006). A Brunswikian evolutionary developmental theory of preparedness and plasticity. *Intelligence, 34*(2), 211–227.

Garcia, J., & Koelling, R. A. (1966). Relationship of cue to consequence in avoidance learning. *Psychonomic Science, 4,* 123–124.

Gonzalo, S. (2010). Epigenetic alterations in aging. *Journal of Applied Physiology, 109*(2), 586–597.

Haidt, J. (2012). *The righteous mind: Why good people are divided by politics and religion.* New York: Pantheon.

Jacobs, W. J., & LoLordo, V. M. (1977). The sensory basis of avoidance responding in the rat: Relative dominance of auditory or visual warning signals and safety signals. *Learning and Motivation, 8,* 448–466.

Jacobs, W. J., & LoLordo, V. M. (1980). Selective association in aversive conditioning: Implications for avoidance learning in the rat. *Learning and Motivation, 11*, 427–455.
Jacobs, W. J., Sisco, M., Hill, D., Malter, F., & Figueredo, A. J. (2012). On the practice of theory-based evaluation: Information, norms, and adherence. *Evaluation and Program Planning, 35*(3), 354–369.
Jerne, N. K. (1955). The natural-selection theory of antibody formation. *Proceedings of the National Academy of Sciences of the United States of America, 41*(11), 849–857.
Johnson, W., Deary, I. J., & Iacono, W. G. (2009). Genetic and environmental transactions underlying educational attainment. *Intelligence, 37*(5), 466–478.
Johnson, W., Turkheimer, E., Gottesman, I. I., & Bouchard Jr., T. J. (2009). Beyond heritability: Twin studies in behavioral research. *Current Directions in Psychological Science, 18*(4), 217–220.
LoLordo, V. M., & Jacobs, W. J. (1983). Constraints on aversive conditioning in the rat: Some theoretical accounts. In P. Harzem & M. Zeiler (Eds.), *Advances in the analysis of behavior* (pp. 325–350). London: Wiley.
Lumsden, C. J., & Wilson, E. O. (1980). Translation of epigenetic rules of individual behavior into ethnographic patterns. *Proceedings of the National Academy of Sciences, 77*(7), 4382–4386.
Lumsden, C. J., Wilson, E. O., Packing, C., & Rare, W. I. E. S. (1981). *The coevolutionary process*. Cambridge, MA: Harvard University Press.
MacDonald, K. (2008). Effortful control, explicit processing and the regulation of human evolved predispositions. *Psychological Review, 115*(4), 1012–1031.
Naour, P. (2009). *E.O. Wilson and B.F. Skinner: A dialogue between sociobiology and radical behaviorism. Developments in primatology: Progress and prospects.* New York: Springer.
Nesse, R. M. (2009). Runaway social selection for displays of partner value and altruism. In J. Verplaetse, J. Schrijver, S. Vanneste, & J. Braeckman (Eds.), *The moral brain* (pp. 211–231). Dordrecht, The Netherlands: Springer.
Piaget, J. (1970). *Psychologie et épistémologie*. Paris: Editions Denoël.
Piaget, J. (1936). *La naissance de l'intelligence chez l'enfant*. Neuchâtel, Switzerland: Delachaux et Niestlé.
Piaget, J. (1945). *La formation du symbole chez l'enfant*. Neuchâtel, Switzerland: Delachaux et Niestlé.
Piaget, J. (1975). *L'équilibration des structures cognitives: problème central du développement (Études d'épistemologie génetique XXXIII)*. Paris: Presses Universitaires de France.

Pigliucci, M. (2007). Do we need an extended evolutionary synthesis? *Evolution, 61*(12). 2743–2749.

Ratel, D., Ravanat, J. L., Berger, F., & Wion, D. (2006). N6-methyladenine: The other methylated base of DNA. *Bioessays, 28*(3), 309–315.

Rozin, P., & Kalat, J. W. (1971). Specific hungers and poison avoidance as adaptive specializations of learning. *Psychological Review, 78,* 459–86.

Sarraf, M., Woodley of Menie, M. A., & Feltham, C. (2018). *Modernity, nihilism and mental health.* London: Routledge.

Seligman M. E. P. (1970). On the generality of the laws of learning. *Psychological Review, 77,* 406–18.

Shapiro, K. L., Jacobs, W. J. & LoLordo, V. M. (1980). Stimulus relevance in Pavlovian conditioning in pigeons. *Animal Learning and Behavior, 8,* 586–594.

Spencer, H. (1855). *The principles of psychology*. London: Longman, Brown, Green and Longmans.

Teif, V. B., Beshnova, D. A., Vainshtein, Y., Marth, C., Mallm, J.-P., Höfer, T., & Rippe, K. (2014). Nucleosome repositioning links DNA (de)methylation and differential CTCF binding during stem cell development. *Genome Research, 24*(8), 1285–1295.

Waddington, C. H. (1953). Genetic assimilation of an acquired character. *Evolution, 7*(2), 118–126.

Waddington, C. H. (1957). *The strategy of the genes. A discussion of some aspects of theoretical biology.* London: Allen and Unwin.

Wahlsten, D. (1990). Insensitivity of the analysis of variance to heredity-environment interaction. *Behavioral and Brain Sciences, 13,* 109–161.

West-Eberhard, M. J. (1979). Sexual selection, social competition, and evolution. *Proceedings of the American Philosophical Society, 123*, 222–234.

West-Eberhard, M. J. (2003). *Developmental plasticity and evolution.* Oxford University Press.

Contents

List of Figures

List of Tables

Introduction

Steven C. Hertler, Aurelio José Figueredo, and Mateo Peñaherrera-Aguirre

A Philosophical Fable

Once upon a time, there lived the *Troglodytes*, people who devolved into selfishness. They ended up living in a state of nature where each abided only by his individual interest. The first step in this descent toward dissolution came when a strong man abducted the wife of a weak man. The weak man's attempts to regain his wife came to naught, as the strong man would not give her over. There was then an appeal to a disinterested third party, but he proved himself uninterested. This third party of appeal forbore to devote his time to adjudicating a matter that had no direct bearing upon him and instead left to cultivate his fields. Frustrated, the man with no wife happened upon a beautiful woman and, actuated by the force of might and whim, took her to his bed. This turned out to be the wife of the man who refused to mediate on the weaker man's behalf, who failed to labor disinterestedly on behalf of any abstract principle of justice. From there, community relations unraveled. Grain could not be reaped by those that planted the seed. Decorum in exchange, even of necessaries, was flouted. One man used his leverage to demand twice the going rate for a commodity, only to witness his trading partner return an even more usurious bargain. In short, all social relations degraded to the detriment of the community at large. Finally, a scourge afflicted the land. A physician of great skill came from a distant land to cure the people, but, upon recovering, the Troglodytes spurned their deliverer, and refused him payment. When this same scourge returned, the physician refused the Troglodytes his services.

Eventually, but two families were left. These families were unrepresentative of the Troglodyte type. They were indeed opposite in every way. None would think of himself, but did only for others. There was consideration only of one's brother, father, mother, wife, or some other. In this way, there grew a society of solidarity: a society that could solve collective action problems and one that was cohesive and cooperative. There were two consequences of this opposite orientation, both redounding to what one might call *group fitness*. First, on its own, the group flourished. They became many and prospered. The second consequence was that they were able to defend themselves. Neighboring persons from distant provinces began to exploit the Troglodytes who were essentially Christian in their altruistic giving. They tolerated some manner of abuse on these grounds. Thereafter, these new Troglodytes were threatened with treachery. They were unified in the face of attack, defending one another, and the community as a whole, instead of fearing for their own lives. In consequence, they closed ranks, avenged one another's deaths, and vanquished the enemy.

The reader may recognize the foregoing yarn as the figment of the imagination of Charles-Louis de Secondat, Baron de La Brède et de Montesquieu. These Troglodytes are described in his allegorical *Persian Letters*,[1] with *Persian* standing in for *Parisian* so as to evade, if not deceive, the censorious *Ancien Régime* of eighteenth-century France. Many of these *Persian Letters* arrest the thought of our long-dead Montesquieu as he contemplated individual interest as it affected the larger whole. Emphasized by George R. Healy in his introduction to the *Persian Letters*, Montesquieu, as always ahead of his time, was something of a proto-utilitarian,[2] developing lines of thought that were thereafter well trodden by John Stuart Mill and Jeremy Bentham. At the same time, Montesquieu thought and wrote as a scion of the early Enlightenment and consequently emphasized its rationalized cosmopolitanism. The confluence of these intellectual developments surfaces in notes and quotes populating his *Lettres Persanes* and his *Pensées.*[3]

Throughout, Montesquieu considers the individual *vis-à-vis* the group. He believed, according to Healy, that the family should be preferred to the individual, the nation to the family, the region to the nation, and the human species to race or region. In doing so, Montesquieu strangely inverted what an evolutionist might see as the natural order of things.[4] Yet, the Troglodytes represent the perils of narrow, individual interest, and, thereafter, the advantage of broad, group altruism. Montesquieu then had an ideal of cooperation for the greater good and was possessed of a nuanced understanding of how that ideal redounded to the benefit of all; but at the same time, he was not privy to the biological processes underpinning cooperation or impeding its development. Be forewarned: gaining that biological understanding comes from reviewing a litany of theories, theorists, concepts, syntheses, discoveries, and developments more than a century in the making; but more than this, it brings one through a series of insights about how evolution operates, which transcends, and encapsulates Montesquieu's more narrowed questions of cooperation. Darwin's original writings contain the first speculations on such questions. It is here where the story starts, even as it then quickly segues through a long interlude, followed by a burst of synthetic work leading right up to the present.

The Present Volume: From Possible to Patent

Mindful that multilevel selection theory has itself evolved into something different altogether, it is unfortunate that so many books, like those of Okasha (2006),[5] Borrello (2010),[6] and Bahar (2017),[7] were necessarily trained rearward, focused on refuting prior vocal detractors of group selection, such as Lack, Williams,[8] and Dawkins. These are excellent works, but they are, in large measure, works of intellectual history. To the extent that such books get beyond intellectual history, they are implicitly on the defensive, attempting to establish the possibility, instead of the reality, of group selection as an actual level of selection within multilevel selection theory. This is not to criticize these uniformly excellent books, as they were necessary apologetics withstanding the torrent of criticism raining down on multilevel selection theory. With the exception of Borrello (2010) which is designedly a work of intellectual history, most books on group selection have enough reserve thrust to *segue* from defense to offense. Extant books provide rarified mathematical models, tortuous intellectual histories,[9] incisive philosophical treatments, and instructive theoretical discussions, though they are all alike improvidently neglectful of evidentiary proofs of human group selection. The present volume represents a first effort to establish the *reality of human multilevel selection on empirical grounds*. That being said, this would neither be practical nor possible, if not for the many miles of track laid down by prior authors, especially those works of D. S. Wilson, which deal most directly with humans, thereby running contrary to the aforementioned generalized tendency to exempt humans from discussions of multilevel selection. Even as we try to condense much of what went before within Part I of this book, many of its three chapters reference prior books, illustrative of the fact that this book is in some sense a chapter in a multivolume history of multilevel selection.

As Michael E. Gilpin[10] rightly states, group selection is not a God hypothesis, but is instead capable of being empirically tested. Still, multilevel selection theory is in some ways a less tractable idea to test than is, for instance, Einstein's *General Theory of Relativity*. There is no hallmark data point or precise prediction by which the predictive validity of multilevel selection theory can be evaluated. One cannot look for starlight bent by the sun's gravitational field. To the contrary, there are several classes of support that are compelling in combination. We propose three classes of support, separately featured in this volume's three main parts, which, together, show human multilevel selection to be *possible*, *probable*, and *patent*. It is in the first of our three parts where we sketch the past of multilevel selection theory; contrast that original formulation with its modern equivalent; introduce theories and theorists; review animal models, experimental bacterial data, and the instructive examples of eusocial insects; and generally intellectually orient the reader to understand the remainder of the book. After a fashion, this first part will therefore be a condensation of all other books that we have read on multilevel selection. By the end of this first part, the reader understands that multilevel selection is empirically demonstrable among other species and at least theoretically possible among humans.

Having established this prerequisite knowledge across Part I's three chapters, Parts II and III are clear to present empirical tests of human multilevel selection. Part II makes the assumption that the evidence of multilevel selection rests in plain sight within the recorded history of our species. Most basically, we focus on intergroup conflicts resulting in significant, directional changes in gene frequencies, as inferred through documented biodemographic change.[11] We specifically emphasize sources demonstrating the frequently genocidal effects of organized warfare,[12] world history,[13] and comparative history.[14] Additionally, we review historical illustrations of nationalism, religion, and ethnicity, among other markers or collective identity delineating biocultural groups from one another. We add to this a review of early and late modern propaganda employed to exaggerate these naturally occurring group differences. Further still, we discuss military drill emphasizing integrated movement, rearguard actions that save main military bodies, and basic training undertaken to activate kinship sentiments and group solidarity. All of this contributes to an understanding as to how groups effectively compete with one another. However, before ending this second part, we also review historical examples of social, legal, and military sanctions designed to maintain internal integrity, especially in the midst of external danger, defection, and desertion. Rules, both formally ensconced in law and informally embodied in custom, mark this struggle to retain sufficient unity to overcome rival groups. By reviewing select portions of the historical record through a multilevel selectionist lens across four chapters, one can see how human biocultural groups form and cleave as they balance external threat against internal dissolution. All such content is carefully assembled to chart civilizational cycles, with four chapters addressing the following four respective phases: aggregation, growth, decline, and collapse.

Both the second and third parts are historical and empirical. However, whereas the second part demonstrates by way of comparative review, the third part demonstrates by way of statistical analyses of primary research data. Our research group has produced what may be the only emerging corpus of peer-reviewed quantitative empirical evidence of human group selection of which we are currently aware. After reviewing appropriate literature, we start by presenting evidence of multilevel selection operating on the number of male chimpanzee patrols. This review of martial behavior in our most closely related cousins is preparatory to studying lethal intergroup conflict among small-scale human societies. Thereafter, we present an empirical examination of Ancient Rome's integration that contrasts the Republic with the Imperium, as it tracks battle frequency, civil wars, monetary indicators, and elite mortality. Through these metrics, a highly cohesive Republic can be seen to expand, only to collapse through its transition to a diffuse and diverse Empire. In a readable form, we also review discussion points from the previously published Nexus 200[15] and Nexus 400[16] analyses that we performed on modern populations. These quantitative analyses, taken together, amass convergent evidence in demonstration of human multilevel selection using biodemographic data, lexicographic analyses, male fluctuating asymmetry, sinistrality prevalence, body mass index, height, cognitive abilities, and brain mass. Thereafter, as it relates to historical biocultural group competition between the early and late modern Gallic and Britannic Empires and their successor states, we review our two most recent companion publications: (1) *War and Peace: A Diachronic Social Biogeography of Climate Change, Life History Strategy, and Evolving Between-Group Relations in Two Western European Populations* and (2) *The Ecology of Empire: The Dynamics of Strategic Differentiation-Integration in Two Competing Western European Biocultural Groups*.

In closing we recall a decades' old conversation about group selection between Wynne-Edwards and Theodosius Dobzhansky, wherein the latter asked the former, "Don't you think you could sort this thing out once and for all?" If this volume does not unequivocally *sort this thing out once and for all*, it will effectually produce what Sober and Wilson[17] refer to as *the smoking gun*, that is, quantitative empirical evidence of human multilevel selection in action. This evidentiary information was previously extant, though it has heretofore remained scattered across diverse peer-reviewed articles, technical monographs, and our private notes. Consolidated in what we hope proves an accessible monograph, however, it may possibly serve as a pivot point, allowing future works on multilevel selection to accrete additional evidence in support (or possible disconfirmation) of human multilevel selection.

Of final note, this volume was written by one integrated team of writers. As with any aggregate subject to the pressures of multilevel selection, however, we reassorted ourselves into different combinations and permutations for the writing of each chapter. Each individual chapter therefore sports the more proximal authorship of its main contributors for optimal apportionment of academic credit. However, this is very different from an edited anthology, as all of us participated, directly or indirectly, in the creation of this entire work as an integrated whole. In addition, we would like to acknowledge the efforts of our five research assistants, Maya Louise Bose, John Michael Jurgensen, Jonathan Revel, Robyn Stea, and Garrett Dien. Also, we would like to express our appreciation for the copy editing services of Marian Hertler. Finally, we would like to thank the co-contributors of the Preface that were not coauthors of the entire book.

Notes

1. Montesquieu, C. (1964). *The Persian letters.* Indianapolis, Indiana: Hackett Publishing Company
2. "It is interesting to note that in 1721 Montesquieu, in this as in so many things at least a generation ahead of his colleagues, was very near the utilitarian position. It will be noted, for example, that in The Persian Letters he attacks Louis XIV's revocation of the Edict of Nantes not on the ground that the revocation denies the natural right of every man to worship as his conscience may dictate, but rather because it weakened and divided the state of France, though the king's intent was to strengthen and unify; because, in a word, it did not fulfill the useful purpose for which it was intended. Montesquieu bitterly deplored John Law's financial manipulations, not as abstractly or inherently bad or "unnatural," but largely for the social disruption they caused. Criminal punishment, he argued, must be scaled reasonably to the magnitude of the crime, not for humanitarian reasons rooted in nature, but because if it is not so proportioned, criminal behavior will increase and further harm society. Montesquieu detested political despotism for various and immediate reasons; his many arguments against it in The Persian Letters, however, most often come down to the entirely pragmatic and utilitarian objection that despotism simply does not work very well" (Montesquieu, 1964; p. xviii; in the introduction by Healy).

3. Montesquieu, C. (2012). *My thoughts.* Indianapolis, Indiana: Liberty Fund This source contains several interesting quotes relative to group selection, of which the following two serve as examples:

   But if it is true that love of Country has in all times been the source of the greatest crimes, because men have sacrificed more general virtues to his particular virtue, it is no less true that, once rectified, it is capable of honoring a whole nation. (Montesquieu, 2012; p. 347)

   "If I knew something useful to me, but prejudicial to my family, I would reject it from my mind," he wrote in his Pensées. "If I knew something useful to my family, but not to my country, I would try to forget it. If I knew something useful to my country, but prejudicial to Europe, or useful to Europe and prejudicial to the human race, I would regard it as criminal."
4. This later quote is cited and discussed in Chap. 2.

   For instance, this is contrary to *inclusive fitness theory* and thereafter *genetic similarity theory*, both of which concepts will be treated subsequently.
5. Okasha's *Evolution and Levels of Selection* is a work of scientific philosophy exploring conceptual implications associated with group selection.
6. M. E. Borrello's *Evolutionary Restraints: The Contentious History of Group Selection*, published by Chicago University Press, is a serviceable, and sometimes inspired, work of intellectual history.
7. Bahar's *The Essential Tension: Competition, Cooperation and Multilevel Selection in Evolution* brings one on the long tour of group selection's intellectual history after the fashion of Borrello (2010).
8. George C. Williams' *Group Selection*, an early and now canonical treatment of the topic, reviews altruistic behavior, social organization, the adaptive regulation of population density, sex, and sex ratios.
9. Certainly, some may want to study the history of defunct scientific postulates like phlogiston and the ether, and there is certainly some scientific value, though for many, the intellectual history of group selection merits study to the extent that early instantiations of group selection have contributed to later instantiations of group selection that are truly informative of a biological reality within our evolutionary past.
10. In his *Group Selection in Predator-Prey Communities*, Gilpin produced a useful though specialized treatment of group selection that begins with introductory information concerning the controversy surrounding group selection.

11. Though we do not attempt an analysis of genetic markers, haplotypes promise the possibility of doing this and are being accessed by members of our research team to, for example, look at conflict and proportional representation of rival genetic clusters within Britain.
12. Laitin, D. (2006). Mann's dark side: Linking democracy and genocide. In, J. A. Hall & R. Schroeder (Eds.) *An anatomy of power: The social theory of Michael Mann* (pp. 328–339). New York: Cambridge University Press.

    Courtois, S., Werth, N., Panné, J., Paczkowski, A., Bartošek, K., & Margolin, J. (1999). *The black book of communism: Crimes, terror, repression.* Cambridge, Massachusetts: Harvard University Press.
    Shaw, M. (2015). *War and genocide: Organised killing in modern society.* John Wiley & Sons.

13. In addition to historians like H. G. Wells, William Durant, Michael Mann, and Oswald Spengler, William H. McNeill is an example of such a world historian; and one we featured in Chap. 9 of our prior publication with Palgrave Macmillan entitled *Life History Evolution: A Biological Meta-Theory for the Social Sciences.* Below is a sampling of sources that would be of relevance to the present project, all of which have already been read and reviewed:

    McNeill, W.H. (1963). *The rise of the West: A history of the human community.* Chicago, Illinois: Chicago University Press.
    McNeill, W.H. (1974). *The shape of European history.* New York: Oxford University Press.
    McNeill, W.H. (1979). *A world history, third edition.* New York: Oxford University Press.
    McNeill, W.H. (1990). *Population and politics since 1750.* Charlottesville, Virginia: University of Virginia Press.
    McNeill, W.H. (1992). *The global condition: Conquerors, catastrophes, and community.* Princeton, New Jersey: Princeton University Press.
    McNeill, W.H. (1998). *Plagues and peoples.* New York: Anchor Books.

14. Arnold Joseph Toynbee in an exemplar of the comparative historian. In reviewing his writings for Chap. 8 of *Life History Evolution: A Biological Meta-Theory for the Social Sciences*, we read and wrote about Toynbee's 12 volume, *A Study of History.* Therein we found and recorded more than 47,000 words worth of notes and quotes, many of which were relevant to group selection.

15. Woodley, M.A. & Figueredo, A.J. (2013). *Historical variability in heritable general intelligence: Its evolutionary origins and socio-cultural consequences*. Buckingham, UK: University of Buckingham Press.
16. Woodley of Menie, M.A., Figueredo, A.J., Sarraf, M.A., Hertler, S. C., Fernandes, H.B.F., & Aguirre, M.P. (2017). *The Rhythm of the West: A Biohistory of the Modern Era, AD 1600 to Present*. Journal of Social, Political and Economic Studies Monograph Series, Volume 37.
17. In *Unto Others: The Evolution and Psychology of Unselfish Behavior*, E. Sober and D. S. Wilson dutifully review the requisite history of group selection, recalling significant names, intellectual arguments, and interactions with related concepts, such as kin selection.

References

Bahar, S. (2017). *The essential tension: Competition, cooperation and multilevel selection in evolution*. New York: Springer.

Borrello, M. E. (2010). *Evolutionary restraints: The contentious history of group selection*. Chicago: Chicago University Press.

Okasha, S. (2006). *Evolution and levels of selection*. New York: Oxford University Press.

1

An Intellectual History of Multilevel Selection from Darwin to Dawkins

Aurelio José Figueredo, Steven C. Hertler, and Mateo Peñaherrera-Aguirre

1 Introduction

Whole books, and sections of books, have been dedicated to reviewing the intellectual history of multilevel selection, some small swath of which are considered in this first chapter. Readers are then made aware of other pertinent publications, acquiring something of their substance in this condensed review. For example, readers will certainly gain knowledge of multilevel selection's prototypical origins as they are present in Darwin's *Descent of Man*, while also being introduced to its reformulation a century later as a measure of population regulation. In addition to preferring main ideas to tortuous detail, we here take such content's subsequent relevance as our litmus test for inclusion. We also review recent bouts of controversy between adherents and detractors. Where historical review is the end of other books, we use the history of multilevel selection instrumentally, with the end of contrasting its original formulation with its

S. C. Hertler et al., *Multilevel Selection*, https://doi.org/10.1007/978-3-030-49520-6_1

present instantiation. Having drawn these distinctions, one can clearly see that most controversies and objections no longer apply, as they do not pertain to modern manifestations of multilevel selection. Even as naysayers continue to criticize it for something it either never was or no longer is, the modern formulation of multilevel selection is becoming known to, and accepted among, most evolutionists. Nevertheless, this is only the first act in the two-part drama, taking us to the lowest ebb of multilevel selection's reputation. The astute reader may notice that we use the expression *multilevel* selection when referring to the more general theoretical framework that is the subject matter of the present volume, while reserving the term *group* selection for describing the positions of past commentators that directed their reflections specifically to this particular component of multilevel selection theory, especially as more narrowly defined by past formulations of this principle. Multilevel selection theory is a more inclusive term that recognizes the operation of selection at multiple levels of biological organization, including both solitary individuals and social groups, as exemplified but never explicitly named by Charles Darwin's thinking on the matter.

2 Darwin and the *Descent of Man*

Let us begin at the beginning. Charles Darwin marks our beginning. Only after Darwin's lifetime of thinking, collecting, traveling, and writing does a modern world unfold wherein blind, bottom-up processes are known to outstrip, in the creation of wonder and complexity, even the most ingenious top-down design. Too many pass by this epoch with a nod and a glance, satisfying themselves with Herbert Spencer's oversimplified characterization: *The Survival of the Fittest*. Yes, evolution is elegantly simple. From a few basic premises, operative over geologic time, come marvelous variation and staggering creation. Still, the natural products of evolution are not invariably "red in tooth and claw" (Tennyson, 1850, In Memoriam *A. H. H.*, Canto 56) and cannot be reduced to Spencer's aphoristic formulation, or to any single phrase whatsoever. From the start, within the original writings of Darwin, much of the complexity of evolution was extant. By contrast, brilliant and capable though he was, Alfred Russel Wallace intuited his understanding of natural selection suddenly in a malarial fever dream and thereafter developed it less fully, whereas Darwin's understanding of natural selection was better elaborated, as well as being better theoretically and empirically supported by the time he published it in 1859. In the process of critically viewing his own theory, with bravery overcoming trepidation, Darwin catalogued potentially contrary facts and held out possible mechanisms by which his theory of evolution could be disproven. It was through this process that Darwin expanded his vision of evolution from a purely organismic view of selfish competition to one encompassing the evolution of groups in altruistic cooperation.[1]

The most pointed examples of apparent evolutionary paradoxes considered by Darwin, which were only solved by transitioning from an individual to a multilevel selectionist paradigm, are found among Darwin's writings on social insects and human tribes. For instance, Darwin understood worker sterility as advantageous to the community. Even as he understood that sterile workers were born capable of work, but not of procreation, Darwin remarked, "I can see no very great difficulty in this being effected by natural selection." The reasoning behind this judgment is explained more fully in the fourth edition of *On the Origin of Species*:

> Thus I believe it has been with social insects: a slight modification of structure, or instinct, correlated with the sterile condition of certain members of the community, has been advantageous to the community: consequently the fertile males and females of the same community flourished, and transmitted to their fertile offspring a tendency to produce sterile members having the same modification. And I believe that this process has been repeated, until that prodigious amount of difference between the fertile and sterile females of the same species has been produced, which we see in many social insects.

With respect to human evolution, Darwin noted that self-sacrifice is necessary for the welfare of a *tribe*. As Darwin (1871, p. 166) realized long ago, "at all times throughout the world tribes have supplanted other tribes." Or as Steven Mirsky notes, Darwin is thought to have said that tribes of moral men have an immense advantage over fractious bands of pirates. Embryonic sympathies with multilevel selection are manifest in Darwin's passages concerning the evolution of altruism, self-sacrifice, intelligence, and obedience. Also evident are considerations as to how outgroup competition and ingroup pressures simultaneously select for group cohesion in opposition to individual self-interest.

Selfishness was punished by death and displacement at the hands of outsiders, while being punished by shame and ostracism at the hands of one's own group. Thus, notwithstanding some few dissenting opinions, Darwin did recognize that multilevel selection was possible and could furthermore result in directional change. In other words, if tribes could supplant other tribes, this recognizes that groups can compete; in turn competition can change gene frequencies even up to the replacement of one group by another.

> Obedience, as Mr. Bagehot has well shewn, is of the highest value, for any form of government is better than none. Selfish and contentious people will not cohere, and without coherence nothing can be effected. A tribe possessing the above qualities in a high degree would spread and be victorious over other tribes; but in the course of time it would, judging from all past history, be in its turn overcome by some other.

Without irresponsibly extrapolating from the direct quotation above, one can infer a positive feedback effect, wherein between-group competition ratchets up cooperation and cohesion over millennia. This is then combined with an acknowledgment that specific traits, in this case cooperation and altruism, decide the contest, with the result that those traits become more common in aggregate. In short, *cooperation and altruism are under directional selection within groups due to competition between groups*. However, cooperation and altruism only extend to the ingroup, with quite a different standard of behavior applicable to relations with outgroups:

> No tribe could hold together if murder, robbery, treachery, &c., were common; consequently such crimes within the limits of the same tribe 'are branded with everlasting infamy;' but excite no such sentiment beyond these limits. A North American Indian is well pleased with himself, and is

> honoured by others, when he scalps a man of another tribe; and a Dyak cuts off the head of an unoffending person and dries it as a trophy. The murder of infants has prevailed on the largest scale throughout the world, and has met with no reproach...

We also see passages more distinctly detailing how altruism might have been selected for within the group via sexual selection and status seeking:

> We may therefore conclude that primeval man, at a very remote period, would have been influenced by the praise and blame of his fellows. It is obvious, that the members of the same tribe would approve of conduct which appeared to them to be for the general good, and would reprobate that which appeared evil. To do good unto others—to do unto others as ye would they should do unto you,—is the foundation-stone of morality. It is, therefore, hardly possible to exaggerate the importance during rude times of the love of praise and the dread of blame. A man who was not impelled by any deep, instinctive feeling, to sacrifice his life for the good of others, yet was roused to such actions by a sense of glory, would by his example excite the same wish for glory in other men, and would strengthen by exercise the noble feeling of admiration. He might thus do far more good to his tribe than by begetting offspring with a tendency to inherit his own high character.

Though it has a larger scope, sections of T. Shanahan's *The Evolution of Darwinism: Selection, Adaptation and Progress in Evolutionary Biology* distill Darwin's writings as they are relevant to multilevel selection. With this astute guide, we locate an excerpt from Darwin's *Descent of Man* that frames the potential problem altruism posed to a nascent evolutionary theory premised on natural selection (Darwin, 1871, vol. 1, p. 163 as quoted in Shanahan, 2004):

> It is extremely doubtful whether the offspring of the more sympathetic and benevolent parents, or of those which were the most faithful to their comrades, would be reared in greater numbers than the children of selfish and treacherous parents of the same tribe. He who was ready to sacrifice his life, as many a savage has been, rather than betray his comrades, would often

> leave no offspring to inherit his noble nature. The bravest men, who were always willing to come to the front in war, and who freely risked their lives for others, would on an average perish in larger numbers than other men. Therefore it seems scarcely possible… that the number of men gifted with such virtues, or that the standard of their excellence, could be increased through natural selection. (p. 30)

Thus, Darwin frames the problem and then provides a multilevel selectionist answer (Darwin, 1871, vol. 1, p. 166, as quoted in Shanahan, 2004):

> It must not be forgotten that although a high standard of morality gives but a slight or no advantage to each individual man and his children over the other men of the same tribe, yet… an advancement in the standard of morality and an increase in the number of well-endowed men will certainly give an immense advantage to one tribe over another. There can be no doubt that a tribe including many members who, from possessing in a high degree the spirit of patriotism, fidelity, obedience, courage, and sympathy, were always ready to give aid to each other and to sacrifice themselves for the common good, would be victorious over most other tribes; and this would be natural selection. (p. 31)

Without using the term *multilevel selection*, the multilevel selectionist reasoning could not be clearer. In fact, some group selectionist reasoning is irrefutably manifest, leaving detractors like Ruse to characterize it as a temporary equivocation with respect to humankind alone. Shanahan believes, as a straightforward interpretation dictates, that the passage above constitutes unequivocal advocacy for group selectionist thought. Furthermore, Shanahan's reading of Darwinian writings suggests the frequency of group selectionist thought, as well as its application, beyond the peculiarities of human morality, to encompass all manner of evolutionary oddities, in "a way that leaves little doubt that he [Darwin] believed that selection can and does operate at the level of communities" (Shanahan, 2004, p. 32). In further support of this interpretation of Darwinian thought, Shanahan provides yet another quotation taken from Darwin (1871, vol. 1, p. 155):

> With strictly social animals, natural selection sometimes acts indirectly on the individual, through the preservation of variations which are beneficial only to the community. A community including a large number of well-endowed individuals increases in number and is victorious over other and less well-endowed communities; although each separate member may gain no advantage over the other members of the same community. (p. 32)

Reflecting on these and other writings, Shanahan (2004) then provides the following summary statement that we will reproduce in full as it cogently communicates our own textual analysis of Darwin's writings (1871, vol. 1, p. 155):

> The best guide to understanding what Darwin actually thought are his actual words, taken at face value if possible, and only reinterpreted if absolutely necessary. When Darwin writes that 'certain mental faculties… have been chiefly, or even exclusively, gained for the benefit of the community,' we should take this as a genuine expression of his thoughts on the matter. When we do this, it becomes evident that although Darwin preferred explanations in terms of selection operating on individual organisms, he was perfectly willing to entertain explanations in terms of selection at the level of groups when the situation warranted it. (p. 32)

As Shanahan reads additional passages within Darwin's fourth edition of *On the Origin of Species*, and his later *The Variation of Animals and Plants under Domestication*, further evidence of prototypical multilevel selectionist thinking is found. Then, reflecting on these chapter and verse citations, as well as other excerpts from Darwin's original writings, Shanahan reflects on their common thread of sociality, writing, "Darwin considered sociality to be a distinct factor in evolution, one that in some cases perhaps licenses (or requires) the postulation of selection operating at a level more inclusive than that of the individual organism" (Shanahan, 2004, p. 33). Further still, Shanahan cites an April 6, 1868 letter from Darwin to Wallace wherein he maintains that selection must redound to the good of the *individual*, "including in this term a social community." So, it could be an individual or a community that selection operates upon so far as Darwin was concerned. This insight is the essence of modern multilevel selection theory.

In exclusively detailing our own and Shanahan's readings of Darwin's writings, we would be remiss not to mention that there exists a degree of dissent, sometimes attaining to vituperative excess, surrounding Darwin's writings, which are subject to intense textual analysis, akin to biblical exegesis.[2] As Timothy Shanahan discusses, Michael Ruse, whom we use to exemplify this species of dissent, maintains that Darwin's writings do not provide evidence of group selectionist sympathies, describing Darwin as a firm and even aggressive proponent of individual selection. As per Ruse (1984, p. 14), Darwin "looked long and hard at group selection and rejected it." After reviewing Darwin's writings on plants and animals, especially as they relate to sterility, Ruse (1984) turns to study Darwin's writings on the possibility of group selection within human beings. Human morality, Ruse is at pains to concede, was believed by Darwin to confer a selective advantage at both the individual and group levels, with the former being signally stronger than the latter, as the following passage reveals (Darwin as quoted by Ruse, 1984):

> It must not be forgotten that although a high standard of morality gives but a slight or no advantage to each individual man and his children over the other men of the same tribe, yet that an advancement in the standard of morality and an increase in the number of well-endowed men will certainly give an immense advantage to one tribe over another. (p. 21)

Bowing to this concession, Ruse attempted to summarize Darwin's thought on group selection, finding him resolutely opposed to group selection with respect to all nonhuman species, and then *weakening,* allowed for group selection only among humans, and then only with respect to human morality. "Apparently," Ruse (1984, p. 22) writes, "at the final point of evolution, Darwin became a group selectionist." Having come shortly after discussing Alfred Russel Wallace's reservations about human cognition, secondary to absorbing spiritualistic sentiments, this review of Darwin's invocation of group selection is suggestive. Ruse seems to cast group selection as a superstition, akin to Wallace's spiritualism, invoked from weakness, stemming ultimately from a failure to imagine how moral sentiments could provide individual selective advantage.

Shanahan himself refutes these as simplistic interpretations of Darwinian thought that erroneously insulates Darwin's achievements from later twentieth-century innovations in evolutionary thought. Contrary to Ruse and Gayon, Borrello, like Shanahan, finds precedent for group selection in both Darwin's *On the Origin of Species* and his *Descent of Man*. In support of his position, Borrello convincingly cites Darwin's comments on social insect caste formation. He finds prior authors, most specifically Ruse, to have constructed an artificially rarified and modernized vision of group selection that was fully cognizant of genetics. According to Borrello (2010), Ruse erroneously denied Darwin's support to group selection on that basis. We fully agree with Borrello on this point. Nevertheless, we herein limit ourselves to providing authors and sources that can be pursued in search of further information for and against our interpretations of Darwin's writings, rather than attempting to fully recreate the able work of Shanahan and Borrello. Having done our due diligence in alerting the reader to countervailing opinion without losing our way and subverting our objective, we move on.

3 From Heterodoxy to Heretical Anathema

Darwin's thoughts on social insect colonies and altruism among hominids were taken up in the twentieth century by Fisher, Wright, and Haldane, each of whom commented on group selection. To be clear, none of these three really advanced or defended group selection: Fisher thought it unimportant as a force in evolution because individuals cycle through faster than groups; Wright focused on demic migration and the resultant importance of genetic material into the group, which ultimately had little bearing on group selection as it is traditionally perceived; and Haldane credited the possibility of group selection, but believed it to come about only under a certain set of highly restricted conditions. Nevertheless, their work is relevant to the study of multilevel selection, both with respect to its intellectual history and its conceptual understanding. Their relevance is not incidentally connected to their tripartite founding of population genetics, which offers a valuable perspective from which to view the evolution of groups.

The later "gene's-eye" thinking we discuss further below is traceable in embryonic form to the 1930s, wherein Sir Ronald Fisher was attempting to reconcile Darwinian evolution and Mendelian genetics in what became known as the *modern synthesis* (Okasha, 2006, pp. 143–144). "Instead of thinking in phenotype space, as Darwin had done," Okasha asserts, "Fisher operated at the level of the underlying genes." Okasha (2006) continues, explaining that Fisher

> thought of natural selection as operating on a large population of genes, or gene pool, altering the pool's allelic composition over time…Evolutionary change, on this view, is simply gene frequency change, and natural selection is a force that leads fitter genes to be substituted for their alleles.[3] (pp. 143–144)

The ideas of Sewall Wright were marginalized by others as a result of his allowance for competition between groups (Bahar, 2017). As Bahar explains, Sewall Wright had also advocated for selective neutrality and drift occurring for certain traits. Such positions were shunned by the majority of the evolutionary biology community, which was then stressing the adaptive value of any and every trait, including many now thought to be arbitrary and random, such as the swirl pattern in snail shells. Yet, it was not just his allowance for drift, but what an emphasis on drift allowed that may well have informed the resistance to his theory. Drift might allow divergence and the prying of a particular population or subsets of a population off an adaptive peak into a valley below. From there, Wright envisioned it as possible for a population to mount a different adaptive peak. All this is a consequence of drift and not particularly reprehensible to Wright's peers. However, once these populations were situated on different adaptive peaks, there came the possibility for intergroup competition, which was formulated in such a way as to smack of group selection. Certainly, there is ambiguity and room for interpretation as to whether or not Wright had group selection in mind. For example, Stephen Jay Gould's recollection of a conversation with the superannuated Sewall Wright represents Wright as characterizing the exclusive focus on individual selection as a major error of the modern synthesis (Bahar, 2017). Like Fisher and Wright, Haldane's contributions to group selectionist thinking were inchoate and oblique, relating to his thoughts on cooperation and altruism as explored through game theory. Haldane's *public goods game* demonstrated the theoretical possibility of a stable equilibrium among altruists existing in a *monomorphic population* (Bergstrom, 2002).

The history of much of the later controversy surrounding the phenomenon of group selection in some ways came to center on the work of a single man: *V. C. Wynne-Edwards*. Indeed, some say that the doctrine of group selection was originally articulated by V. C. Wynne-Edwards[4] in the 1960s, a time wherein population increase trained intellectual attention on Paul Ehrlich's *The Population Bomb*, John Calhoun's *Behavioral Sink*, and related Neo-Malthusian eschatological narratives. Wynne-Edwards himself understood group selection as a population regulation mechanism, which would impose reproductive limits ahead of Malthusian constraints. In this sense, it is antithetical to the competitive foundations upon which Darwinian evolution rests. Wynne-Edwards posited altruistic restraint at the highest levels of population aggregation without clear mechanisms by which individual interests could be suppressed by group interests. Richard Dawkins, David Lack, George C. Williams, and F. W. Braestrup were among a chorus of critics that savaged *Animal Dispersion in Relation to Social Behaviour*, the book wherein Wynne-Edwards' views on group selection first appeared prominently in print.

Bahar sympathizes with Wynne-Edwards, implying that his views were distorted by his critics. Lack, for instance, says that Wynne-Edwards argued for control of group-level reproductive rate via group selection, a charge which Bahar believes was not sufficiently substantiated by Lack. In that passage written in 1966, Lack rather candidly admits that the rival explanatory hypothesis that he used to smother group selection was itself only obliquely supported by observational and experimental data. Bahar goes on to note that, with this want of evidence, Lack bases his rejection of Wynne-Edwards largely on logical grounds. Stephen Jay Gould, and later Bahar herself, trace this rejection to Lack's distortion of Occam's razor. Lack took Occam's razor to mean that the *smallest explanation or most pared-down explanation is always best*, rather than that *one should not unnecessarily complicate an explanatory hypothesis*. G. C. Williams doubled down on Lack's critique of

Wynne-Edwards, going further and more dogmatically down the same path. Later, in 1966, Williams would argue that group selection was both "inherently weak and theoretically superfluous" (Okasha, 2006, p. 176). Then John Maynard Smith took a similar tack, with the combination pushing toward a consensus wherein group selection was thought theoretically possible in principle, even while it was purported to be exceedingly rare in fact. In other words, it's not that group selection couldn't happen, just that it doesn't happen very often. Subsequently, Hamilton's kin selection and Trivers' reciprocal altruism theories had the effect of strengthening this view, as both were thought to obviate the perceived need to invoke group selection. While a more judicious critique might have shaped Wynne-Edwards' vision of group selection into one reflective of reality, one should not lose sight of the larger point: that group selection, as operationalized by Wynne-Edwards, is not a viable hypothesis. In recognition of this, it has been termed *naïve group selection*, to contrast it with the modern formulation of multilevel selection from which it substantively differs and which eschews any notion of cooperative population regulation at the species level. Nevertheless, learning more about this straw man version of group selection and how it was dismembered fails to advance toward a modern operationalization of multilevel selection. Citing Borrello as a source for the reader interested in learning more about this *lacuna* of intellectual history, we now move on.

4 Kin Selection and Inclusive Fitness Theory

J. B. S. Haldane was famously "prepared to lay down his life for eight cousins or two brothers." This calculus illustrates the logic of what later came to be known as *kin selection*, wherein one's own genes are proportionally distributed among kin, allowing an alternate route to traditional reproductive success. When obituaries say that one is survived by family members, this is evolutionarily accurate from a kin selectionist perspective. The theory of kin selection, later introduced by W. D. Hamilton (1964), and originally equated to *inclusive fitness*, figured high among the lists when Brandon and Burian (1984) explained contentious relations between proponents of group and individual selection. Kin selection and inclusive fitness theory both entail preferentially directing altruistic behaviors toward conspecifics that share a greater proportion of genes with the altruist. As originally introduced by Hamilton (1964), kin selection and inclusive fitness theory were one and the same and were therefore treated as virtually synonymous expressions. As a simplifying assumption, both kin selection and inclusive fitness theory were based on the presupposition that any shared genes were inherited by recent common descent. This quantitative theory can be summarized within a single inequality:

$$rB > C$$

In this expression, *r* is the coefficient of relatedness (proportion of shared genes) between any two individuals, *B* is the benefit gained by the recipient of the altruistic act, and *C* is the cost to the altruistic individual performing the act. Both the benefits (*B*) and costs (*C*) in this expression are assessed in terms of reproductive fitness.

Aside from mathematics, Hamilton (1964) based his main narrative on the case of the eusocial insects, notably the social wasp. From the very introduction of the theory of evolution by natural selection, extreme forms of altruism found among eusocial insects posed a special problem. Darwin (1859) characterized it as one that

> first appeared to me insuperable, and actually fatal to my whole theory. I allude to the neuters or sterile females in insect communities: for these neuters often differ widely in instinct and in structure from both the males and fertile females, and yet, from being sterile, they cannot propagate their kind. (p. 269)

Darwin solved this theoretical problem by proposing that selection must be operating at the level of the *family*, so that the specialized characteristics of sterile morphs (e.g., "workers," "soldiers") are indirectly propagated by the reproduction of the fertile morph ("queen") within each family ("colony"). The problem with interpreting this scenario in modern terms is that a eusocial insect colony is at once a *family* and an entire *society*. Does this process therefore represent what is now known as *kin* selection or as *group* selection?

Just over a century after Darwin's seminal discussion, Hamilton (1964) came down solidly on the side of attributing the evolution of eusociality to the dynamics of *inclusive fitness theory*. As his paramount example of this process, Hamilton alluded to the peculiar system of sexual reproduction, *haplodiploidy*, which characterizes one of the main taxonomic groups, the Hymenoptera, containing about 14,000 eusocial species, including all of the eusocial Aculeata or stinging insects (ants, bees, and wasps). This iconic illustration relates how a female wasp will, by virtue of the dynamics of haplodiploid sex determination, be more genetically related to her full sisters (by a coefficient of relatedness of 0.75) than to her own daughters (by a coefficient of relatedness of 0.50). It is thus more advantageous from the standpoint of genetic replication, all else being equal, to assist her mother in producing more full sisters than in producing daughters of her own. This dynamic presumably led to the evolution of a "sterile" worker caste, which is sterile only insofar as they cannot produce diploid daughters (due to the lack of a *spermatheca* in which to store the necessary sperm, as females can develop only from fertilized eggs) while they can in many cases still produce haploid sons (which develop from unfertilized eggs) parthenogenetically. In most cases, however, the laying of male eggs by workers is largely suppressed by either other workers or the queen herself by the eating of such eggs upon detection. Although the theory of inclusive fitness does not depend on haplodiploidy, as Hamilton's rule can be generalized to any coefficient of relatedness provided the balance of costs and benefits to the altruistic behavior is favorable, the extreme special case of Hymenopteran full sisters became a widely accepted explanation for eusociality in Hymenoptera generally. As time went by, however, more and more problems with that story became recognized. Among them is the fact that, although all Hymenoptera are haplodiploid, most Hymenoptera are not eusocial, but remain stubbornly solitary.

Another problem is that the Hymenoptera represent only one of the four orders of insects in which eusociality is found, the others being the Isoptera (including all the termites), the Homoptera (including some eusocial aphids), and the Thysanoptera (including some eusocial thrips). Of these remaining three orders of insects, only the Thysanoptera are also haplodiploid, whereas the Isoptera and Homoptera are not. For example, the entire order of Isoptera is eusocial (there are no extant solitary termites); and all 3000 or so species of termite are as fully diplodiploid as humans. Termites are a much more ancient group and are believed to have originated in the Permian, whereas the social Hymenoptera probably did not evolve until the Cretaceous. The case of termites is particularly interesting. Although they are not haplodiploid, many researchers continue to attribute the origins of termite eusociality to kin selection, or inclusive fitness theory, arguing that "If relatedness between helpers and reproductives produced averages of 50%, as is the case in monogamous families, only slight ecological benefits are required to favor helping alleles" (Howard, Johns, Breisch, & Thorne, 2013, p. 1). Thus, as noted above, the ecologically determined balance of costs and benefits may favor altruistic behavior even in the absence of unusually elevated coefficients of relatedness. In spite of this, some researchers have continued to search for functional analogues of haplodiploidy in termites. For example, translocation complexes of sex-linked chromosomes have been predicted to produce higher within-sex than between-sex coefficients of relatedness (Lacy, 1980). Other researchers have invoked substantial inbreeding and yet others have invoked within-colony and between-colony inbreeding-outbreeding cycles, expected to produce higher within-generation than between-generation coefficients of relatedness. Nevertheless, studies using multilocus DNA fingerprinting (Husseneder, Brandl, Epplen, Epplen, & Kaib, 1998, p. 1046) have failed to find substantial support for any of these hypotheses, concluding instead that "ecological factors and constraints must be considered a major selective force."

What this means is that although unusually elevated coefficients of relatedness such as 0.75 might be in some cases sufficient to favor indirectly producing full sisters over directly producing daughters, *ceteris paribus*, normal diplodiploid coefficients of relatedness of about 0.5 and even lower can be adequate to promote advanced eusociality if other ecological conditions favor it, as all else is rarely ever equal. Hamilton's famous rule implies as much mathematically, but comparisons across eusocial taxa substantiate this principle empirically. Coming full circle, how does one apply this reasoning to the social Hymenoptera? Taking the broader perspective, one can see how having the elevated coefficients of relatedness sometimes produced by haplodiploidy might nonetheless remain one of many factors influencing the evolution of eusociality by lowering the threshold of adaptiveness for altruistic behavior, in terms of fitness costs and benefits. However, this perspective suggests that such elevated levels of relatedness are not absolutely necessary for eusocial evolution even in the Hymenoptera.

This is a fortuitous theoretical conclusion in light of the evidence that the most primitively social wasps do not generally found new colonies based on intergenerational mother-daughter bonds, as the classical haplodiploidy narrative would have it, but instead form intragenerational *foundress associations* among individual fully fertile females of variable degrees of genetic relatedness (West-Eberhard, 1967). The adaptive problems faced by such foundress associations, including the exigencies of collective nest construction, nest defense, and offspring provisioning, tip the balance of ecological factors toward facilitating the operation of kin selection dynamics among even "quite distant relatives" (West-Eberhard, 1975, p. 1). More advanced eusocial wasps (as well as many bees and ants) generally do conform to the pattern of subordinate "worker" daughters becoming "helpers at the nest" in the service of a sole despotic foundress queen mother, but that *monogynous* condition is probably derived and not ancestral (at least in the social wasps, which are the most evolutionarily ancient of the eusocial Hymenoptera).

This does not mean that higher-than-average genetic relatedness is irrelevant to the evolution of eusociality, but merely that it is one of the many factors that promote it (e.g., Bourke, 2011, 2014; Kapheim, Nonacs, Smith, Wayne, & Wcislo, 2015; Waibel, Floreano, & Keller, 2011). In addition to contemporary ecological factors influencing the balance of costs and benefits entailed in kin-selected altruism, for example, there are the often-ignored factors of *phylogeny* or evolutionary history that constitute one of Tinbergen's (1963) *Four Questions* that we must always ask in constructing a complete account of behavioral evolution. Evans and West-Eberhard (1970) explain this evolutionary progression as follows:

> A brief summary of the apparent major steps in the evolution of nesting behavior of solitary wasps may assist in an understanding of the origin of sociality. These steps are progressive in the sense that they must occur in approximately this order and that they lead from simple to more complex behavior. They may be thought of as "rungs in the social ladder" in the sense that the social wasps must once have "climbed" them (Fig. 75); but of course each rung must have its own adaptive value, since many wasps inhabit each rung successfully and the evolution of sociality was by no means preordained. (pp. 114–115)

These authors afterward detail seven major steps, some with multiple subtypes, which set the stage for social evolution within erstwhile solitary wasps. These are too involved to reproduce in full within the present context, so we advance to the major milestones believed to have been achieved by the ancestors of eusocial Hymenoptera:

> We considered four developments to be preadaptations to sociality: (a) the possession of a nest to which the female returns repeatedly; (b) placement of numerous cells at one site; (c) provisioning more than one cell at a time; and (d) increased longevity of females. All of these would contribute to the likelihood of nesting in groups and, in particular, in groups of close relatives. (p. 203)

Thus, although all Hymenoptera are haplodiploid, only some clades within that order have undergone this particular evolutionary progression leading to the acquisition of behavioral preadaptations for sociality. That is why most species of Hymenoptera have remained solitary in spite of having the additional preadaptation of haplodiploidy, which is neither necessary nor sufficient by itself for eusocial evolution. When examining the evidence across taxa, there is no statistically significant relation between haplodiploidy and eusociality in general (Wilson & Hölldobler, 2005).

Furthermore, this line of reasoning is not limited to the eusocial Hymenoptera, as termites are known to possess other preadaptations which bias them toward sociality. These include the need to reinfect themselves with symbiotic gut protozoa (which are necessary to digest wood) from the anal secretions of conspecifics every time they molt (as the internal gut lining is shed each time with the discarded exoskeleton and the essential protozoa are thus periodically lost). This biases termite evolution toward gregariousness, at the very least, and that gregariousness no doubt preadapts them to evolve the collective defense of nest sites, offspring, and nutritional resources. Unfortunately, there is no evolutionary progression of extant solitary termites, each at different stages, with which to fully reconstruct the elaborate phylogeny that we have for social wasps. All existing termites are fully eusocial. The only surviving models for their possible primitive sociality are the Cryptocerid wood-feeding cockroaches, which have many attributes that appear to be homologous to those of the lower Isoptera (Klass, Nalepa, & Lo, 2008).

The case of the eusocial aphids is even more mysterious. These aphids generally live in *galls*, which are tumorous growths produced on their host plants. Aphids live in clonal colonies of parthenogenetically produced females, and some species have evolved a self-sacrificing sterile "soldier" caste to help defend the galls in which they live from predators and parasites. Thus, the coefficient of relatedness among individual aphids within a monoclonal colony is 1.0, which is even higher than that for Hymenopteran full sisters, which might be interpreted as predisposing them to greater kin-selected altruism. Nevertheless, there are many species of aphid that live in such genetically homogeneous monoclonal colonies that are *not* eusocial and it is not known what other factors might preadapt some species and not others to eusociality (Abbot, 2009).

Finally, some species of gall-inducing thrips have also crossed the threshold to eusociality, having evolved a "soldier" caste with the same defensive functions as in gall-dwelling aphids. Unlike the clonal colonies of aphids, however, these thrips are haplodiploid like the Hymenoptera and thus generate the same pattern of elevated coefficients of relatedness as the latter, while falling short of full clonality. The theoretical problem, as with the Hymenoptera, is that *all* Thysanoptera are haplodiploid, whereas only *some* species of gall-inducing thrips are eusocial (Gadagkar, 1993). The lineages of thrips in which eusociality is believed to have originated, however, appear to have high degrees of within-group relatedness attributable to high inbreeding coefficients (Chapman, Crespi, Kranz, & Schwarz, 2000), as was hypothesized, but not empirically supported for termites. That, at least, is consistent with the theory of kin-selected altruism.

In light of these considerations, we return to the question of whether colony-level selection in eusocial insects is best interpreted as *kin* selected or as *group* selected. The answer can be *either* or *both*, as this has been shown to be an unnecessary dichotomy. Early proponents of kin selection thus understood group selection as a rival explanation, leaving one or the other true or false as if the two explanatory views were rivals in a zero-sum game (Bahar, 2017). To some, kin selection was somehow separate from group selection, with many going so far as to cast group selection as untrue or unnecessary. On the other hand, group selectionists simply saw kin selection as a specific instantiation of group selection that was aided by genetic relatedness.

As Gintis describes, even as proponents of group selection gave credence to kin selection, such recognition was not reciprocated. D. S. Wilson speaks to this point:

> An important development in the history of thinking on kin and group selection is called equivalence—the possibility that the two theories do not invoke different causal processes and are inter-translatable, thereby deserving to co-exist rather than one replacing the other.[5]

Interestingly, Hamilton himself modified his earlier position, even as he is lastingly remembered for promulgating kin selection at the expense of group selection. Sober and Wilson (1998) recall Hamilton's autobiographical essays wherein he converses with George Price, perhaps for the first time. After participating in the Manhattan Project and dabbling in foreign policy, Price abandoned a position with IBM to study evolutionary biology, during which time he developed the *Price equation*. Before turning ascetic and dying as a squatter after having given over all his possessions to the homeless, Price worked out relevant theorems and mathematics on first principles, the result of which was a reinterpretation of kin selection that brought multilevel selection theories back into the fold. Price did this by focusing his mathematical argument on gene frequencies within and between groups, rather than the sharing of genes among specific pairs of individuals (as discussed further in Chap. 9).

Price (1970, 1972) integrated inclusive fitness theory and multilevel selection theory into a common mathematical framework, reducing theoretical differences to semantic framing of common evolutionary processes (e.g., Frank, 1995, 1997; Gardner, 2008; Gardner, West, & Wild, 2011; Hamilton, 1975; Kramer & Meunier, 2016; Korb & Heinze, 2004; Leigh, 2010; Lion, Jansen, & Day, 2011; Marshall, 2011). The *Price equation* accomplishes this by partitioning selection into within-group and between-group components of covariance, making it the first true quantitative model of multilevel selection. Wilson and Wilson (2007) have written:

> Hamilton's key insight about the importance of genetic relatedness remained valid, but his previous interpretation of inclusive fitness theory *as an alternative to group selection* was wrong, as he freely acknowledged. (p. 335)

The culmination of Hamilton's brief collaboration with Price is the following paper: *Innate Social Aptitudes of Man: An Approach from Evolutionary Genetics* (1975), wherein "Hamilton re-described inclusive fitness theory as representing a multilevel selection process" (Sober & Wilson, 1998, p. 75). Sober and Wilson ultimately see this paper as nearly aligning Hamilton's view with that of Darwin. Sober and Wilson (1998) summarize this shift in Hamilton's understanding:

> Hamilton now saw the multi-group nature of social interactions, regardless of when and where breeding occurs. Social interactions among genetic relatives correspond to the nonrandom formation of groups. The significance of relatedness for the evolution of altruism is that it increases genetic variation among groups, thereby increasing the importance of group selection. Furthermore, any process that increases variation among groups will accomplish the same job. As Hamilton observed, "it obviously makes no difference if altruists settle with altruists because they are related … or because they recognize fellow altruists as such. Or settle together because of some pleiotropic effect of the gene on habitat preference." Genetic relatedness loses its status as the exclusive factor responsible for the evolution of altruism and becomes one of many factors that can promote group selection. Hamilton (1975) left no doubt that the difference between inclusive

> fitness theory and group selection theory is a matter of perspective, not process. Earlier (1963) he had said that group selection should be treated with reserve so long as it remains unsupported by theoretical models. Now those models were available and the major so-called competing theory had vanished. The only process to explain the evolution of altruism was the one that Darwin identified long ago. (pp. 76–77)

Sober and Wilson then cite Frank's (1995) biography of Price, which remarks that it is strange how a man of Price's age could enter a field and produce such a unique contribution. Frank then remarks how it is equally strange for Hamilton to happily accept these findings even as they rendered inclusive fitness just another way of looking at multilevel selectionist explanations.[6] This was not revolutionary science in the Khun (1962) sense. One might have otherwise expected Hamilton to metaphorically go down with the ship, as Sober and Wilson rightly observe; to defend the original interpretation of inclusive fitness as it was originally formulated. Far from it: Hamilton seemed to delight in the *post-Price synthesis.* Nevertheless, as per Sober and Wilson, Hamilton's papers acknowledging this new understanding were accepted and cited among a much smaller group of theorists, while the larger contingent of biologists continued to cast kin selection and group selection as competing evolutionary explanations. In Hamilton's 1996 reflections, he notes that his 1975 paper is not often cited and, when cited at all, is apt to be erroneously cited, selectively read, and inaccurately interpreted. For instance, Dawkins invokes Hamilton's paper to describe evolutionary biology's hard-won eradication of group selection in spite of its widespread allure. Sober and Wilson puzzle not only at how the H. M. S. Kin Selection was understood to be afloat as before, but, more interestingly, how, after this 1975 paper and his contact and collaboration with George Price, partisans could think Hamilton still at the helm.

Little over a decade after the introduction of his theory, Hamilton (1975) drew a finer-grained distinction between kin selection and inclusive fitness theory based on whether or not genes were shared by recent common descent. As Hamilton then explained, recent common descent for shared genes was nothing more than a simplifying assumption that he had introduced to make the mathematical model more tractable, not a necessary condition for the evolution of altruism. Kin selection theory was thus distinguished by retaining that assumption, consistent with the vernacular meaning of the word kin; inclusive fitness theory was thereafter not required to rest on that assumption in that shared genes would operate in the same way regardless of their provenance.

In spite of this, Dawkins (1976) and certain similar theorists continued to insist that the assumption of recent common descent for shared genes was a necessary one for the evolution of kin-selected altruism. Other theorists, notably Rushton (1989, 1998), came to different conclusions. As described in his *genetic similarity theory*, Rushton posited that individuals enter into marriage, friendship, social relationships, and like mutualistic endeavors based on their ability to perceive shared genetic variance among genealogical kin and nonkin alike. While both kin selection and genetic similarity theory suggest shared genetic variance, only kin selection (*sensu stricto*) requires shared genetic variance to derive from common descent, whereas genetic similarity theory does not. Genetic similarity theory was a natural extension of kin selection theory and did not suffer from its previously restrictive assumptions, thereby establishing the requisite theoretical link to multilevel selection. Thus, it is unclear what distinguishes Hamilton's (1975) updated definition of inclusive fitness theory from Rushton's (1989, 1998) genetic similarity theory. One can go so far as to cast genetic similarity theory as redundant given the aforementioned post-Price synthesis. Functionally speaking, it was nevertheless not so because few evolutionary thinkers assimilated the post-Price synthesis; instead, at the instigation of Dawkins (1976) and similarly vocal critics, many continued to stipulate the necessity for shared common descent, and this simplifying assumption has subsequently become engrained in the general understanding of inclusive fitness theory.

5 Reciprocal Altruism

Without requiring any particular degree of genetic relatedness, *reciprocal altruism* (Trivers, 1971) theoretically explains altruistic behavior as selected in situations where the benefactor is eventually repaid by the beneficiary of the altruistic act. Reciprocity is distinguished from *mutualism* in that there is a delay between the initial act and its repayment. As a result of this delay, there is consequently some uncertainty, as there is always some probability that the beneficiary may *defect* or fail to return the altruistic favor. The main obstacle to the evolution of reciprocity is thus the possibility that the recipient might defect, whether intentionally or unintentionally (as by the premature death of the recipient).

Reciprocity is therefore said to depend on certain cognitive abilities, presumably evolved for this purpose. For example, there must be mutual recognition of individuals, so that the beneficiary can identify who to repay and the benefactor can track from whom to expect repayment. There must also be sufficient memory capacity on both sides to recall what services were exchanged and the magnitude of their fitness value. In establishing a relationship of repeated and mutual reciprocity, these abilities allow a potential benefactor to discriminate among potential partners to select only those who will reliably reciprocate. It is important for the theory that defectors, also known as *cheaters*, be discriminated against in future altruistic interactions, by means of tactics ranging from mere exclusion to actual punishment. Punishment of cheaters, however, is a problematical behavior as it may incur costs to the punishing altruist, while conferring unmerited benefits to surrounding non-punishing altruists, and might thus fail to evolve within purely dyadic interactions, barring other supporting conditions such as indirect reciprocation via social reputational effects (e.g., Figueredo, Tal, McNeill, & Guillén, 2004).

In spite of the theory's popularity, reciprocal altruism has been difficult to establish unequivocally outside of the human species due to the stringent conditions associated with the process. For example, alarm calling has been offered as an illustration of this principle but is confounded with the possibility of *predator signaling* of antipredator vigilance (Bergstrom & Lachmann, 2001; Zuberbühler, Jenny, & Bshary, 1999), as well as with kin-selected altruism toward relatives residing nearby (e.g., Cheney & Seyfarth, 1985). In other words, what appears to some as motivated by reciprocation might be either an individually adaptative communication to a predator that the element of surprise has been lost or, alternatively, a behavior warning nearby kin and thereby indirectly preserving the genes of the individuals producing the signal. Altruistic blood regurgitation in vampire bats has been offered as another example (Wilkinson, 1984), but it has yet to be established that non-reciprocators are either discriminated against or actually punished in repeated interactions. In other oft-cited cases, such as those of cleaner fish (Grutter, 2002), reciprocity has been confused with mutualistic symbioses, in which there is no delay between the benefits accruing to the beneficiary and those simultaneously accruing to the benefactor, who immediately obtains a meal by removing ectoparasites from the beneficiary.

Even in the case of humans, for which evolutionary psychologists have presented an impressive experimental corpus of research with respect to the innateness and modularity of *cheater detection* (e.g., Cosmides, 1989; Cosmides, Barrett, & Tooby, 2010; Cosmides & Tooby, 2013), the phenomenon of reciprocal altruism has possibly been confounded with that of innate *deontic reasoning* (e.g., Cummins, 1996). The latter refers to the rule-following behavior to be expected of a species that has evolved within the context of rule-governed dominance hierarchies. Many of the most notable examples of cheater detection reported actually involve identifying infractions of implicit or explicit social norms, in the absence of any exchange of obvious fitness costs or benefits. As a result of this and other difficulties, reciprocal altruism has not gained much empirical traction as a general theory for explaining altruistic behavior except within the rarefied *demimonde* of abstract mathematical modeling and computer simulations.

In response to some of these criticisms, De Waal and Brosnan (2006) developed a theoretical taxonomy with three mechanisms involved in the distribution of benefits during reciprocal exchanges between cooperators. For the authors, symmetry-based reciprocity emerges based on shared features between cooperative partners (e.g., kinship, affiliation, sex, or age). Given that this mechanism relies on individuals exhibiting similar characteristics, De Waal and Brosnan claimed symmetry-based reciprocity did not rely on the preexistence of complex cognitive phenomena. In contrast to symmetry-based interactions, attitudinal reciprocity occurs when cooperators modify their strategy based on their partner's attitude during previous encounters. Hence, the contingent nature of every interaction influences the way partners will behave in the future. Once again, De Waal and Brosnan argued that attitudinal reciprocity did not impose considerable cognitive demands on cooperators. Finally, calculated reciprocity fits the description of traditional reciprocal altruism, wherein following each interaction partners did not interact for a prolonged period. This temporal gap required cooperators to recall the outcome of past encounters and classify each companion based on these contingencies and the ability to respond and punish free riders (De Waal & Brosnan, 2006). Consequently, only calculated reciprocity relies on the presence of complex cognitive mechanisms (e.g., scorekeeping).

Traditionally, the combination of inclusive fitness theory and reciprocal altruism theory has been used to try to explain all altruistic and cooperative behavior without recourse to any theories of multilevel selection. Inclusive fitness theory was considered to apply to interactions among genetically *related* individuals, whether specifically by common genealogical descent or more generally by genetic similarity; reciprocal altruism theory was considered to apply to interactions among genetically *unrelated* individuals. All proverbial bases were thus presumably covered. However, this formulation failed to take into account the observation that many human and nonhuman animal societies are kin structured, meaning that the degree of genetic relatedness is higher within groups than between groups, but that most pairings of individuals within them do not formally meet the stringent criteria for altruism imposed by Hamilton's rule. That implies that most reciprocal exchanges within ancestral human societies were probably occurring among individuals that had many widely varying but generally nonzero degrees of shared genes. We propose that such generally elevated but not necessarily very close degrees of relatedness function to facilitate the evolution of reciprocal altruism among members of kin-structured groups by raising the expected benefits and lowering the expected costs of such interactions.

The model of reciprocal altruism has been conceptualized as the following equation, where c is the cost accrued by the helper when assisting the recipient, b is the benefit acquired by the recipient, and w is the helpers' expectation of future repayment based on the cost of aiding its partner (Patton, 2000):

$$c < wb$$

This inequality, however, conflates certain model parameters which we will have to discriminate among more clearly in order to properly integrate it with Hamilton's rule. We therefore introduce the following more comprehensive set of variable definitions:

c = reduction of donor's fitness by the altruistic act
b = increase in recipient's fitness by the altruistic act
r = coefficient of relatedness between donor and recipient
i = increase in donor's fitness by reciprocation of the altruistic act (*indemnification benefit*)
j = decrease in recipient's fitness by reciprocation of the altruistic act (*indemnification cost*)
w = probability of reciprocation by recipient of the altruistic act

We then use these symbols to restate Hamilton's rule in conventional terms:

$$c < rb$$

We then reformulate the equations for reciprocal altruism using this expanded set of terms, defining the expected fitness costs to the donor and the expected fitness benefits to the recipient over a large number of interactions:

$$E(c) = c - wi$$
$$E(b) = b - wj$$

This means that the expected fitness cost to the donor is the initial and immediate fitness cost of the act (c) minus the eventual fitness benefit to the donor of indemnification (i), by reciprocation on the part of the recipient, times the probability of such indemnification actually occurring. Similarly, the expected fitness benefit to the recipient is the initial and immediate fitness benefit of the act (b) minus the eventual fitness cost to the recipient of the same indemnification (j), by reciprocation to the original donor, times the probability of such indemnification actually occurring.

We further now stipulate that for both the initial altruistic act and the eventual reciprocation to be adaptive among genetically unrelated individuals, the following two conditions must be met: (1) the expected fitness cost to the donor of the initial altruistic act, after the expected reciprocation, must be less than (or at most equal to) zero and (2) the expected fitness benefit to the recipient of the initial altruistic act, after the expected reciprocation, must be greater than (or at least equal to) zero. This can be expressed as follows:

$$E(c) < 0$$

$$E(b) > 0$$

And it therefore follows by substitution that

$$c < wi$$

$$b > wj$$

Note that the traditional inequality for reciprocal altruism ($c < wb$) assumes that b is equal to i, meaning that the fitness benefit of the initial altruistic act by the donor to the recipient (b) is equal to the fitness benefit of the eventual reciprocation by the recipient to the donor (i), which is not assumed by the present more generalized model. However, as stipulated above, these inequalities only apply to reciprocal altruism interactions among individuals that are entirely genetically unrelated.

To synthesize this reasoning with that of Hamilton's rule, we now presume that the expected fitness cost to the donor over a large number of such altruistic interactions must be less than the expected fitness benefit to the recipient over the same large number of interactions, multiplied by their coefficient of relatedness:

$$[E(c)] < r[E(b)]$$

And it therefore follows by substitution that:

$$[c - wi] < r[b - wj]$$
$$c - wi < rb - rwj$$

And finally, this inequality implies that even a lower probability (w) of reciprocation can be compensated for by a higher coefficient of relatedness (r) and vice versa. These derivations therefore demonstrate that an average within-group coefficient of relatedness that is somewhat higher than the average between-group coefficients of relatedness, even if it does not meet the more stringent threshold for kin-selected altruism, is sufficient to facilitate the evolution of within-group reciprocal altruism by proportionally increasing the expected benefits and reducing the expected costs of altruistic interactions among individuals. This simplified hybrid model, however, is a purely stochastic one and does not assume any assortative sociality among the interacting individuals. There are more complex models that do so and show even greater expected fitness benefits that might accrue, conditional upon the exercise of competence-based discrimination among potential cooperative partners.

Kin-biased cooperation occurs due to the additive nature of benefits obtained by related individuals during a collective action (Chapais, 2006). Hence, in addition to acquiring direct fitness gains from the collective activity, cooperators indirectly improve their fitness by helping their relatives (depending on their coefficient of relatedness). This dynamic differs from instances of cooperation between unrelated partners, wherein cooperators only attain direct fitness benefits from the activity. Although this distinction has led some authors to claim that individuals are prone to cooperate with kin over nonkin, in previous publications, Chapais (2001) emphasized the difficulties of disentangling kin selection from reciprocal altruism between kin:

> It may be extremely difficult to differentiate between bilateral [kin selection], on the one hand, and reciprocal altruism between kin, on the other. In this context, it might be tempting to simply deny the operation of reciprocal altruism among kin, accounting for bilateral altruism among kin in

> terms of [kin selection], and among nonkin in terms of reciprocal altruism. But this would amount to a sort of double-standard reasoning. If a single mechanism, in this case, reciprocal altruism, can account for a given behavior both between kin and between nonkin, one cannot eliminate the possibility of its operating among kin. The latter possibility is rendered even more likely in view of the idea that it takes some degree of relatedness between individuals to set reciprocal altruism in motion. (p. 214)

Chapais (2006) also argued that traditional distinctions between kin selection and reciprocal altruism often ignore the role of individual competences moderating the likelihood organisms choose cooperative partners. Hence, if the collaborative activity is independent of the actor's qualifications, then individuals should prefer to cooperate with kin. According to the author, in nonhuman primates, activities, such as collective thermoregulation, receiving grooming, and gaining maternal experience, exemplify instances of cooperation relatively independent of the partner's qualification (i.e., low competence). However, if the action relies on the partner's abilities, organisms should carefully choose their cooperative partner above and beyond the influence of preexisting kin biases. In nonhuman primates, some examples of competence-dependent cooperation include gaining access to resources, rising in rank, catching prey, and practicing social and motor skills (Chapais, 2006).

Chapais (2006) estimated that individuals should opt to cooperate with an unrelated partner when:

$$\mathrm{qB} > \mathrm{B} + \mathrm{Br}$$

where q indicates the ratio between the competence of kin relative to nonkin, B is the direct fitness benefit of cooperating with kin, and r is the coefficient of relatedness. Following the latter equation, the author calculated that individuals would cooperate with nonkin if q remains higher than 1.5. Even though competence-dependent cooperation often occurs between nonkin, under certain conditions relatives can also be the most qualified individuals in the group. For instance, Chapais mentions members of high-ranking lineages (e.g., matrilines in cercopithecine primates) preferring to cooperate with their high-ranking relatives. Primate field observations support Chapais' predictions. Cheney and Seyfarth (2008) discovered that low-ranking Chacma baboons face the dilemma of either affiliating with their low-status relatives or instead attempting to establish coalitions and alliances with high-ranking females. Consequently, low-ranking matrilines are less cohesive relative to dominant lineages.

Thus far, this chapter has broached the interrelations between genes and relatedness on the one hand and cooperation and group formation on the other. We began with an extreme and ended with an opposite extreme, reviewing more moderate positions in between. On one side we saw group selection dismissed in favor of genetic relatedness, whereas on the other side we saw genetic relatedness denied a causal role. As can be seen through the evolution of social insects, since the middle 1960s, primarily through the work of Hamilton, much has been made of haplodiploid genetic relatedness, which became a primary explanation of eusociality in hymenoptera. By the mid-1970s with E. O. Wilson's *sociobiology*, all degrees of relatedness were incorporated, not simply the especially high relatedness among Hymenoptera. In this latter view, any degree of relatedness might influence the calculus of cooperation. That being said, Hamilton himself would agree with this formulation, as he was using Hymenoptera as an extreme case to illustrate a point, rather than as some kind of exception in kind to disallow the operation of relatedness at lesser levels.

6 The Gene's-Eye View

Traditional Neo-Darwinian models, view the organism as the unit of selection and, thus, labor to explain phenomena such as altruism. By contrast, altruistic behavior is perfectly understandable when we see altruism as perpetuating an organism's genes, which reside also in related group members instead of residing solely within the altruistic organism. At length, a "gene's-eye view" perspective was most famously articulated by Richard Dawkins and has lastingly been associated with his 1976 book, *The Selfish Gene*. Of course, there were intermediate stages within the development of this concept. In moving toward selfish gene theory, both G. C. Williams (1966) and thereafter Dawkins (1976) were influenced by their respective studies of genetic distorters and transposons, both of which are illustrative of genes causing behaviors detrimental to the organisms in which they are housed. The gene's-eye view perspective gains credence when considering the processes of *interference*, *over-replication*, and *gonotaxis*, all ways in which selfish genetic elements gain untoward representation as they pass from generation to generation (Burt & Trivers, 2006). Burt and Trivers (2006) also review selfish genetic elements, such as maternal effect killers, genomic imprinting, and *B* chromosomes, the signature of which began to be implicitly documented in the late nineteenth century. Relevant to the present volume on multilevel selection, Burt and Trivers, after characterizing the operation of selfish genetic elements across hundreds of pages, provide the following reflection: "The disunity of the organism is manifest once more" (Burt & Trivers, 2006, pp. 420–421). The weight of evidence and example leads to the insight that what appear as unitary wholes are no more than tightly associated federations of cells, cohering as a complex whole with the aid of occasional coercion. In this way, the organism may just be the most tightly associated of all such groupings.

Among the most incongruous ironies of science, vehement objections to multilevel selection theories originated from proponents of kin selection. The irony extends from the recognition that families and extended kin networks are themselves groups. Similarly ironic were the vocal objections to group selection that came from Richard Dawkins and others advocating the so-called gene's-eye view of selection, which frames genes as opposed to organisms as the target of evolutionary selection. This is ironic because group selection is not comprehensible except from such a population genetics lens. Borrello (2010, p. 140) puts it succinctly in saying that some thought "*The Selfish Gene* served as the nail in group selection's coffin." To that end, Sober and Wilson (1998, p. 50) quote Dawkins who writes what purports to be a sort of *epitaph for group selection*. In it, Dawkins compares group selection to a perpetual motion machine, in having the allure of the impossible. More precisely, and unlike the perpetual motion machine, group selection was not deemed technically impossible but only exceedingly improbable. So group selection theory continues to be perpetuated, in Dawkins' view, from some romantic notion of accomplishing what some say cannot be accomplished.

Instead, we believe that rather the opposite is taking place. A gene's-eye view of selection is complementary to multilevel selection theory. Indeed, it is exactly what is necessary for multilevel selection to be possible because it makes conceivable evolving populations of aggregate gene frequencies competing against other such populations. To understand this reasoning, one first has to more closely consider the relationship between genes, organisms, and groups; it is from this distinction that Dawkins extrapolates, erroneously in our view, to the eventual rejection of multilevel selection. Dawkins (1984) distinguishes between *replicators* and *vehicles*, better known, respectively, as *genes* and *organisms*. Allowing Dawkins to speak for himself, we highlight the following passage:

> At first sight, it appears that 'the individual' is intermediate in some nested hierarchy between the group and the gene. This paper shows, however, that the argument over 'group selection' versus 'individual selection' is a different kind of argument from that between 'individual selection' and 'gene selection.' The latter is really an argument about what we ought to mean by a unit of selection, a 'replicator' or a 'vehicle'. (p. 176)

As can be seen, groups of organisms are classed alongside organisms themselves as vehicles. In a later publication, Dawkins (1984) most pointedly addresses groups as they are criticized within his vision of genetic evolution:

> Evolution results from the differential survival of replicators. Genes are replicators; organisms and groups of organisms are not replicators, they are vehicles in which replicators travel about. Vehicle selection is the process by which some vehicles are more successful than other vehicles in ensuring the survival of their replicators. The controversy about group selection versus individual selection is a controversy about whether, when we talk about a unit of selection, we ought to mean a vehicle at all, or a replicator. In any case, as I shall later argue, there may be little usefulness in talking about discrete vehicles at all. (p. 162)

Here there is a sharper distinction between replicators and vehicles, delineating arguments over levels of selection so that one avoids classing genes alongside organisms and groups. In other words, one should *not* place the organism in the middle and then look up to the group and down to the gene as though these were linear differences in aggregation. As Dawkins informs us, this is not proper and fitting because one is attempting to forge a hierarchy of different things: (1) the replicators and (2) the vehicles that carry those replicators.

We have come this far without taking issue with the veracity of any of these claims. Having cited the special nature of genes as replicators, Dawkins goes on to depict individuals[7] and groups alike as excessively large and too temporally unstable to constitute a *unit of natural selection*: "Genetically speaking, individuals and groups are like clouds in the sky or dust storms in the desert. They are temporary aggregations or federations" (Dawkins, 1976, p. 36).[8] All this, too, is true; and yet, it does not follow that these implications undermine multilevel selection theory. Such would be the implication given the excessive gene flow through the human species. However, groups were so often isolated by distance, with oceans, seas, channels, and rivers, like deserts, fens, mountains, and tundra, serving as physical barriers to gene flow. Moreover, adjacent peoples maintained differences as they warred and defended territories. Propinquity was certainly then (Jonason, Nolland, & Tyler, 2017), if not also now, a decisive determinant in mate choice (Haandrikman, 2019; Marches & Turbeville, 1953). So one can ask: Are groups mixing and recombining with *relative* within group fidelity? To the extent that one can answer this question in the affirmative, Dawkins' premise could be correct without being at odds with multilevel selection. Even as individuals die, groups endure. Individual members can be replaced with later generations comprising similar genetic profiles cycling through generations of aggregation and disaggregation as members marry and mate, are born and die. As Dawkins (1976) proposed, this phenomenon would be akin to clouds or storms forming and breaking apart, only to form once again into new clouds comprising the same water vapor.

7 Conclusions

Before the publication of *On the Origin of Species*, *gemmules* populated the thoughts of Charles Darwin as he labored to articulate his theory of natural selection. A sort of inchoate unit of inheritance, the Lamarckian-inspired gemmule, was replaced by the Mendelian-inspired gene with the advent of the *modern synthesis*, the aforementioned milestone in scientific integration wherein Darwinian theory was reconciled with modern genetics. Several decades later, one major implication of the modern synthesis became apparent, namely, that *genes are ultimately the things that are selected*. As Dawkins asserts, theoretical insights at length following from the modern synthesis relegated organisms to the role of vehicles for the propagation of genes. With that insight gained, it followed that evolution progressed via intergenerational changes in gene frequencies. From thence it was understood that genes reside in other *vehicles*: not just in *individual* organisms but in *groups* of organisms at varying degrees of social aggregation. Up until this point, there is no meaningful divergence in opinion between multilevel selection's adherents and its opponents. Proponents of kin selection, like proponents of a gene's-eye perspective, fabricated a controversy from failing to see kin selection theory as merely a specific, familial instantiation of multilevel selection. Putting aside its overstatement and poor empirical moorings, we should not fail to note that reciprocal altruism exists in much the same relationship to multilevel selection as the aforementioned theories, as we have demonstrated mathematically by means of hybrid models. Though framed as a superior alternative explanation of altruistic acts, reciprocal altruism is more productively viewed as a complementary mechanism facilitating group cohesion where genetic relatedness is absent or weakly present.

Having seen how, in the mid- to late twentieth century, controversy surrounding individual selection and group selection repeatedly arose from perceiving controversy where there was only complementarity, in the chapter that follows, we will be charting the evolution of theoretical ideas regarding a resurgent group selection theory based on this broader framework of thinking, which came to be known as multilevel selection (Keller, 1999).

Notes

1. Darwin did not use the same lexicon as found in modern evolutionary theory, so it is in vain to look for *multilevel* selection, *group* selection, or related terms. Nevertheless, the rudiments of the latent ideas represented by those terms are evident through careful review of Darwin's original writings.
2. The interested reader can pursue the representative writings of Michael Ruse and Jean Gayon.
3. To this Wright argued, as do modern critics of this position, that evolution acts on organisms which are clusters of genes and that genes combined in different ways in different bodies do not generally have the same fitness implications. More generally, following from this observation, the reader should be aware of debates centering on whether a gene's-eye perspective is inappropriately reductionist (Okasha, 2006, p, 172).
4. *Evolution Through Group Selection* consolidates decades of Wynne-Edwards' thinking on group selection, arguing against narrow visions of individual, adaptationist models of evolution.
5. https://evolution-institute.org/blog/the-tide-of-opinion-on-group-selection-has-turned/
6. See Okasha (2015) for a more recent examination regarding the relation between kin selection and multilevel selection.
7. It is not the province of this chapter or this book, but it should be noted that the selfish gene was controversial with respect to individual selection, as illustrated by the following quote from Okasha: "Despite the prominence Dawkins attached to this argument, arguably it confuses the unit of inheritance with the unit of selection" (Okasha, 2006, p. 145). Additionally, consider that Dawkins co-opted objections to group selection and applied them also to individual selection, as seen in the following quotation:

   > Dawkins' work on evolutionary theory has elicited comment primarily because of his heavy emphasis on selection at the level of particular genes. He has taken the arguments which organism selectionists have used against group selection and turned them on the organism selectionists themselves. (Hull, 1984, p. 147)

8. Also of note is the fact that this argument is confined to sexually reproducing organisms wherein genes recombine such that offspring are geneti-

cally dissimilar by half from their parents. The same does not follow for asexually reproducing organisms as the following passage by Hull details:

> In asexual reproduction, the entire genome might be the unit of selection. In cases in which no crossover occurs between homologous chromosomes, entire chromosomes might function as units of selection. But sexually reproducing organisms and anything that might be considered a group can never function as units of selection. In the vast majority of cases, genes are selected and everything else goes along for the ride. (Hull, 1984, p. 142)

References

Abbot, P. (2009). On the evolution of dispersal and altruism in aphids. *Evolution: International Journal of Organic Evolution, 63*(10), 2687–2696.

Bahar, S. (2017). *The essential tension: Competition, cooperation and multilevel selection in evolution*. New York: Springer.

Bergstrom, C. T., & Lachmann, M. (2001). Alarm calls as costly signals of antipredator vigilance: The watchful babbler game. *Animal Behaviour, 61*(3), 535–543.

Bergstrom, T. C. (2002). Evolution of social behavior: Individual and group selection. *Journal of Economic Perspectives, 16*(2), 67–88.

Borrello, M. E. (2010). *Evolutionary restraints: The contentious history of group selection*. Chicago: Chicago University Press.

Bourke, A. F. (2011). The validity and value of inclusive fitness theory. *Proceedings of the Royal Society B: Biological Sciences, 278*(1723), 3313–3320.

Bourke, A. F. (2014). Hamilton's rule and the causes of social evolution. *Philosophical Transactions of the Royal Society B: Biological Sciences, 369*(1642), 20130362.

Brandon, R. N., & Burian, R. M. (1984). *Genes, organisms, populations: Controversies over the units of selection*. Cambridge, MA: The MIT Press.

Burt, A., & Trivers, R. (2006). *Genes in conflict: The biology of selfish genetic elements*. Cambridge, MA: The Belknap Press of Harvard University Press.

Chapais, B. (2001). Primate nepotism: What is the explanatory value of kin selection? *International Journal of Primatology, 22*(2), 203–229.

Chapais, B. (2006). Kinship, competence and cooperation in primates. In P. M. Kappeler & C. P. Van Schaik (Eds.), *Cooperation in primates and humans* (pp. 47–64). Berlin, Germany: Springer.

Chapman, T. W., Crespi, B. J., Kranz, B. D., & Schwarz, M. P. (2000). High relatedness and inbreeding at the origin of eusociality in gall-inducing thrips. *Proceedings of the National Academy of Sciences, 97*(4), 1648–1650.

Cheney, D. L., & Seyfarth, R. M. (1985). Vervet monkey alarm calls: Manipulation through shared information? *Behaviour, 94*(1/2), 150–166.

Cheney, D. L., & Seyfarth, R. M. (2008). *Baboon metaphysics: The evolution of a social mind.* Chicago: University of Chicago Press.

Cosmides, L. (1989). The logic of social exchange: Has natural selection shaped how humans reason? Studies with the Wason selection task. *Cognition, 31*, 187–276.

Cosmides, L., Barrett, H. C., & Tooby, J. (2010). Adaptive specializations, social exchange, and the evolution of human intelligence. *PNAS, 107*, 9007–9014.

Cosmides, L., & Tooby, J. (2013). Evolutionary psychology: New perspectives on cognition and motivation. *Annual Review of Psychology, 64*, 201–229.

Cummins, D. D. (1996). Evidence for the innateness of deontic reasoning. *Mind & Language, 11*, 160–190.

Darwin, C. (1859). *On the origin of species by means of natural selection or the preservation of favoured races in the struggle for life.* H. Milford: Oxford University Press.

Darwin, C. (1871). *The descent of man and selection in relation to sex.* London: John Murray.

Dawkins, R. (1976). *The selfish gene*. London: Oxford University Press.

Dawkins, R. (1984). Replicators and vehicles. In R. N. Brandon & R. M. Burian (Eds.), *Genes, organisms, populations: Controversies over the units of selection* (pp. 161–180). Cambridge, MA: The MIT Press.

De Waal, F. B., & Brosnan, S. F. (2006). Simple and complex reciprocity in primates. In P. M. Kappeler & C. P. Van Schaik (Eds.), *Cooperation in primates and humans* (pp. 85–105). Berlin, Germany: Springer.

Evans, H. E., & West-Eberhard, M. J. (1970). *The wasps.* Ann Arbor, MI: University of Michigan Press.

Figueredo, A. J., Tal, I. R., McNeill, P., & Guillén, A. (2004). Farmers, herders, and fishers: The ecology of revenge. *Evolution and Human Behavior, 25*(5), 336–353.

Frank, S. A. (1995). George Price's contributions to evolutionary genetics. *Journal of Theoretical Biology, 175*(3), 373–388.

Frank, S. A. (1997). The Price equation, Fisher's fundamental theorem, kin selection, and causal analysis. *Evolution, 51*(6), 1712–1729.

Gadagkar, R. (1993). And now… eusocial thrips! *Current Science, 64*(4), 215–216.

Gardner, A. (2008). The price equation. *Current Biology, 18*(5), R198–R202.

Gardner, A., West, S. A., & Wild, G. (2011). The genetical theory of kin selection. *Journal of Evolutionary Biology, 24*(5), 1020–1043.

Grutter, A. S. (2002). Cleaning symbioses from the parasites' perspective. *Parasitology, 124*(7), S65–S81.

Haandrikman, K. (2019). Partner choice in Sweden: How distance still matters. *Environment and Planning A: Economy and Space, 51*(2), 440–460.

Hamilton, W. D. (1964). The genetical evolution of social behaviour. II. *Journal of Theoretical Biology, 7*(1), 17–52.

Hamilton, W. D. (1975). Innate social aptitudes of man: An approach from evolutionary genetics. *Biosocial Anthropology, 133*, 115–132.

Hamilton, W. D. (1996). *Narrow roads of gene land: Evolution of social behaviour* (Vol. 1). Oxford, UK: Oxford University Press on Demand.

Howard, K. J., Johns, P. M., Breisch, N. L., & Thorne, B. L. (2013). Frequent colony fusions provide opportunities for helpers to become reproductives in the termite *Zootermopsis nevadensis*. *Behavioral Ecology and Sociobiology, 67*(10), 1575–1585.

Hull, D. (1984). Units of evolution: A metaphysical essay. In R. N. Brandon & R. M. Burian (Eds.), *Genes, organisms, populations: Controversies over the units of selection* (pp. 142–160). Cambridge, MA: The MIT Press.

Husseneder, C., Brandl, R., Epplen, C., Epplen, J. T., & Kaib, M. (1998). Variation between and within colonies in the termite: Morphology, genomic DNA, and behaviour. *Molecular Ecology, 7*(8), 983–990.

Jonason, P. K., Nolland, M., & Tyler, M. D. (2017). Incorporating geographic distance into mate preference research: Necessities and luxuries, 2.0. *Personal Relationships, 24*(3), 585–597.

Kapheim, K. M., Nonacs, P., Smith, A. R., Wayne, R. K., & Wcislo, W. T. (2015). Kinship, parental manipulation and evolutionary origins of eusociality. *Proceedings of the Royal Society B: Biological Sciences, 282*(1803), 20142886.

Keller, L. (1999). *Levels of selection in evolution*. Princeton, NJ: Princeton University Press.

Khun, T. (1962). *Paradigm shift*. Cambridge, Massechusetts: Harvard University Press.

Klass, K. D., Nalepa, C., & Lo, N. (2008). Wood-feeding cockroaches as models for termite evolution (Insecta: Dictyoptera): Cryptocercus vs. Parasphaeria boleiriana. *Molecular Phylogenetics and Evolution, 46*(3), 809–817.

Korb, J., & Heinze, J. (2004). Multilevel selection and social evolution of insect societies. *Naturwissenschaften, 91*(6), 291–304.

Kramer, J., & Meunier, J. (2016). Kin and multilevel selection in social evolution: A never-ending controversy?. *F1000Research, 5*, F1000 Faculty Rev-776. https://doi.org/10.12688/f1000research.8018.1

Lacy, R. C. (1980). The evolution of eusociality in termites: A haplodiploid analogy? *The American Naturalist, 116*(3), 449–451.

Leigh Jr., E. G. (2010). The group selection controversy. *Journal of Evolutionary Biology, 23*(1), 6–19.

Lion, S., Jansen, V. A., & Day, T. (2011). Evolution in structured populations: Beyond the kin versus group debate. *Trends in Ecology & Evolution, 26*(4), 193–201.

Marches, J. R., & Turbeville, G. (1953). The effect of residential propinquity on marriage selection. *American Journal of Sociology, 58*(6), 592–595.

Marshall, J. A. (2011). Group selection and kin selection: Formally equivalent approaches. *Trends in Ecology & Evolution, 26*(7), 325–332.

Okasha, S. (2006). *Evolution and levels of selection.* New York: Oxford University Press.

Okasha, S. (2015). The relation between kin and multilevel selection: An approach using causal graphs. *The British Journal for the Philosophy of Science, 67*(2), 435–470.

Patton, J. Q. (2000). Reciprocal altruism and warfare: A case from the Ecuadorian amazon. In L. Cronk, N. Chagnon, & W. Irons (Eds.), *Adaptation and human behavior: An anthropological perspective* (pp. 417–436). New York: Aldine de Gruyter.

Price, G. R. (1970). Selection and covariance. *Nature, 227*(5257), 520–521.

Price, G. R. (1972). Fisher's 'fundamental theorem' made clear. *Annals of Human Genetics, 36*(2), 129–140.

Ruse, M. (1984). Charles Darwin and group selection. In R. N. Brandon & R. M. Burian (Eds.), *Genes, organisms, populations: Controversies over the units of selection* (pp. 9–28). Cambridge, MA: The MIT Press.

Rushton, J. P. (1989). Genetic similarity, human altruism, and group selection. *Behavioral and Brain Sciences, 12*, 503–559.

Rushton, P. (1998). Genetic similarity theory, ethnocentrism, and group selection. In *Ethnic conflict and indoctrination: Altruism and identity in evolutionary perspective*. New York: Berghahn Books.
Shanahan, T. (2004). *The evolution of Darwinism: Selection, adaptation and progress in evolutionary biology*. New York: Cambridge University Press.
Sober, E., & Wilson, D. S. (1998). *Unto others: The evolution and psychology of unselfish behavior*. Cambridge, MA: Harvard University Press.
Tennyson, A. (1850). *In memoriam*. Boston: Knight & Millet.
Tinbergen, N. (1963). On aims and methods of ethology. *Zeitschrift für Tierpsychologie, 20*(4), 410–433.
Trivers, R. L. (1971). The evolution of reciprocal altruism. *Quarterly Review of Biology, 46*, 35–57.
Waibel, M., Floreano, D., & Keller, L. (2011). A quantitative test of Hamilton's rule for the evolution of altruism. *PLoS Biology, 9*(5), e1000615.
West-Eberhard, M. J. (1967). Dominance hierarchies and the evolution of social behaviour. *Science, 157*, 1585–1595.
West-Eberhard, M. J. (1975). The evolution of social behavior by kin selection. *The Quarterly Review of Biology, 50*(1), 1–33.
Wilkinson, G. (1984). Reciprocal food sharing in the vampire bat. *Nature, 308*(5955), 181–184.
Williams, G. C. (1966). Natural selection, the costs of reproduction, and a refinement of Lack's principle. *The American Naturalist, 100*(916), 687–690.
Wilson, D. S., & Wilson, E. O. (2007). Rethinking the theoretical foundation of Sociobiology. *The Quarterly Review of Biology, 82*(4), 327–348.
Wilson, E. O., & Hölldobler, B. (2005). Eusociality: Origin and consequences. *Proceedings of the National Academy of Sciences, 102*(38), 13367–13371.
Zuberbühler, K., Jenny, D., & Bshary, R. (1999). The predator deterrence function of primate alarm calls. *Ethology, 105*(6), 477–490.

2

An Intellectual History of Multilevel Selection: Reformulation and Resuscitation

Steven C. Hertler, Aurelio José Figueredo, and Mateo Peñaherrera-Aguirre

1 The Resurgence of Group Selection

The previous chapter depicted a rising chorus of consensus starting in the 1970s. Sober and Wilson describe how group selection was buried in the 1960s and 1970s and treated with utter contempt. It was so reviled that it was not forgotten, but recalled as an example of how *not* to think. Even in the 1980s, as Sober and Wilson recount, an unidentified, distinguished biologist once advised a younger colleague that there are three things that one does not defend in biology: *group selection*, *phlogiston theory*, and *Lamarckian evolution*. Indeed, discussion of group selection, at certain points in the history of evolutionary biology, evoked criticism and even ostracism, as illustrated by the following reflection shared by the still skeptical professor Detlef Fetchenhauer (2009):

S. C. Hertler et al., *Multilevel Selection*, https://doi.org/10.1007/978-3-030-49520-6_2

> Arguing with group-selection ideas in biology was for a long time like arguing with psychoanalytic ideas in psychology, namely an embarrassing signal of being old-fashioned and outdated. When some years ago I was giving a course on the evolution of human altruism at the department of biology at the University of Groningen I was told that I should not even mention group selection. The idea was regarded as so absurd it was not even worth[y] of being refuted. (p. 283)

David Sloan Wilson defended group selection unremittingly through these decades of withering criticism. Honing his reasoning on the whetstone of contrarian opinion, Wilson produced convincing arguments applicable to humans and human societies. Together, Sober and Wilson offer unique perspectives on group selection using comparative religion, field data, reviews of experimental work, and descriptions of parasitism and sex ratio. Comprehensive and various, their writings invariably wend their way back to human group selection, which they understand to be established and evident. "We claim," Sober and Wilson write, "that human social groups are so well designed at the group level that they must have evolved by group selection." Sober and Wilson, in short, see the stamp of evolutionary design on human groups, just as Darwin saw the stamp of evolutionary design on individual organisms. Within human populations, religion, language, and ideology are layered onto ecological and biological determinants, which jointly contribute to the formation of cooperating and competing groups. D. S. Wilson's *Darwin's Cathedral* and Sober and Wilson's *Unto Others* are among a short list of works establishing the theoretical probability of human multilevel selection.

Sober and Wilson (1998, p. 51) invoke Kuhn (1962) to explain the contrast between group selection's real explanatory power and its attenuated impact. In the rejection of group selection, Sober and Wilson see science not progressing in a straight line toward truth, as Kuhn warned that it would not. William's work on sex ratio and Lewontin's work on virulence are cited by Sober and Wilson as empirical examples of group selection functioning as a "significant evolutionary force." Nevertheless, these and other avenues of extant evidence have remained relatively unassimilated. "Normal science," Sober and Wilson write, "did its job, but somehow it failed to have the right impact." These and other supportive empirical studies have not "forced a general reassessment of group selection theory." Accordingly:

> Many evolutionary biologists continue to play the 'group selection is dead' song ... Little wonder, then, that scholars from other disciplines who are interested in evolution have heard almost nothing about these scientific developments. (p. 51)

Modern formulations of group selection as one among other levels of selection within multilevel selection theory remain conflated with what has been called *naïve group selection*, perhaps equally so within the minds of those opposing and those neglecting group selection. Extending back to the aforementioned work of Wynne-Edwards (1962), naïve group selection posited Panglossian circumstances, whereby species husbanded resources by limiting population growth for the good of the whole population. The *naïve* aspect of naïve group selection derives from its failure to incorporate plausible opposing forces to those of individual selection. Naïve group selection might then be compared to idealized conceptions of communism in its theoretically final utopian state. Both theories promote the good of the whole, but both lack any viable controls on individual group members that might undermine the collective good to benefit their selfish ends. This is why one is actually absent in nature and the other is impracticable in human sociopolitical affairs. By way of contrast, modern formulations of group selection within multilevel selection theory are akin to governmental forms employing checks and balances, keeping opposing forces in dynamic tension. Illustrating the point, Wilson (2016) speaks thus of the opposing forces between individual and group selection:

> Cooperators might be at a selective disadvantage compared to free riders and exploiters within the same group, but groups composed primarily of cooperators are at a selective advantage compared to groups crippled by free riders and exploiters. Natural selection takes place both within and between groups. Group-beneficial adaptations can evolve if between-group selection is strong enough to oppose disruptive within-group selection. (pp. 33–34)

Competition from rival groups is the implicitly stated counterweight to selfishness within the group. In addition to external competition from rival groups, selfishness is societally controlled by a system of reward and punishment meted out within societies to coerce noncooperators, as will be discussed in Sects. 5 and 6 of this chapter as well as in Part II of this book.

As implausibly posited in the case of naïve group selection, groups never compass whole species. Instead, groups can range from immediate families to extended kinship bands and to nation-states but ineluctably fracture before progressing toward the whole of a species. Wilson (2016) recently made this point: "Social interactions almost always take place in groups that are small, compared to the total population" (pp. 33–44).

Writing a subsequent chapter in Wilson's edited volume, *Complexity and Evolution: Toward a New Synthesis for Economics*, Gowdy, Mazzucato, van den Bergh, van der Leeuw, and Wilson (2016) follow the theme, providing the following maxim: "Adaptation at any level of a multitier hierarchy requires a process of selection at that level and tends to be undermined by selection at lower levels" (p. 336). Envisioning a hierarchical view of society with ascendant levels of aggregation, Gowdy and colleagues understand the tendency for *disruptive self-serving behaviors* to serve as a brake in the progression from one rung to the next on the ladder of aggregation. In doing so, Gowdy et al. (2016) plainly illustrate the nature of the conflict: "What's good for me can be bad for my family. What's good for my family can be bad for my clan. All the way up to what's good for my nation can be bad for the global village" (p. 336). Gowdy and colleagues' rendering recalls a well-known Arabian proverb: "It was me against my brother; me and my brother against our father; my family against my cousins and the clan; the clan against the tribe; and the tribe against the world" (Uris, 1984, p. 14).

This heady realism, grounded in an evolutionary truism, is precisely what Enlightenment philosophers attempted to work against as they promoted cosmopolitanism, as seen through the writings of Montesquieu (1964): "If I knew something useful to me, but prejudicial to my family, I would reject it from my mind. If I knew something useful to my family, but not to my country, I would try to forget it" (p. xviii; in the introduction by Healy). Following the chain of reasoning to its logical conclusion,

Montesquieu (1964) concludes: "If I knew something useful to my country, but prejudicial to Europe, or useful to Europe and prejudicial to the human race, I would regard it as criminal" (p. xviii; in the introduction by Healy).

Even as the opposing forces of selfish individualism are obviously operative, it does not follow that the processes of social selection are invariably pulled all the way down the ladder of aggregation such that evolution never takes place at the group level. Unfortunately, something of an ideological consensus emerged wherein individual selection was thought to be *invariably* stronger than group selection. Instead of imagining a dynamic interplay of vying forces, sometimes stronger, sometimes weaker, and sometimes equal in either synergy or opposition, critics of group selection admitted the theoretical possibility of an interplay so weak as to nearly never be able to overbalance individual selection's countervailing forces. Nevertheless, opponents of group selection saw the stable formation of aggregations in families, clans, tribes, states, and nations and, perforce, had to explain their existence in alternative ways. Pinioned between the necessity to explain reality and the unwillingness to consider group selection as a factor came *inclusive fitness theory*, *evolutionary game theory*, *selfish gene theory*, and indeed "most of the theories of social evolution that emerged during the second half of the twentieth century," which were nothing other than efforts to "explain apparent group-advantageous behaviors without invoking group selection" (Wilson, 2016, p. 34). According to Wilson's (2016) and later Okasha's (2006) reading of the history of the controversy surrounding group selection, the emerging synthesis cannibalizes rather than discounts alternative theories, such that they become incorporated into multilevel selection theory:

> In retrospect, the theories developed to explain apparent group-advantageous behaviors without invoking group selection can be seen to have the logic of multilevel selection embedded within their own structures. They offer different perspectives on a single causal process rather than invoke a separate causal process, a topic discussed under such terms as pluralism and equivalence. (p. 34)

Notwithstanding opposition, and bolstered by decades of thought, theory, and comparative analysis, Wilson has moved on, understanding selective pressures to impart directional selection in response to group competition over and above the drag of individual selection. More than this, Wilson extends his claims to insist, not only on the possibility that group selection can be strong enough to exist but that it can sometimes come to dominate individual selection, especially in instances of evolutionary transition.

2 Neo-group Selectionism

The idea of multilevel selection, as outlined above, changed incrementally so as to have become a very different concept when compared to its original formulation; yet, this distinction is not always evident to critics of group selection. Even Borrello does not pointedly differentiate group selection as understood by Wynne-Edwards from group selection as understood within modern multilevel selection theory. He comes closest to doing this toward the end of his book; and it is certain that the author himself understands this distinction. Still, there is no real discussion of how group selection operates in conjunction with individual selection or, for example, how groups are simultaneously challenged with free-riding exploiters from within and cohesive rival groups from without. From that time till this, group selection has been supported by the research underpinning all the bulleted points below, in addition to becoming embedded in a matrix of multilevel selection theory wherein individual and group selection are known to simultaneously operate and dynamically interact.

Slowly, as if by erosion, consensus was selectively undermined in the decades thereafter, with fewer and fewer evolutionists stalwartly denying the possibility of group selection in the present (Okasha, 2006). Critics have historically harbored notions of group selection that are antiqued or otherwise inadequately operationalized, rendering the derived critiques irrelevant. Common to the intellectual history of group selection, skeptics have based their deconstructions from false premises, starting from an understanding of group selection that no longer is, or never was. Now accepted among most evolutionists, the modern formulation of group selection within the broader framework of multilevel selection is becoming better known.[1] Group selection, as the most controversial level of selection within multilevel selection theory, has been established on new grounds, having been clarified in the following ways (Grueter, Chapais, & Zinner, 2012; Mayr, 1970; Richerson et al., 2016; Wilson & Sober, 1989, 1994):

- Groups were operationally defined
- Groups were understood as fluctuating between different levels of aggregation and kin relatedness[2]
- Group selection does not always end in the extinction between one of the two groups. Instead, there can be *character displacement* wherein one group is marginalized or enters a different ecological zone at the point at which it comes into contact with the other group
- Group selection can still occur even while migration and gene flow take place; this happens when migration and gene flow are strategic or directional, such that gene flow drives, rather than undermines, between-group differences
- Group selection does not start where individual selection stops; rather, both processes can occur simultaneously[3]
- Group selection provides an emergent group-level advantage even if the trait in question was neither self-sacrificial (it was actually under positive individual-level selection) nor social in any sense initially
- Traits that are selected for at the group level may well have been initially generated via individual level selection

Even when properly differentiated from naïve group selection, group selection has remained controversial mostly because it has been theorized to require certain preconditions, such as the punishment of cheaters, free riders, social loafers, and related individually selected opportunists. Take the example of *altruism*: Altruism might be selected for, with the effect that the altruistic group becomes a superior competitive unit relative to less altruistic groups. The population of the altruistic group may therefore expand to the detriment of competitor populations. Staying with the example, that altruistic group, however, is prone to exploitation from population members who do not add to aggregate altruism or who actively undermine it. Within the altruistic group, if free riders and cheaters then reproduce to the detriment of their altruistic neighbors, mean altruism will be brought down to the level found among other populations, eliminating the group's prior advantage. Consequently, to maintain high mean altruism, the proliferation of free riders and cheaters must be counteracted. This is a fair criticism, but one that has been duly answered. For instance, in eusocial insect colonies, group-selected colony cohesion is enforced by selectively consuming any eggs laid by workers, attacking ovulating workers,[4] and policing by other means. In historical and contemporary human populations, we find analogous mechanisms that thwart free riders, for instance, shaming, incarceration, ostracism, sanctions, fines, and related mechanisms, reviewed in Chap. 5's fourth section. Selfish individuals do indeed attempt to undermine groups and group selection, but these selfish individuals may suffer depressed fitness in consequence of group punishment. In the context of such group punishment, altruism can become individually adaptive, leading to the evolution of an altruistic group. Humans are especially capable of this form of self-domestication wrought by imposing socially selective pressures on one another. An additional point made by Wilson and Sober (1998) is that cheater detection and punishment may be more common mechanisms among highly intelligent humans, making human group selection more plausible.

One can then see how multilevel selection recovered theoretical respectability as it differentiated itself from naïve group selection, after which multilevel selection demanded thoroughgoing consideration for having undergone successful operationalization and theoretical defense. The work of Okasha (2006) furthered multilevel selection's renaissance. As seen in reading Okasha, social groups have emergent properties, meaning that the *within-group* component of selection is not necessarily a simple additive function of the within-group fitness of constituent individuals, because the interactions among individuals produce a nonadditive component to the mix. There are thus two types of multilevel selection characteristics hypothesized by Okasha (2006): Type 1 (*MLS1*) multilevel selection characteristics are individual difference traits, and their corresponding aggregates, that affect both the individual's relative fitness and the aggregate fitness of individuals within the group, and Type

2 (*MLS2*) multilevel selection characteristics are traits of the collectives that are irreducible to the phenotypes of the constituent individuals, wherein the collectives' fitness is defined as that of parent groups multiplying into offspring groups. The Type 2 group selection characteristics can at times become so pronounced, presumably as reflected by the between-group component of selection in the Price equation, that some have characterized the entire insect societies as *superorganisms*, with constituent individuals assuming the role of specialized parts selected to function as integral components of a discrete, collective entity (e.g., Korb & Heinze, 2004; Leigh, 2010; Reeve & Hölldobler, 2007; Wheeler, 1910, 1911; Wilson & Sober, 1989; Wilson & Wilson, 2008; Wilson & Hölldobler, 2005).

Another way that modern multilevel selection theory has distinguished itself from naïve group selection theory is in gaining greater clarity regarding the level of biological organization of the objects at which altruistic behavior is directed. Many otherwise respectable theorists previously made what would now be considered somewhat extravagant claims regarding the motivating principles behind much observed animal behavior. Dawkins (1976), for example, points out that

> Konrad Lorenz, in *On Aggression,* speaks of the 'species preserving' functions of aggressive behaviour, one of these functions being to make sure that only the fittest individuals are allowed to breed. (p. 8)

Thus, the dominant wolf is depicted as not pursuing dominance for his own selfish reproductive benefit, but instead, for the higher purpose of this principle of racial hygiene; with equal implausability, the subordinate wolf is ostensibly accepting of his nonreproductive status as consistent with this self-effacing eugenic strategy for the good of the species. Although an argument might be made from modern multilevel selection theory that acceptance of dominance hierarchies might have evolved for the "good of the group," the idea that these behaviors evolved for the benefit of the *entire species* now sounds to us as highly improbable. Unlike the artificially created groups of captive wolves that Lorenz (1966) studied, we now know that naturally formed wolf packs in the wild are extended families that accrete around a core group of parents and their adult offspring. The mechanism that ensures the generally amicable relations among dominants and subordinates is not loyalty to some eugenicist ideals for their species, but the fact that the individuals are genetically related and thus subject to the conflict-moderating forces of kin selection.

Such ideas from naïve group selection theory nonetheless disseminated widely into the popular culture and found their way into the narratives of many nature documentaries and educational resources. To this day, the website of the *International Wolf Center* (2019) makes the following rather extravagant claim with equally cavalier disregard for any conceptual plausibility or support from evolutionary theory:

> All of these ungulates have adaptations for defense against wolves, including a great sense of smell, good hearing, agility, speed, and sharp hooves. As these prey are so well adapted to protecting themselves, wolves feed upon vulnerable individuals, such as weak, sick, old, or young animals, or healthy animals hindered by deep snow. By killing the inferior animals, wolves help increase the health of their prey population a tiny bit at a time. When inferior animals are removed, the prey population is kept at a lower level and there is more food for the healthy animals to eat. Such "culling" also ensures that the animals which reproduce most often are healthy and well suited for their environment. Over many generations, this selection helps the prey become better adapted for survival.[5]

The evident implication here is that wolves do not prey upon the weak and the sick merely because they are much easier to capture and subdue, but instead out of eugenical regard for the genetic health of the prey species and the provision of sufficient resources and *lebensraum* for the fittest among them to survive. One might reasonably wonder why wolves would have evolved any desire to improve the antipredator adaptations of their prey populations, making them more difficult to hunt in the future. Given our current understanding of the principle of natural selection, this narrative appears implausible to the point of absurdity.

In contrast to the gist of these narratives, modern multilevel selection theory requires one to specify exactly at what level(s) of biological organization any adaptation is selected and by what mechanism. Furthermore, all this theoretical debate regarding the sometimes conflicting interests of "individuals" and "groups" does little to clarify precisely what one means by a "group." Does one mean a bounded and kin-structured social group, a local population, an entire species? These questions must be answered for the theory to have any coherence whatsoever. For example, some group selection theories use smaller units than the deme to model "trait-groups," which is a collection of individuals defined on the basis of common properties, or heritable traits. Wilson (1975) explains this usage as follows:

> Evolution's most easily conceived population unit is the deme, and it is determined by the movement occurring during the dispersal phase. Yet most ecological interactions, in terms of competition, mating, feeding and predation are carried out during the nondispersal stages in the smaller subdivisions, which I term "trait-groups." In some cases the trait-groups are discrete and easily recognized, such as for vessel-inhabiting mosquitoes and dung insects. In other cases they are continuous and each individual forms the center of its own trait group, interacting only with its immediate neighbors, which comprise a small proportion of the deme. (p. 143)

3 Pressing the Offensive

Only more recently have Wilson, Sober, and other *neo-group selectionists* (Okasha, 2006) have gotten beyond arguing for group selection's existence and have gone on to insist on its theoretical superiority. Co-opting kin selection, casting it as a form of group selection, is only one of several tenets undergirding Wilson and Sober's conclusion that group selection is not only plausible but also powerful (Okasha 2006, p. 177 et seq.). Staying with the example of kin selection, we find that Nowak, Tarnita, and Wilson (2010) insist that it is kin selection, more than group selection, that "requires stringent assumptions, which are unlikely to be fulfilled by any given empirical system." These include the assumption that interactions between organisms are additive and pairwise, thus excluding any situation with synergistic effects or where more than two organisms interact. Moreover, inclusive fitness is relevant only to a limited set of population structures (Bahar, 2017, p. 277). Along with Tarnita and Wilson,

Nowak finds inclusive fitness to provide no special biological insight that could not be otherwise better accounted for using the more generalized theory of group selection. Again, this is just one example of how neo-group selectionists have taken the offensive, which can be seen as a qualitatively different level of defense, one that looks a lot more like offense.

In the view of neo-group selectionists, ecological factors (such as the distribution of food) are thought more relevant than genetic relatedness to group formation and cooperation. This conclusion is evidently based on observations of eusocial insects. For example, observation and experimentation witnessed unrelated termite colonies merge into super colonies, both from naturally occurring and artificially imposed ecological pressures (Howard, Johns, Breisch, & Thorne, 2013). Indeed, when the comparative literature is systematically searched, eusociality and genetic relatedness are demonstrably uncoupled, thus breaking the association between one and the other, resulting in the following claim (Nowak, Tarnita, & Wilson, 2010):

> Relatedness is better explained as the consequence rather than the cause of eusociality. Grouping by family can hasten the spread of eusocial alleles, but it is not a causative agent. The causative agent is the advantage of a defensible nest, especially one both expensive to make and within reach of adequate food.

Nowak and colleagues presume that the causal arrow assumed in inclusive fitness theory should be reversed. In other words, ecological conditions pressing toward eusociality sometimes allow high levels of relatedness, rather than high levels of relatedness allowing eusociality (Bahar, 2017).

In attempting this coup, D. S. Wilson, E. O. Wilson, Martin Nowak, and others have placed the pole more fully outside the bounds of what Jerry Coyne, Richard Dawkins,[6] and other detractors find acceptable. The *Overton window* has historically been bounded on one side by positing that relatedness was the sole requirement for eusociality and on the other by positing that relatedness was necessary but not sufficient for the evolution of eusociality. By replacing relatedness with ecology as the prime determinant of eusociality, and then further relegating relatedness to the role of a by-product rather than a driver of eusociality, Wilson, Wilson, and Nowak have fully inverted the relationship as it has been understood by many evolutionists. Attempting a dispassionate analysis of this partisan strife, H. Gintis, an economist and evolutionist, insists on the *analytical validity* and *ultimate importance* of kin selection. At the same time, Gintis does not understand kin selection to be a driving force in the "formation and evolution" of sociality. Nowak and Wilson, Gintis writes, were warranted in exposing kin selection's limitations and thereby circumscribing its scope and influence, even as they went a step too far by "questioning its validity and in understating its [kin selection's] contribution to sociobiology." In turn those kin selectionists on the other side of the divide "err in claiming that organisms in a social species maximize their inclusive fitness and that inclusive fitness theory explains social structure" (Gintis, 2017, p. 184).

The preceding sections document a range of views on genes and relatedness as they are applicable to multilevel selection theory and group formation. Hamilton, following Price, came to see the debate as semantic, some product of perception that disappears with perspective. Others, like Nowak, have relegated relatedness to a product of cooperation, thereby inverting the traditionally assumed directionality of the causal arrow, as has been seen. However, most see genetic relatedness as a driver of cooperation, even as the ultimate importance of relatedness varies proportionally. Both Hamilton and Wilson consider relatedness necessary, though the Wilsonian model more readily recognizes extra-genetic, ecological variables, thereby assigning proportionally less influence to relatedness. Relatedness then finds its place as a necessary variable in sociality, though it is a precondition, rather than a lone cause. With relatedness effectively circumscribed came clearer and more comprehensive descriptions of the ways in which relatedness may weigh in favor of cooperation amid a broader multiplicity of causes, such as nest defense, territoriality, ecology, and life history traits. This is illustrated by those species with high relatedness that are not at all social, such as certain species of armadillo birthing septuplets, which, though genetically identical, go on to live more or less solitary lives (Greulich, 1938). In sum, with the possible exception of Nowak and some few other theorists, the spectrum of opinion centers on how heavily weighted is relatedness, compared to other factors, with some degree of relatedness being necessary.

The purely theoretical literature on relative degrees of relatedness within and between social groups is mixed. Some mathematical simulations predict that high population *viscosity*, generated by the slow movement of individuals from their places of birth (also known as *philopatry*), tends to promote more local *cooperative* interactions among genetic relatives due to proximity but simultaneously tends to promote more local *competitive* interactions among genetic relatives for the same reason, with these two effects purportedly cancelling each other out (Mitteldorf & Wilson, 2000; Taylor, 1992; Wilson, Pollock, & Dugatkin, 1992). Other purely mathematical models have instead predicted the opposite (Schonmann, Vicente, & Caticha, 2013):

> We conclude that contingent forms of strong altruism that benefits equally all group members, regardless of kinship and without greenbeard effects, can spread when rare under realistic group sizes and levels of migration, due to the assortment of genes resulting only from population viscosity. (p. 1)

The data from social insects on local genetic relatedness, however, is generally less equivocal than this body of theory. For example, much of the research on the evolution of sociality in primitive wasps has abandoned the strict kin selection model of Hamilton (1964), where shared genes must be identical by recent common descent, and their proportions inflated to elevated levels by the mechanism of *haplodiploidy*. This is largely because the most primitively social wasp colonies are not typically created by mother-daughter bonds, as envisioned by Hamilton's simplified model, but instead by *foundress associations* among fully fertile female wasps of the same generation coalescing to construct communal nests (e.g., West-Eberhard, 1967, 1969, 1975). Although their genealogies of origin are typically unknown to researchers, these foundresses have not been generally found by molecular genetic methods to conform to the levels of relatedness seemingly required by Hamilton's original model for kin-selected altruism to evolve. Nevertheless, they have often been found to be more closely related to each other within local groups than to the general surrounding population outside of these groups (e.g., Bluher, 2018; Wehren Gaspar, López-Uribe, & Del Lama, 2007). This is not a purely accidental outcome of population viscosity, but is behaviorally mediated, at least in part, as foundresses have been observed to migrate among nests to maximize inclusive fitness: "A clear motivation for moving to new nests was high genetic relatedness; by the end of the foundress period all females were on nests with full sisters" (Seppa, Queller, & Strassmann, 2012, p. 1). Genetic relatedness among foundresses has also been proposed to play a contributory role in the degree of reproductive skew, or differential dominance, emerging among nestmates (e.g., Bolton, Sumner, Shreeves, Casiraghi, & Field, 2006; Sumner, Casiraghi, Foster, & Field, 2002). Such reproductive skew is believed to set the stage for more advanced stages of eusocial evolution, such as the evolution of the purportedly *sterile* worker caste.

With the understanding that genetic relatedness exists on a continuum and thus can proportionally influence cooperation, group selection becomes possible for distantly related human kin and even supranational human groupings that share genetic variance but also only for close hymenopteran or human kin. In consequence, we can see the logical extension of kin selection in *genetic similarity theory* (Rushton, 1998; Rushton & Nicholson, 1988; Rushton, Russell, & Wells, 1984), a theory which has been critiqued (Mealey, 1985) and thereafter successfully defended (Rushton, 2005). As with Hamilton's (1975) updated inclusive fitness theory, genetic similarity theory subsumes relatedness among kin within a broader view of relatedness wherein trait similarities can foster cooperation and favoritism even when the genes underlying those traits reside in more distantly related conspecifics. In Rushton's theory, genetically based favoritism and cooperation can thus operate in the province of mate preference, friendship, and *ethnic nepotism*. One can then cooperate with a group in competition with another as a result of having more shared genetic variance with one group than another and not necessarily based on any strong relatedness as found among siblings.

4 Commentary on the Controversy

We proceed to a summary statement of the levels of selection controversy; in doing so, we begin with an instructive quote from Gilpin (1975), orienting us in time and viewing this controversy as recapitulating that which surrounded evolution generally:

> Group selection is thus an old concept that is believed in by many but has not been demonstrated to the satisfaction of anyone. The history of group selection perhaps resembles that of individual (Darwinian) selection up to about 1910. Both forms of selection were discovered by 'economists' working on human populations (Malthus and Carr-Saunders). For both, these economic ideas were shown to apply to animals (Darwin and Wynne-Edwards), which introduced a form of selection. But for both, the method of inheritance was not clearly defined. And for both, mathematical models of the evolutionary behavior had to await later development. (p. 8)

We select a second quote, like the one above, for its ability to stand aloof from the debate, giving perspective unavailable to partisans of either side (Hull, 1984):

> One reason that the controversy over the levels at which selection takes place has remained so intractable is that some of the issues are basically metaphysical: what sorts of things are organisms in contrast to groups, what general characteristics must an entity have to be selected, can entities which have what it takes to be selected also evolve or are the requisite characteristics mutually exclusive, etc.? (p. 144)

In commenting on the levels of selection debate, Gintis (2017) notes that there is

> a certain asymmetry in the mutual criticism of the two schools of thought. Few supporters of group selection deny the importance of inclusive fitness theory, while virtually all its opponents regularly deny the importance of group selection theory. (p. 192)

In contrast to these tendentious critics, Gintis (2017) accurately observes that "[t]he correct way of thinking is to embrace both atomistic [inclusive fitness] and structural [group selection] approaches and analyse the corresponding interplay of forces" (p. 192).

Evolution can operate on any entity whose heritable phenotypic variation results in differential fitness. Traditionally, that entity is understood to be the individual organism, and, of course, organisms evolve. Notwithstanding, groups also evolve, as they, too, are aggregates of heritable genetic material; they, too, display phenotypic variation; and they, too, evince differential fitness. In other words, *genes create phenotypes that exist at multiple levels of aggregation*: some at the level of cells, some at the level of individuals, and some at the level of groups of varying levels of complexity.[7] Any of these levels can be the target of selection or, in other words, drive changes in gene frequencies. Thus situated as one level of selection by which gene frequencies within a species change, it should then be recognized that groups may be as small as an immediate family, or extend to larger aggregates, such as human tribes, states, nations, or continental populations. Accordingly, together with cells and organisms, these groups, both small and large, each comprise levels on which selection can operate; hence the term *multilevel selection*.

The various levels at which evolution operates within multilevel selection theory range from the obvious and long accepted to the contentious and vigorously debated. All understand that populations of organisms evolve over time. At the level of the family, selection can rest upon a basis of genetic relatedness as per *inclusive fitness theory*. However, beyond the confines of the family, or extended kinship networks, selection among larger aggregations of individuals, referred to as *group selection*, remains controversial in some quarters, as we have seen. Therefore, even as all see individuals as targets of selection, and most see families as targets of selection, there remains a fair degree of resistance to viewing populations, demes, tribes, nations, states, or continental populations as potential targets of selection. Group selection is part of multilevel selection. Multilevel selection theory, articulated later, did not replace as much as incorporate group selection. While always understanding evolutionary processes to proceed within the overarching framework of multilevel selection, we nonetheless focus on large, non-closely related groups and group selection. *We attempt to level empirical data precisely at the group selection controversy, which remains the most contentious aspect of multilevel selection theory.* So yes, we are arguing for the validity of multilevel selection theory, but specifically doing so by attempting to buttress the "weakest" leg or level on which it partially stands, which is group selection.

Thus, with proper perspective, objections to naïve group selection have been co-opted as support for multilevel selection theory. Within this context, readers are brought to the understanding of the larger point, namely, that genes resident in groups of individuals can precipitate cooperation and cohesion such that lines of competition and conflict are often drawn along the fault line of genetic difference. At still a higher level of abstraction, we see colonial organisms, from siphonophores to slime molds, as cooperative, group-selected ventures. Moreover, eusocial insect colonies, with their high genetic relatedness, blur the boundary between individual and collective. Even multicellular life is in some ways a feat of group selection, in that it presupposes the cooperation of many genes to perpetuate the survival and reproduction of the entire organism via the suppression of selfish genetic elements.

5 Multiple Levels of Aggregation: A Brief Illustrative Survey

There is a growing body of literature on selfish genetic elements (Okasha, 2006, p. 145), of which cancer may be held out as the most well-known example (Fishman & Jainike, 2014).[8] Looking to the case of cancer, we see that, if a cancerous uprising can beat the odds by mutating within a long-lived stem cell, and continue without repair in such a way that meaningfully and harmfully alters protein production, while also significantly boosting replication above and beyond the rate of other somatic cells, it can then metastasize (Greaves, 2000). Immunosurveillance (Schreiber, Old, & Smyth, 2011; Waldhauer & Steinle, 2008), inflammatory response (Grivennikov, Greten, & Karin, 2010), apoptosis (Kerr, Winterford, & Harmon, 1994), killer *T* cells (Krijgsman, Hokland, & Kuppen, 2018), natural killer cells (Morvan & Lanier, 2016; Wu &

Lanier, 2003), macrophages (Mills, Lenz, & Harris, 2016), and dendritic cells (Palucka & Banchereau, 2012) act to suppress mutinous cancer cells, just as federal, state, and local law enforcement work with prosecutors, judges, jailers, and executioners to suppress criminal, revolutionary, or treasonous citizens.

We learn more about multilevel selection when we observe obligate symbiotes intermediate a continuum of cooperation anchored on one extreme by complex multicellular bodies and on the other by solitary organisms. Lewontin (1970), Sober and Wilson assert, was the first to connect group selection to the evolution of virulence in parasites. Lewontin studiously differentiated group selection or *population selection* as he seems to have termed it in his 1970 publication *The Units of Selection*, from Wynne-Edwards's *species selection*. Though Lewontin believed the conditions were rare and the requirements strict, he countenanced group selection, providing two examples, one of which related to the virus *myxoma* and the evolution of its virulence. The myxoma virus was introduced purposefully to Australia to control the accidentally[9] introduced rabbit population, the explosion of which was displacing native fauna and taxing native flora. The lethality of myxoma was legion. It killed almost 100% of infected rabbits, though the surviving few seemed to eventually have evolved resistance, which was confirmed via laboratory testing. Resistance within the rabbits is fully explicable and expected as an individually selected evolutionary outcome. Testing, however, was also performed upon free-ranging myxoma, which had evolved toward reduced virulence, a finding explicable through the lens of group selection. As Lewontin explained, myxoma was spread via mosquitoes into a single rabbit, and were thereafter trapped within that rabbit. The fate of host and parasite was one. As Lewontin states, a host rabbit then amounted to a deme from myxoma's perspective. Especially virulent groups or populations of myxoma extinguished themselves as they extinguished their hosts, making less virulent strains of myxoma comparatively more prevalent as they existed alongside the hosts they spared.

Sober and Wilson (1998, p. 50) then cite Nesse and Williams, who note that the evolution of virulence is now an obvious example wherein group selection and individual selection are opposing forces in selecting for altruism and selfishness. The contest will be decided based on the "relative strengths of within-host and between-host competition in pathogen evolution." Suggesting that between-host competition sometimes prevails, Miralles, Moya, and Elena (1997) found group selection sufficiently powerful to attenuate virulence, even as individual selection acted to augment virulence. In addition to group selection acting to decrease virulence in *horizontal transmission* of parasites to hosts, group selection can attenuate virulence in certain forms of *vertical transmission*, wherein parasites are transmitted from a host to that host's offspring (Ferdy, 2009).

Mitochondria were free-living prokaryotes that merged with the ancestors of contemporary eukaryotes 1.45 billion years ago, giving rise to extant eukaryotic life (Gray, 2017; Sagan, 1967). Reflecting on these capabilities, Bahar (2017, p. 170) writes, "even for prokaryotes, then, the transition from individual to collective, with its delicately negotiated balance between competition and cooperation, occurs with comparative ease."

In this vein, slime molds are of great interest. Illustrative of the miracle of slime mold formation, some amoebae form bricks in the stalk, which will never reproduce, allowing other amoebae to ably disperse from the heights gained by virtue of those below. Field studies of slime molds find aggregate slug formation from genetically distinct amoebae. Even as different amoebae species do not combine, slugs contain considerable intraspecific genetic diversity, which nevertheless does not preclude slug formation, even as some individual amoebae must come to comprise the stalk of the fruiting body, thus allowing others to persist at their expense (Bahar, 2017, p. 185).[10]

Coral is an association of animal-like polyps and photosynthetic algae that serves as a common example of obligate symbiosis among other species. Still further, siphonophores are of certain relevance. With more than one hundred species of these sea creatures being classed into three distinct suborders, one observes specialization in buoyancy, propulsion, and digestion (Kirkpatrick & Pugh, 1984). The Portuguese man o' war, a famed representative of the order commonly mistaken for a jellyfish, displays remarkable division of labor and unitary integration. We learn still more about multilevel selection when we observe those organisms facultatively switching between solitary existence and colonial cooperation in response to environmental demands. To this point are Bahar's extensive reviews of bacterial biofilms, secreted matrices within which individual bacteria become embedded so as to resist attack and transfer nutrients. Complex processes of chemical communication and genetic switching allow quorum sensing and coordinated responses to environmental changes of which the creation of biofilms is only one example.[11]

Sober and Wilson review self-sacrificial altruism among multicellular parasites. They specifically use the example of the lancet liver fluke (*Dicrocoelium dendriticum*), a parasite that creates a *brain worm* within ants, which induces parasitized ants to rise high and lock their jaws around a grass blade, making it more likely that they will be eaten by a ruminant, such as a sheep. From thence, the parasite can infect the ruminant liver, and exit as feces, which are eaten by snails. Thereafter, the parasite comes out in a mucus envelope, which is then eaten by ants; and so, the life of these parasites cycle between ants, ruminants, snails, and back again. The ant phase is focused on by many ecologists and evolutionists, for it is an account of behavioral change induced by a parasite on its host. From a group selectionist perspective, however, the point of interest is not the manipulation of the host, but the self-sacrificial process by which that manipulation is accomplished. The *brain worm* is, in effect, an altruistic volunteer member of the parasitic population, which generates the ant's behavioral change at the cost of its own reproductive potential. As per an individual selectionist view, brain worms should decrease and disappear. However, the key to understanding why this does not happen is to consider the individual ant with its population of fifty odd parasites inside, vying in competition against conspecific parasite populations sequestered in other ants. Within the ant, brain worms have lower fitness. However, parasitic populations with brain worms, because they manipulate their host ants into being reliably eaten by ruminants, have higher fitness than parasitic populations without brain worms. Thus, there is an individual selection pressure opposing a group selection pressure. The very existence of brain worms, in this sense, may be taken as evidence of group selection's ability to overcome individual selection. What looks to be something impossible to explain, Sober and Wilson (1998) conclude, becomes easy to explain.[12]

Though ecological exigencies are among the other factors inducing cooperation, competition and cooperation may have been its main drivers. This is evident even in the laboratory. Release a predatory small mouth ciliate into a population of single-celled algae (*Chlorella*), and witness those algae form into eight-celled units (Bahar, 2017), with predation inducing de novo multicellular clustering.

Think of what is happening—a predator creates a pressure to which there is an adaptive response toward cooperative aggregation. We witness the rudiments of complexity through cooperation among organisms against other organisms—a point we return to in Chap. 4 when describing how large-scale cooperative societies begin ratcheting toward complexity.

Evolution is famously directionless. It is a branching bush not a ladder. Lay descriptions, especially those perverting evolutionary science to the ends of ideology, often conceive of a teleological evolution replete with levels of hierarchal organization. Such levels then are understood as improvements on prior forms all progressing to a platonic ideal of perfection. Evolution, of course, is nothing of the sort. It is, rather, a blind bottom-up process of continual adaptation. However, that is not to say that evolution is absent trends. There is a robust trend toward complexity, when taking the overarching view of life as a whole (Bonner, 1988; Wilson & Kirman, 2016; Yaeger, Griffith, & Sporns, 2008). Most simply, time affords the evolution of complex forms (McShea, 1994). Yet, trends toward complexity are not simply a matter of endless iterations of integrational evolutionary cycles. Complexity comes of competition (Robson, 2005). Biotic competition, life struggling against life, both within and between species, tends, all else being equal, to augment complexity (Brockhurst et al., 2014). To be sure, the Cretaceous-Tertiary extinction ending eons of evolutionary complexity is only one of many examples of long-evolving and slowly won complexity being suddenly stricken from the evolutionary record by a sharp reversal in the prevailing selective regime. Nevertheless, complexity has a way of reasserting itself and will do so as long as complexity is a viable mode of competition (Benton, 1987). Thus, prokaryotes were joined by eukaryotes, multicellularity evolved, organisms developed lungs rather than relying on diffusion, and competition for light caused plants to invest in costly trunks and the complex plumbing known as xylem and phloem. That is to say, complexity is ratcheted up by competition via a red queen effect in an evolutionary arms race between life forms. To competition must be added cooperation, an equally powerful spur toward complexity, as seen among the many intricate mutualisms (Thrall, Hochberg, Burdon, & Bever, 2007) ranging from multicellularity to eusociality (Thorne, Breisch, & Muscedere, 2003).

The foregoing examples all illustrate that the dynamics of multilevel selection generalize across a wide array of different levels of biological organization as well as different taxa.

6 Conclusions

Together, Chaps. 1 and 2, traversing selfish genes and kin selection, and thereafter colonial organisms and eusocial insects, were so organized as to alert readers from an essentialist torpor wherein clear boundaries separate organisms and their constituent cells. To the contrary, as we have seen, aggregations exist at various hierarchical levels of biological organization, forming temporary federations, lasting associations, or permanent mutualisms. Aggregation at any of these levels represents selection favoring group formation in reaction to some selective pressure. When we see associations ranging from complex multicellularity, to colonial organisms, to eusocial colonies, we are observing, in some sense, degrees of aggregation reflective of the levels of selection within multilevel selection theory. Genetic relatedness among group members, rather than being an alternative explanatory framework, is simply an auxiliary adjunct, perhaps necessary but not sufficient for group formation. This principle is illustrated by the aforementioned uncoupling of eusociality and relatedness wherein the co-occurrence of eusocial insect communities comprising individuals that are not closely related exist alongside populations of closely related individual insects that are not eusocial. Genetic information briefly occupies and animates individual organisms. With the death of the organism comes the dissolution of the genetic aggregation. Yet, genetic information recombines generation after generation, allowing for stable aggregation at the group level.

Notes

1. https://evolution-institute.org/blog/the-tide-of-opinion-on-group-selection-has-turned/
2. *Haystack Model*: Hamilton's theory of inclusive fitness was brought forth and framed as a competitor to group selection, a view reinforced by John Maynard Smith's *Haystack Model.* The Haystack Model makes assumptions that maximize the force of individual selection and minimize the force of group selection, in addition to confusing some concepts, as discussed by Sober and Wilson (1998, p. 71).
3. Sober and Wilson (1998) explain *Simpson's paradox* when discussing altruism and group selection. Simpson's paradox, or the *Yule-Simpson effect*, is a phenomenon in probability and statistics wherein a trend appears in several different groups of data but disappears or reverses when these groups are combined, which is also why some additionally use the terms *reversal paradox* or *amalgamation paradox* synonymously. From one factor pulling in this direction and another factor pulling in that direction can come a cancellation of effects. So, Simpson's paradox is a myopic focus on outcome that fails to appreciate how that outcome came about. This is relevant in a multilevel selection model as individual selection and group selection are often thought to be working at odds. Take the example of a tug of war where the flag marking the rope's center hovers in the middle, not because it is at rest but because pull on one side is correspondingly countered by pull on the other.
4. Bourke, A. F. (2007). Social evolution: Community policing in insects. *Current Biology, 17* (13), R519–R520.
5. (https://www.wolf.org/wolf-info/basic-wolf-info/biology-and-behavior/hunting-feeding-behavior/)
6. "Richard Dawkins agreed, writing that the Nowak, Tarnita, and Wilson paper was "'no surprise' since 'Edward Wilson was misunderstanding kin selection as far back as this seminal 1975 work, *Sociobiology*.' David Sloan Wilson leapt into the mix, writing an 'open letter to Richard Dawkins' titled 'Why Are You Still In Denial about Group Selection'? Mutual allegations of ignorance of the literature are a common motif. 'Your view is essentially pre-1975', wrote Wilson, 'a date that is notable not only for the publication of *Sociobiology* but also a paper by W. D. Hamilton, one of your heroes, who correctly saw the relationship between kin selection and group selection thanks to the work of George Price.'"

7. These three characteristics were defined by Richard Lewontin and were described by Okasha on page 13 of the following work:
   Okasha, S. (2006). *Evolution and levels of selection*. New York: Oxford University Press.
8. Fishman and Jainike (2014) also provide the example of selfish genetic elements among stalk-eyed flies. In this case selfish genetic elements bias toward female production, so that when males and females mate, they produce a preponderance of females. As males become rare, it becomes extremely advantageous to rid oneself of the burden of these distorting elements. Stalk eyes are actually associated with not having these selfish genetic elements; they are a marker of being free of them; therefore they give rise to female preference for stalk eyes. Thus, you have sexual selection driven by selfish genetic elements.
9. It seems that rabbits were brought in cages on the First Fleet from England, and so it was their escape into the wild that was accidental.
10. See Bahar (2017, p. 188) for an interesting image depicting the stages of amoeba collectivization: growth, aggregation, differentiation, migration, and culmination.
11. Interestingly, in the course of this presentation, Bahar describes *persisters*, which might first sound like antibiotic-resistant bacteria. However, these persisters are more common where the biofilms are denser, and thus their presence and numbers appear to be density dependent. They are quite good at founding the colony anew where it has been decimated. They have a slower metabolism allowing them to exist in a sort of stasis or diapause so that they are not ingesting toxins to lethal rates.
12. It should be noted that, when first explaining the significance of the aforementioned parasite, Sober and Wilson do not mechanistically explain how the selfishness in the group does not entirely displace the altruistic. In other words, how within-group selfishness does not undermine between-group fitness. By the end of page 31, these authors begin to broach this subject. Sober and Wilson essentially argue that there will be a stable polymorphism and then, without using the word, at least just then, make an argument about an evolutionarily stable strategy maintained by negative frequency-dependent balancing selection. However, there is no actual evidence of the presence of this polymorphism; it is only posited.

References

Bahar, S. (2017). *The essential tension: Competition, cooperation and multilevel selection in evolution*. New York: Springer.

Benton, M. J. (1987). Progress and competition in macroevolution. *Biological Reviews, 62*(3), 305–338.

Bluher, S. (2018). *Demographic history and population structure of Polistes fuscatus paper wasps*. Master of science thesis, Cornell University.

Bolton, A., Sumner, S., Shreeves, G., Casiraghi, M., & Field, J. (2006). Colony genetic structure in a facultatively eusocial hover wasp. *Behavioral Ecology, 17*(6), 873–880.

Bonner, J. T. (1988). *The evolution of complexity by means of natural selection*. Princeton, NJ: Princeton University Press.

Brockhurst, M. A., Chapman, T., King, K. C., Mank, J. E., Paterson, S., & Hurst, G. D. (2014). Running with the Red Queen: The role of biotic conflicts in evolution. *Proceedings of the Royal Society B: Biological Sciences, 281*(1797), 20141382.

Dawkins, R. (1976). *The selfish gene*. Oxford, UK: Oxford University Press.

Ferdy, J. B. (2009). Virulence and transmission modes in metapopulations: When group selection increases virulence. *Journal of Theoretical Biology, 256*(2), 286–296.

Fetchenhauer, D. (2009). Evolutionary perspectives on religion–what they can and what they cannot explain (yet). In E. Voland & W. Schiefenhövel (Eds.), *The biological evolution of religious mind and behavior* (pp. 275–291). Waldhilsbach, Germany: Springer-Verlag.

Fishman, L., & Jainike, J. (2014). Selfish genetic elements and genetic conflict. In J. B. Losos (Ed.), *The Princeton guide to evolution* (pp. 347–355). Princeton, NJ: Princeton University Press.

Gilpin, M. E. (1975). *Group selection in predator-prey communities: Monographs in population biology 9*. Princeton, NJ: Princeton University Press.

Gintis, H. (2017). *Individuality and entanglement: The moral and material bases of social life*. Princeton, NJ: Princeton University Press.

Gowdy, J., Mazzucato, M., van den Bergh, J. C. J. M., van der Leeuw, S. E., & Wilson, D. S. (2016). Shaping the evolution of complex societies. In D. S. Wilson & A. Kirman (Eds.), *Complexity and evolution: Toward a new synthesis for economics* (pp. 185–199). Cambridge, MA: The MIT Press.

Gray, M. W. (2017). Lynn Margulis and the endosymbiont hypothesis: 50 years later. *Molecular Biology of the Cell, 28*(10), 1285–1287.

Greaves, M. F. (2000). *The deepest legacy: The evolutionary origins of cancer*. New York: Oxford University Press.

Greulich, W. W. (1938). The birth of six pairs of fraternal twins to the same parents: A discussion of the possible significance of such cases in the light of some recent observations. *Journal of the American Medical Association, 110*(8), 559–563.

Grivennikov, S. I., Greten, F. R., & Karin, M. (2010). Immunity, inflammation, and cancer. *Cell, 140*(6), 883–899.

Grueter, C. C., Chapais, B., & Zinner, D. (2012). Evolution of multilevel social systems in nonhuman primates and humans. *International Journal of Primatology, 33*, 1002–1037.

Hamilton, W. D. (1964). The genetical evolution of social behaviour. II. *Journal of Theoretical Biology, 7*(1), 17–52.

Hamilton, W. D. (1975). Innate social aptitudes of man: An approach from evolutionary genetics. *Bios, 133*, 115–132.

Howard, K. J., Johns, P. M., Breisch, N. L., & Thorne, B. L. (2013). Frequent colony fusions provide opportunities for helpers to become reproductives in the termite. *Zootermopsis nevadensis*. *Behavioral Ecology and Sociobiology, 67*(10), 1575–1585.

Hull, D. (1984). Units of evolution: A metaphysical essay. In R. N. Brandon & R. M. Burian (Eds.), *Genes, organisms, populations: Controversies over the units of selection* (pp. 142–160). Cambridge, MA: The MIT Press.

Kerr, J. F., Winterford, C. M., & Harmon, B. V. (1994). Apoptosis. Its significance in cancer and cancer therapy. *Cancer, 73*(8), 2013–2026.

Kirkpatrick, P. A., & Pugh, P. R. (Eds.). (1984). *Siphonophores and velellids: Keys and notes for the identification of the species* (Vol. 29). Bath, England: The Pittman Press.

Korb, J., & Heinze, J. (2004). Multilevel selection and social evolution of insect societies. *Naturwissenschaften, 91*(6), 291–304.

Krijgsman, D., Hokland, M., & Kuppen, P. J. (2018). The role of natural killer T cells in cancer—A phenotypical and functional approach. *Frontiers in Immunology, 9*, 367.

Kuhn, T. (1962). *Paradigm shift*. Cambridge, Masssechusetts: Harvard University Press.

Leigh, E. G. (2010). The group selection controversy. *Journal of Evolutionary Biology, 23*, 6–19.

Lewontin, R. C. (1970). The units of selection. *Annual Review of Ecology and Systematics, 1*(1), 1–18.

Lorenz, K. (1966). *On aggression*. London: Methuen.
Mayr, E. (1970). *Populations, species, and evolution: An abridgment of animal species and evolution* (Vol. 19). Cambridge, MA: Harvard University Press.
McShea, D. W. (1994). Mechanisms of large-scale evolutionary trends. *Evolution, 48*(6), 1747–1763.
Mealey, L. (1985). Comment on genetic similarity theory. *Behavior Genetics, 15*(6), 571–574.
Mills, C. D., Lenz, L. L., & Harris, R. A. (2016). A breakthrough: Macrophage-directed cancer immunotherapy. *Cancer Research, 76*(3), 513–516.
Miralles, R., Moya, A., & Elena, S. F. (1997). Is group selection a factor modulating the virulence of RNA viruses? *Genetics Research, 69*(3), 165–172.
Mitteldorf, J. J., & Wilson, D. S. (2000). Population viscosity and the evolution of altruism. *Journal of Theoretical Biology, 204*, 481–496.
Montesquieu, C. (1964). *The Persian letters*. Indianapolis, IN: Hackett Publishing Company.
Morvan, M. G., & Lanier, L. L. (2016). NK cells and cancer: You can teach innate cells new tricks. *Nature Reviews Cancer, 16*(1), 7.
Nowak, M. A., Tarnita, C. E., & Wilson, E. O. (2010). The evolution of eusociality. *Nature, 466*(7310), 1057.
Okasha, S. (2006). *Evolution and levels of selection*. New York: Oxford University Press.
Palucka, K., & Banchereau, J. (2012). Cancer immunotherapy via dendritic cells. *Nature Reviews Cancer, 12*(4), 265.
Reeve, H. K., & Hölldobler, B. (2007). The emergence of a superorganism through intergroup competition. *Proceedings of the National Academy of Sciences, 104*(23), 9736–9740.
Richerson, P., Baldini, R., Bell, A. V., Demps, K., Frost, K., Hillis, V., et al. (2016). Cultural group selection plays an essential role in explaining human cooperation: A sketch of the evidence. *Behavioral and Brain Sciences, 39*, e30.
Robson, A. J. (2005). Complex evolutionary systems and the Red Queen. *The Economic Journal, 115*(504), F211–F224.
Rushton, J. P. (1998). Genetic similarity theory, ethnocentrism, and group selection. In I. Eibl-Eibesfeldt & F. K. Salter (Eds.), *Indoctrinabilitv. Warfare, and ideology: Evolutionary perspectives* (pp. 369–388). Oxford, UK: Berghahn Books.
Rushton, J. P. (2005). Ethnic nationalism, evolutionary psychology and genetic similarity theory. *Nations and Nationalism, 11*(4), 489–507.
Rushton, J. P., & Nicholson, I. R. (1988). Genetic similarity theory, intelligence, and human mate choice. *Ethology and Sociobiology, 9*(1), 45–58.

Rushton, J. P., Russell, R. J., & Wells, P. A. (1984). Genetic similarity theory: Beyond kin selection. *Behavior Genetics, 14*(3), 179–193.

Sagan, L. (1967). On the origin of mitosing cells. *Journal of Theoretical Biology, 14*(3), 225–IN6.

Schonmann, R. H., Vicente, R., & Caticha, N. (2013). Altruism can proliferate through population viscosity despite high random gene flow. *PLoS One, 8*(8), e72043.

Schreiber, R. D., Old, L. J., & Smyth, M. J. (2011). Cancer immunoediting: Integrating immunity's roles in cancer suppression and promotion. *Science, 331*(6024), 1565–1570.

Seppa, P., Queller, D. C., & Strassmann, J. E. (2012). Why wasp foundresses change nests: Relatedness, dominance, and nest quality. *PLoS One, 7*(9), e45386.

Sober, E., & Wilson, D. S. (1998). *Unto others: The evolution and psychology of unselfish behavior*. Cambridge, MA: Harvard University Press.

Sumner, S., Casiraghi, M., Foster, W., & Field, J. (2002). High reproductive skew in tropical hover wasps. *Proceedings of the Royal Society, B, 269*(1487), 179–186.

Taylor, P. D. (1992). Altruism in viscous populations—An inclusive fitness model. *Evolutionary Ecology, 6*(4), 352–356.

Thorne, B. L., Breisch, N. L., & Muscedere, M. L. (2003). Evolution of eusociality and the soldier caste in termites: Influence of intraspecific competition and accelerated inheritance. *Proceedings of the National Academy of Sciences, 100*(22), 12808–12813.

Thrall, P. H., Hochberg, M. E., Burdon, J. J., & Bever, J. D. (2007). Coevolution of symbiotic mutualists and parasites in a community context. *Trends in Ecology & Evolution, 22*(3), 120–126.

Uris, L. (1984). *The haj*. New York: Bantam.

Waldhauer, I., & Steinle, A. (2008). NK cells and cancer immunosurveillance. *Oncogene, 27*(45), 5932.

Wehren Gaspar, J. C. W., López-Uribe, M. M., & Del Lama, M. A. (2007). Allozyme variation and sociogenetic structure of Polistes satan Bequaert 1940 colonies (Hymenoptera: Vespidae). *Genetics and Molecular Biology, 30*, 470–474.

West-Eberhard, M. J. (1967). Foundress associations in polistine wasps: Dominance hierarchies and the evolution of social behavior. *Science, 157*(3796), 1584–1585.

West-Eberhard, M. J. (1969). The social biology of Polistine wasps. *Miscellaneous Publications of the University of Michigan Museum of Zoology, 140*, 1–101.

West-Eberhard, M. J. (1975). The evolution of social behavior by kin selection. *Quarterly Review of Biology, 50*, 1–33.
Wheeler, W. M. (1910). *Ants: Their structure, development and behavior* (Vol. 9). New York: Columbia University Press.
Wheeler, W. M. (1911). The ant-colony as an organism. *Journal of Morphology, 22*(2), 307–325.
Wilson, D. S. (1975). A general theory of group selection. *Proceedings of the National Academy of Sciences, 72*, 143–146.
Wilson, D. S. (2016). Two meanings of complex adaptive systems. In D. S. Wilson & A. Kirman (Eds.), *Complexity and evolution: Toward a new synthesis for economics* (pp. 31–46). Cambridge, MA: The MIT Press.
Wilson, D. S., & Kirman, A. (2016). *Complexity and evolution: Toward a new synthesis for economics*. Cambridge, MA: The MIT Press.
Wilson, D. S., Pollock, G. B., & Dugatkin, L. A. (1992). Can altruism evolve in a purely viscous population? *Evolutionary Ecology, 6*, 331–341.
Wilson, D. S., & Sober, E. (1989). Reviving the superorganism. *Journal of Theoretical Biology, 136*(3), 337–356.
Wilson, D. S., & Sober, E. (1994). Group selection: The theory replaces the bogey man. *Behavioral and Brain Sciences, 17*(4), 639–654.
Wilson, D. S., & Sober, E. (1998). Multilevel selection and the return of group-level functionalism. *Behavioral and Brain Sciences, 21*(2), 305–306.
Wilson, D. S., & Wilson, E. O. (2008). Evolution "for the Good of the Group": The process known as group selection was once accepted unthinkingly, then was widely discredited; it's time for a more discriminating assessment. *American Scientist, 96*(5), 380–389.
Wilson, E. O., & Hölldobler, B. (2005). The rise of the ants: A phylogenetic and ecological explanation. *Proceedings of the National Academy of Sciences, 102*(21), 7411–7414.
Wu, J., & Lanier, L. L. (2003). Natural killer cells and cancer. *Advances in Cancer Research, 90*(1), 127–156.
Wynne-Edwards, V. C. (1962). *Animal dispersion in relation to social behavior*. Edinburgh, UK: Oliver and Boyd.
Yaeger, L. S., Griffith, V., & Sporns, O. (2008). Passive and driven trends in the evolution of complexity. *Proceedings of Artificial Life, 11*, 725–732.

3

Theoretical Foundations of Multilevel Selection Among Humans

Steven C. Hertler, Mateo Peñaherrera-Aguirre, and Aurelio José Figueredo

1 Introduction

The content of the previous two chapters described mathematical models and presented relevant empirical data pertaining to multilevel selection as a proposed biological universal within the general framework of evolutionary theory. The present chapter turns to phenomena that are believed to apply more specifically to humans. Consistent with the Darwinian *principle of continuity*, we are not claiming that humans stand alone as somehow separate from the rest of animal nature, given that the differences between human and nonhuman animals are most often differences in degree and not in kind. Nevertheless, there is also a case to be made that *all* species are to some extent unique and distinguishable from each other based on *species-typical characteristics*. Following from the principle of continuity, humans are not excepted from the forces of multilevel selection. Nevertheless, humans are unique by virtue of our species-typical characteristics, and so have a unique relationship to multilevel

S. C. Hertler et al., *Multilevel Selection*, https://doi.org/10.1007/978-3-030-49520-6_3

selection deriving from our unique evolutionary history. If we infer correctly, Wilson (2015) concurs with this assessment, figuring among those few authors who recognize the human species as having been particularly susceptible to multilevel selection throughout our evolutionary history. As an explanatory framework, multilevel selection might therefore be most interesting, elaborate, and probable among human populations precisely for the many complex qualities that qualify as human.[1] This section, and all the sections that follow within this chapter, can then be understood to explore the unique properties of humans, both as they were shaped by multilevel selection and as they allowed multilevel selection to assume unprecedented effects and directions. In sum, when simultaneously considering the aforementioned principle of continuity alongside species-typical human universals, one finds certain principles of multilevel selection uniquely applicable to our species and not many others. To fulfill this mandate, we provide prerequisite knowledge of *cultural evolution theories*, *gene-culture coevolution*, and *cultural group selection* before closing with an integrated section embedding group selection within the larger framework of multilevel selection theory.

2 Cultural Evolution Theories

Before addressing the underlying dynamics associated with cultural group selection, it is essential to provide a brief overview of the pertinent forces involved in the more generalized concept of cultural evolution. There has been a long history of theorizing on the evolution of human culture, dating back at least to the classic work of Morgan (1877). This early theory proposed a deterministic progression of social organization from more primitive to more advanced phrases, from *savagery* through *barbarism* to *civilization*. These theories were heavily criticized by the Boasian School of cultural anthropology, which introduced the alternative concept of *historical particularism* (e.g., Boas, 1896), advocating that cultures be studied instead on case-by-case bases, using *idiographic* rather than *nomothetic* approaches (Harris, 1968; Langness, 1974). Historical particularism eschewed the notion that some cultures were more evolved than others. As a result, general theories of cultural evolution fell into general disrepute in anthropology during the mid-twentieth century and were widely viewed as either implicitly or explicitly ethnocentric, if not actually racist.

The revolution in sociobiology of the 1970s and 1980s, however, saw a much renewed interest in evolutionary theories of culture. Having adopted the *gene's-eye view*, Dawkins (1976) proposed that culture could be envisioned as being comprised of discrete pieces of information, calling these elements of cultural transmission *memes*, by analogy with *genes*. This neologism was derived from the Greek *mimesis* or the process of imitation. Like genes, memes feature three properties of replicators: (1) *fidelity*, which concerns the degree to which copies of the meme remain unaltered even after numerous replicating events; (2) *fecundity*, which corresponds to the replicators' capability of generating multiple copies; and (3) *longevity*, which refers to the replicators' survivability (Dawkins, 1999).

As they concern only units of cultural replication, memes are not content-specific. Thus, memes not only serve as vehicles of teaching and productive cultural preservation, they can equally propagate propaganda, vice, or countercultural currents (Dawkins, 1981). As these memes can only exist and propagate by inhabiting human bodies, Dawkins (1976) imagined that memes could stand in any of various forms of symbiotic relation with genes, ranging from mutualism to parasitism. In a mutualistic symbiosis, memes would function to promote the fitness of the host organism's genes; in a parasitic symbiosis, memes would instead function to degrade the fitness of the host organism's genes. In either case, memes would be selected to serve their own self-interest, meaning their own *memetic* replication, and might only secondarily foster *genetic* replication instrumentally within mutualistic symbioses, as a mere means to an end.

For example, Dawkins considers religion to be a pernicious (meaning parasitic) meme (Dawkins, 2016). In the *God Delusion*, as elsewhere, Dawkins laments religion's "intolerance, blind faith, cruelty, extremism, abuse, and prejudice" (Wilson, 2008). Interestingly, Dawkins' extreme opposition to religion directly reflects his equally extreme opposition to multilevel selection; for it is multilevel selection that explains religious ideology, while also explaining religion's prosocial and antisocial values, respectively deriving from its within-group and between-group application.

With religious memes, as with political and moral memes, the value of a meme is dependent on the level of selection considered. Certainly, religious memes may divide populations, potentiating conflict and a sense of otherness *between* religious groups by lowering the threshold for suspicion, xenophobia, and violent confrontation. Yet, if one considers the influence of religion *within* religious groups, prosocial memes predominate, such as *love thy neighbor*, along with conflict-suppressing memes, such as *turn the other cheek*. Dawkins assumes a cosmopolitan perspective, criticizing religion from the species level, rather than at the group level. Even from that perspective, his criticism lacks a counterbalancing tally of prosocial goods produced by all religions collectively. Additionally, even from a non-evolutionary perspective of social criticism, religion should be considered in the context of coexisting modes of group definition, such as race, ethnicity, kinship, and language. At large, religion and religious memes are thrice mistreated, first from being judged at the species level, and next through an emphasis on its negative contributions to humanity's legacy, and thereafter in shouldering the blame for human suffering sometimes better attributed to overlapping group differences.

Thus, Dawkins tended to overemphasize the potentially parasitic role of memes with respect to genes, perhaps to emphasize their separable processes of replication. This emphasis was based on the premise that memes could replicate themselves independently of the genes of the host organism, using the latter as a vehicle analogously to the way in which viruses co-opt the reproductive machinery of cells to replicate themselves at the expense of their host. This idea of memes running amok and spreading like epidemic diseases to the detriment of host genes came to dominate what became known as the field of *memetics*. This asymmetrical treatment of gene-meme host-parasite relations, however, did not go unanswered. In 1981, Lumsden and Wilson published an influential book to counter some of these misconceptions, entitled *Genes, Mind, and Culture*. In this view, one had to consider both sides of the coevolutionary process between genes and memes, which they instead called *culturgens*, perhaps to set their theory apart from that of Dawkins. Essentially, just as a biological host is expected, as per evolutionary theory, to be selected for resistance to a biological parasite, one can expect the genes of the host organism to be selected for resistance to potentially fitness-damaging culturgens. This resistance came in the form of evolved *epigenetic rules of development*, which bias organismic learning processes to better serve the interests of genes. As Wilson (2004) infamously put it:

> The genes hold culture on a leash. The leash is very long, but inevitably values will be constrained in accordance with their effects on the human gene pool. The brain is a product of evolution. Human behavior—like the deepest capacities for emotional response which drive and guide it—is the circuitous technique by which human genetic material has been and will be kept intact. (p. 167)

Lumsden and Wilson (1985) understand cultural evolution to proceed from the interaction between overt behaviors and cultural creations, as well as the covert semantic memory representations that enable them. In other words, cultural evolution involves observable outward change alongside unobservable inward change. In this view, the study of cultural evolution involves the examination of those covert or unobservable semantic memory representations, both as they exist within individuals and as they are transmitted among them.

As we have seen, memes, in addition to their adaptive value being contextually dependent as with genes, can be indirectly relevant to multilevel selection in that they relate to group identity, cohesion, and differentiation. Traditional memes propagated with high fidelity can create cohesion and intergenerational continuity, which might simultaneously augment group solidity and distinctiveness (Sterelny, 2006). Memes might be legitimately said to promote individual or group-selected ideas. For example, *stoicism* and *hedonism* are examples of contrasting philosophies, which themselves can be considered memes relevant, respectively, to group and individual selection, as can be their constituent ideas.

3 Gene-Culture Coevolution

In further developing these themes, the concept of *gene-culture coevolution* is of foremost significance. Humans are bequeathed a *dual inheritance*, one genetic and one cultural (Henrich & McElreath, 2007;

Houkes, 2012; Richerson & Boyd, 1978). Gene-culture coevolution arises from the dynamic interaction of these bequests (Boyd & Richerson, 1976; Durham, 1979; Smith, 1990). Genes act through the humans they construct to create culture, but culture then acts directly on the selection of genes (Hare, Wobber, & Wrangham, 2012; Leach, Groves, O'Connor, Pearson, & Zeder, 2003). Dairying and lactase persistence are customarily invoked as proof of concept (Beja-Pereira et al., 2003). Calories from milk contribute to survival and successful reproduction, but there was no selection pressure driving the conservation past infancy of lactose-digesting enzymes prior to herding and dairying, both of which are cultural innovations (Gerbault et al., 2011; Itan, Powell, Beaumont, Burger, & Thomas, 2009). Incest taboos (Lumsden & Wilson, 1985) and the reduction in digestive organ mass consequent to cooking (Wrangham et al., 1999) are additional examples of gene-culture coevolution. So, human genes ultimately created memes, a culturally driven selective pressure to which they themselves responded via further biological evolution (Gerbault, Moret, Currat, & Sanchez-Mazas, 2009; Gerbault et al., 2011; Hollox, 2004).

It is doubtful that humans ever exclusively reacted passively to environmental pressures. Judging from archeological records, *Homo sapiens* have been living with the effects of *cumulative cultural adaptations* prior to diverging from other hominids. Empathy, guilt, and shame are among the complex emotional states predictive of prosocial behavior and within group cooperation. Such *social instincts* developed via natural selection, then inaugurated a coevolutionary process as they were expressed in tribal societies to the more radical reshaping of our *social psychology* (Boyd & Richerson, 2005). However, agriculture, writing, and thereafter the technological and mechanical outgrowths of the scientific and industrial revolutions, accelerated the evolutionary process, making plausible the advanced evolutionary pace featured in works such as the *10,000 Year Explosion* (Cochran & Harpending, 2009).

The concept of gene-culture coevolution also invokes a variety of ecological and biological terms, such as *niche construction* and the *extended phenotype*. A niche-constructing organism alters the environmental selective pressures to which it is subject and in this way shares in one component of gene-culture coevolution (Odling-Smee, Laland, & Feldman, 2000.)

Shared variance between these concepts increases to the extent that the products and processes of niche construction are, respectively, transmitted and taught (Odling-Smee and Laland, 2000; Peterson et al., 2011; Schoener, 2009). Bowers, damns, nests, scatter hordes, and middens are all examples of extended phenotypes (Dawkins, 1982), which afford their creators an altered, and favorable, selective regime (Weiss & Buchanan, 2009). To the extent that intergenerational learning and transmission supplements heritable instinct, extended phenotypes have some of the elements and consequences of culture (Gintis, 2011; Laland & Hoppitt, 2003). In relation to these terms, gene-culture coevolution remains hierarchically above them, being a more generalized and inclusive process; it is also a process especially applicable to humans. This is because culture, though evident in animals, attains to baroque complexity in humans alone. The life histories, group sizes, cognitive powers, and linguistical abilities are among those unique properties allowing multilevel selection expression in novel avenues and different manners, as compared with the more limited scope of action multilevel selection assumes among most nonhuman animals outside the eusocial insects. And so one can certainly consider potato washing (Nakamichi et al., 1998) and tool use among primates (Van Lawick-Goodall, 1970; Van Schaik, Deaner, & Merrill, 1999), aquatic animals (Krützen et al., 2005; Mann & Patterson, 2013), and birds (Hunt, 1996) as behaviors that are probably or mostly learned, though in this respect humans are so different in degree as to approach a difference in kind. Flexibility, modes of transmission, and sheer quantity of information passed on vertically and horizontally are unprecedented in humans. Consequently, there are unprecedented outcomes, among which cooperative norms can be counted.

A key contribution to gene-culture coevolution theory came with the more detailed specification of the epigenetic rules of development governing the propagation of cultural traits, redubbed *cultural transmission biases* (Boyd & Richerson, 1983, 1985; Richerson & Boyd, 1978, 1984). After making the case for a Darwinian approach to cultural evolution, Richerson and Boyd (2005) proceeded to classify the various forces required for cultural evolution, including the pertinent mechanisms generating variance among cultural items. The authors distinguish between *random* and *decision-making* forces. Random influences can create cultural variation from incorrectly storing or recalling a cultural item and transmitting this *cultural mutation* to others, or otherwise variation can extend from *cultural drift*, a population-level stochastic progression influencing the likelihood a cultural variant will persist across generations analogous to Sewall Wright's *genetic drift*. In contrast to random influences, decision-making forces are characterized by the level of active engagement of models and imitators transmitting and modifying cultural information. Richerson and Boyd (2005) further classify these forces into guided variation and biased transmission. According to the authors, guided variation is evidenced when individuals actively modify cultural variants by means of social learning or invention. Therefore, the rate of cultural evolution is dependent on the degree of cultural variability within the group.

Richerson and Boyd (2005) further classified biased transmission into three types of social learning: *content-based*, *frequency-based*, and *model-based biases*. Regarding content-based biases, learners weigh the costs and benefits of adopting a cultural variant instead of an available alternative. As mentioned before, it is not infrequent for cultural items to compete for cognitive and behavioral resources required for their storage and transmission (Richerson & Boyd, 2005). It follows then that learners could be inclined to adopt cultural items that do not demand a significant cognitive load to memorize, perform, or transmit to other individuals. This preference, however, does not preclude more taxing cultural variants from persisting in a population. For example, in small-scale and state societies, individuals able to master challenging skills often attain gains ranging from wealth to prestige. However, the frequency of these cultural items is expected to be lower relative to less burdensome variants. Frequency-based biases exist as two versions: Adopters can either copy the most common behavior or instead acquire the rarest. The benefits attained from each alternative are frequency dependent. Finally, model-based biases are evidenced when individuals prefer to copy others based on the models' attributes. For instance, adopters may emulate people with whom they share a feature, such as speaking the same language, having a similar ethnicity, or based on kinship; alternatively, learners may choose to imitate prestigious individuals. This theory also specifies the environmental conditions under which these different transmission biases might be favored by selection. Relatively stable environments are thought to indicate that the most frequent cultural traits are well adapted to the local environment and thus favor the evolution of conformity bias (Creanza, Kolodny, & Feldman, 2017).

In their influential book, *Not by Genes Alone*, Richerson and Boyd (2005) maintained that culture follows three Darwinian principles. First, inter-individual variation rises due to the differential acquisition of beliefs and behaviors by means of social learning; second, this variation influences the likelihood a cultural item will be transmitted to other members in the group; and, third, cultural variants compete for cognitive resources; hence, the number of cultural items within a population cannot grow indefinitely. Richerson and Boyd develop this theory without explicitly premising it on the presence of particulate cultural replicators (memes or culturgens), instead casting these more generally as *cultural traits* or *cultural variants*.

Models of Darwinian cultural evolution also considered the role of genetic selection as a necessary and complementary force for cultural evolution. Furthermore, this perspective described the interplay between genetic selection and cultural variation occurring at the level of individuals as well as between groups (Richerson & Boyd, 2005). Cultural variants, for example, can affect the fitness of genes. Given these features, one may interpret biased transmission as a selective process. Richerson and Boyd (2005) acknowledged that terms such as *cultural selection* provide a similar approximation to the persistence and transmission of cultural items due to individuals' genetically evolved preferences. However, the imitation of cultural items (i.e., diffusion) is not equivalent to an organism's genetic fitness (Henrich, Boyd & Richerson, 2008; Richerson & Boyd, 2005). Genetic selection and social learning biases guide complementary and interacting rather than equivalent and parallel aspects of cultural evolution. For example, the assimilation of cultural information rests on underlying cognitive substrates, such as semantic and episodic memory, evolved by means of genetic selection (Lumsden & Wilson, 1981).

In terms of the transmission of cultural information, learners thus seem to gravitate toward *cognitive attractors*, concentrating most of the cultural variation (Henrich, Boyd & Richerson, 2008). This notion predicts that the presence of multiple attractors generates the conditions wherein even weak selective forces increase the likelihood that individuals will be driven to an attractor (Henrich et al., 2008). Contrary to the critics' position, where cultural evolution requires the preexistence of random variation, Darwinian models of cultural evolution presume that selection can operate even though the observed cultural variance is the outcome of nonrandom forces (Henrich et al., 2008).

Even though the notion that genes and memes (or culturgens) share several features was initially well received, this position has faced recurrent criticisms over the last three decades. At the center of this debate were the properties of fidelity and particulate transmission, wherein each unit retains its integrity and it is not recombined during replication. Cognitive and social psychologists, such as Sperber, questioned the argument that memes, or mental representations, existed as clearly defined units of information stored in brains and transmitted with precision. Instead, Sperber described mental representations as a continuum. Other authors noticed that, differently than genes in which allelic integrity is preserved, cultural items were often recombined by learners in a manner contrary to the presumed particularity exhibited by cultural variants. These theoretical pitfalls, among others, encouraged critics to ponder whether evolutionary examinations of cultures were theoretically misguided. Although Richerson and Boyd (2005) concurred with critics of meme theory in that representations and cultural items are not discrete pieces of information, and are often modified during transmission, the authors disagreed with positions arguing against Darwinian models of cultural evolution based on the restricted level of fidelity and particulate transmission of cultural variants. As per Richerson and Boyd, while cultural items and their corresponding mental representations are subject to copying errors, the replication of average cultural traits at the population level enables cumulative cultural evolution (Boyd & Richerson, 2005; Henrich et al., 2008; Richerson & Boyd, 2005).

Differences of opinion regarding these exact details of mechanism aside, the basic take-home message is that genes and cultural traits (memes, culturgens, cultural variants, or whatever one chooses to call them) coevolve, exerting reciprocal selective effects upon each other over evolutionary time. The "nature" and "nurture" aspects of our evolved adaptive heritage are therefore intimately intertwined and have never been completely independent of each other. Human culture is as much of an evolved adaptation as the human genome, shaped by natural, social, and sexual selection. While genes do hold culture on a leash, culture also influences the social environments in which genes subsequently evolve. To extend this metaphor, the cultural dog often pulls back on the genetic leash. Having reviewed some of gene-culture coevolution's content and controversies, we can apply this prerequisite information to a more precise review of cultural group selection, first in relation to morality and thereafter in relation to religion.

4 Cultural Group Selection: The Case of Morality and Moralistic Punishment

In addition to developing a taxonomy of evolutionary forces necessary for cultural evolution, evolutionary researchers proceeded to generalize these tenets to address the evolution of human cooperation and between-group selection. In contrast to other multilevel selection perspectives, cultural group selection theory maintains that selection has a greater influence on intergroup cultural variation rather than genetic variation (Boyd, Richerson, & Henrich, 2011a, 2011b). This approach emphasized the role of social learning biases and the moralistic enforcement of social norms in sustaining between-group variance. Supporters of cultural group selection theory acknowledged that, even though kin selection and reciprocal altruism provide cogent explanations for the persistence of cooperation in small groups, these perspectives were insufficient for understanding the evolution of prosocial behaviors in large-scale societies. Differently than in small groups, where people encounter other members of the community repeatedly, inhabitants of cities or nation-states engage not only with kin and kith but also with strangers. Anonymity and the unlikely possibility of re-encountering a stranger in the future increase the risk of defection in large-scale societies. Hence, collective strategic action and group cohesion depend on an array of cognitive mechanisms and emotions (such as guilt and shame; Richerson & Boyd, 2005) constituting the biological foundations of morality.

In that it interfaces with altruism, morality has obvious connections to multilevel selection. Morality constrains behavior in an informal, secular manner, sharing the burden of within-group social regulation with formal legal systems and sacred religious systems. Many theories of the evolution of morality, such as Darwin's (1871) theory of *moral sentiments*, do not depend on cultural transmission, but are implicitly nativist and are presumed to be evolved directly as heritable traits. Nevertheless, although moral and altruistic impulses might be evolved as general tendencies, it is likely that the specific *content* of moral regulations might be culturally variable and transmissible.

One such implicitly nativist theory is that of Jonathan Haidt (2013a), who has advanced a theory of *moral foundations*—purportedly human universals that may vary cross-culturally only in degree and not in kind. Consistent with our position, Haidt acknowledges group selection's cooperative benefits expressed within groups, contrasting them with their conflict-promoting tendencies when expressed between groups. It is not surprising that he is an adherent of multilevel selectionism because his system of moral emotions lends itself to that perspective. Haidt distinguishes between other-condemning emotions, such as *contempt*, *anger*, and *disgust*, and those relating to oneself, such as *fairness* and *harm avoidance*. The first class of moral emotions consists of judgmental dispositions aimed outward at others and in some way contributing toward regulating group behavior. These other-condemning emotions motivate rule creation, enforcement, and moralistic punishment, all tending to direct and order society, with implications for its subsequent cultural evolution and prevailing selective regime. Alternatively, fairness and harm avoidance are self-referential, demanding that oneself be treated equitably and allowed to pursue life, liberty, and happiness (Haidt, 2001, 2003, 2013b). For instance, those at once governed predominately by self-referential dispositions and relatively free of other-condemning emotions are apt to agree with Jefferson,[2] who opined, "…it does me no injury for my neighbor to say there are twenty gods or no God. It neither picks my pocket nor breaks my leg" (Kaminski, 2006). Self-referential moral emotions relating to fairness and harm avoidance appear to be universal, showing themselves ubiquitous in cross-cultural ethnographic studies. On the other hand, the latter three forms of morality, being consistently found only among conservatives and emphasizing the group and its cohesion and strength (Haidt, 2007), are more variably expressed. Revisited in the empirical reviews of Part II, moral emotions cluster along a single broad axis, which we can array, in turn, along a group to individual selection spectrum.

The evolution of such moral sentiments presupposes sufficient group-selective pressures. Recall that in the group selection debates detailed in Chaps. 1 and 2, even the staunchest critics of multilevel selection generally allowed that group selection was possible, but only so exceedingly improbable as to rarely if ever exist. Group selection's improbability extended from the difficulty of suppressing individually selected selfishness. Indeed, even the staunchest supporters (Wilson, 2015) recognize that suppressing individual selectionist tendencies is a serious problem that must be surmounted if group selection is to operate:

> The suppression of disruptive forms of lower-level selection is only partial and by no means complete. Everyday life and the annals of history are replete with examples of individuals and factions that succeed at the expense of their groups, despite the arsenal of social control mechanisms designed to thwart them. (pp. 49–50)

Though both supporters and detractors recognize individual selection's ability to undermine group selection, only the former have fully thought through the fullness of effort with which the aforementioned "arsenal of social control mechanisms" have suppressed individual selection, so as to allow group selection to take hold. Social control mechanisms are manifold and will be discussed throughout Part II as we provide original reviews of historical examples. Here, consistent with the scope of this first section, we note that other-referential sets of moral emotions motivate behaviors extending beyond self-interested motivations, to induce cooperation and cohesion at the highest level of group formation. Contempt, anger, and disgust are among the moral emotions that induce some to regulate the behaviors of others. These emotions relate to what might be called the regulation of one's social niche or the regulation of the social community. Each of these emotions contributes to *moralistic punishment*, which coerces group cohesion.

Moralistic punishment is explained as follows by Boyd and Richerson (2005):

> To prevent a defector from eating, somebody has to intervene when he reaches into the pot. That someone has to undertake a (perhaps) costly action that reduces the payoff of the defector and thus produces a benefit to the group as a whole. This is an example of what Trivers called "moralistic punishment" and applies to a much wider range of problems than excluding factors from the fruits of cooperation. Even if the defectors cannot be excluded, punishment can create incentives for them to cooperate. Cowards may get the benefits of group defense, but they may also be shunned, beaten, or banished. The real question is under what conditions can selection favor moralistic punishment? (p. 138)

Relative to reciprocal altruism, moralistic punishment operates with greater efficiency providing enough social stability for the evolution of cooperation in sizable groups (Boyd & Richerson, 1992). This difference arises from the type of penalty imposed on free riders. Reciprocal altruism models merely presuppose that cooperators will respond to defectors by refusing to cooperate with them in the future. Alternatively, mathematical models of moralistic punishment consider the effect of cooperators' retribution on free riders (Boyd & Richerson, 1992). Active retribution will inevitably be superior to noncooperation in suppressing defection and free-riding, and thus moralistic punishment provides a more plausible pathway to cooperative civilization, as compared with reciprocal altruism. Moreover, since noncooperators are targeted on a case-by-case basis, moralistic punishment circumvents the risk of initiating a chain of defection (Boyd & Richerson, 1992). Thus, punishment, through social learning, can deter further defection, free-riding, and noncooperation.

Boyd and Richerson's (1992) mathematical model determined that retribution enables the evolution of cooperation through two main avenues. First, if moralistic individuals can obtain long-term gains from cooperation, even if it implies accruing a personal cost when punishing noncooperators, it follows that moralists, non-punishing cooperators, and individuals that cooperate only to avoid being punished may co-occur within a population. Second, moralistic strategies (e.g., ready to cooperate, determined to punish defectors, and keen to punish individuals unwilling to retaliate against noncooperators) can become evolutionarily stable if the costs of experiencing any punishment are sufficiently onerous (Boyd & Richerson, 1992). As opposed to reciprocal altruism, the size of the group does not influence the severity of the punishment inflicted on noncooperators (Richerson & Boyd, 2005). Critics claim punishment itself does not solve the collective action problem, but instead generates a new dilemma: Because punishers incur costs, such as facing the defector's reactive aggression, punishing defectors itself constitutes an altruistic behavior. As such, altruistic or moralistic punishers might exclusively incur costs, while other individuals in the group attain the benefits associated with the resultant prosociality. Experimental data, however, reveal that many human punishers are willing to accept a cost to themselves if the defector is punished without falling into a second-order cooperation dilemma. Furthermore, as discussed in Sect. 4 of Chap. 5, there are ways in which mature societies equitably apportion the costs of meting out punishment.

Although cultural group selection theory offers several predictions regarding the evolution of cooperation, the authors did not develop this approach as a substitute for perspectives examining the role of reciprocal altruism, or kin selection, relative to the persistence of prosocial behaviors (Richerson & Boyd, 2001). Hence, for cultural group selection theory, social tendencies, such as punishing defectors and bias toward conformity (tribal social instincts; Richerson & Boyd, 2001), are laid over more ancestral prosocial tendencies. Although under some circumstances, kin selection, reciprocal altruism, and tribal social instincts can act synergistically, it is not uncommon for these dynamics to eventually clash (Richerson & Boyd, 2001). For instance, individuals must comply with social norms that guarantee the provision of goods and services to the group, above and beyond their personal preference to allocate these resources to their relatives and allies. Cultural group selection must curb self-oriented behaviors to sustain cooperation in sizable groups, such as tribal organizations (Richerson & Boyd, 2001).

5 Cultural Group Selection: The Case of Religion

As with moral sentiments, *religiosity* is a highly heritable trait (Ludeke, Johnson, & Bouchard, 2013), and thus presumably subject to direct natural and sexual selection. Religiosity, however, determines the *degree* to which one is religious, without specifying the particular religious *content* to which one adheres. As with specific moral injunctions, specific religions are culturally transmitted belief systems subject to cultural group selection. Both with respect to heritable religiosity and culturally evolved religious content, religion is an especially important instance of cultural group selection because, as the works of Norenzayan and colleagues (2006; 2010; 2012; 2013; 2016) demonstrates, religion is a "psychological adhesive, promoting coordinated action and altruistic self-sacrifice" (Hertler, 2017). Norenzayan[3] (2013) celebrates religion's ability to induce "unprecedented cooperation within ever expanding groups,"[4] even as he acknowledges the ways in which religion can serve as "source of potential conflict between competing groups." Beyond the bounds of kin and kith, religion imparts cooperation and coordination, as only more lately, and less effectually, accomplished by nationalism and nation-states (Hertler, 2017). Consider that martyrdom and celibacy, for example, are conceivable as extreme forms of altruistic religiosity, but might have insufficient evolutionary rationale from an individual-level perspective. One has to turn to group selection to evolutionarily explain why the religious martyr and the celibate clergyman would respectively forfeit their survival and reproduction for an abstraction (Vaas, 2009).

D. S. Wilson (1980, 1983, 2010) and E. O. Wilson are one among a growing cadre of authors interpreting religion, not simply sociobiologically, but explicitly from a multilevel selectionist perspective. As previously reviewed (Hertler, 2017), E. O. Wilson believes:

> …that religious dogma is a consequence of evolution. Religious belief and the firm adherence to it–and the intense dislike of apostates, people who abandon it–has a very important biologic origin, probably through natural selection, namely the cohesion of the group and the persuasion of people to be more altruistic.[5]

D. S. Wilson's *Darwin's Cathedral* opens with the observation that churchgoers often compare their congregations to social insect colonies. Wilson takes the comparison seriously as a scientific hypothesis, which is then explored in the remainder of his book. Religious mores, rules, regulations, laws, threats, and punishments are deployed as mechanisms of social control to shape social behavior, not exempting seemingly innocuous behaviors, such as dancing and dress. Sometimes social control wrought conformity, while other times it countenanced controlled dissent within the group, without tolerating outright subversion. Here, Wilson (2002) uses Calvinism as an exemplar of religious purpose, arguing that Calvin's words and deeds jointly erected a pattern for imitation:

> Calvin's passion for setting a moral example of selflessness is well known, culminating in his burial in an unmarked grave, but more important are the many features of the religion that are clearly designed to constrain the self-will of the leaders as effectively as the rank-and-file. (p. 119)

"Calvinism," Wilson insists, "is designed to make a human community function as an adaptive unit." Even as he cautions scientists from becoming carried away in their analytical perspective, as did many scions of the Enlightenment in expressing contempt for religion, Wilson frames religious belief as an outgrowth of the adapted mind, which is the product of, and also promotes, multilevel selection. When replacing rationality with evolution as the standard of evaluation, the adaptive significance of religion becomes manifest. Religions impart bids for control, banishment, persecution, and other forms of religious coercion that, on their face, seem at odds with religion—at least those religions like Catholicism emphasizing acceptance and love. "If *Amnesty International* existed in the sixteenth century," Wilson quipped, "every religious and political organization in Europe would be on its list." In support of his conjecture, Wilson cites the use of ostracism, persecution, and capital punishment historically employed to impart social control and root out heresy. Again, through the prism of multilevel selection, the rationale is obvious: Altruism and cooperation strengthen groups, whereas selfish and disruptive behaviors weaken groups. Religion, understood as a boon to group selection, rewards and punishes so as to create optimal levels of cooperation (Wilson, 2015, p. 89). Thus, "most enduring religions are impressively designed to motivate altruism at the level of action by promoting behaviors that are for the good of the group and suppressing disruptive self-serving behaviors within the group" (Wilson, 2015, p. 89). Consequently, religious dogmas and doctrines implicitly or explicitly distinguish between the *ingroup* and the *outgroup*, variously referred to as the *chosen people* and the *infidels*.

If we might condense and order Wilson's thoughts, religion can be said to: (1) promote within-group harmony; (2) impose within-group conformity; and (3) differentiate those within the group from those without. These functions are relayed in the order in which they might be explicitly acknowledged. In other words, a religious adherent would readily acknowledge acceptance and harmony, community, and belonging as tacit values or goals of religious participation, while the second function may be reluctantly seen as a rare and reluctantly resorted to necessity, with the third function apt to be denied altogether. Nevertheless, from a group selectionist perspective, religions create competitive advantage, the successful pursuit of which relies on cohesive groups, clearly delineated from rival groups. Truly, all three functions promote success in competitive bids between groups. In sum, the "successful" religion, altogether, promotes cooperation, suppresses dissent, and separates *us* from *them*.

By way of contrast, Fetchenhauer (2009) is representative of a divergent voice audible among evolutionists. Fetchenhauer is unconvinced, believing multilevel selection an inadequate explanation of religion. However, an analysis of his objections suggests that they are not grounded in a sufficient knowledge of multilevel selection. For instance, speaking of D. S. Wilson, and then generalizing the charge to multilevel selectionists at large, Fetchenhauer alleges ignorance of intragroup conflict, stating that, "humans will be under selection pressure not to let themselves be exploited for 'the good of the group.'" In answer, one can first find the charge unfounded, as D. S. Wilson and other multilevel selectionists are ever aware of individual selection as a counterweight to group-selected evolution. Religion is a cultural mechanism enlisted to act against that very threat. Religion itself deploys excommunication, threats of ostracism, costly signals, social monitoring, and other mechanisms of social control leveled at individual selfishness. Extrapolating from religion to the chapter theme at large, we see that gene-culture coevolution and cultural group selection afford many avenues wherein behaviors transition from maladaptive to adaptive based on their expression inside or outside complex culture.

Before moving on, we take up another point of contention registered by Fetchenhauer (2009). After highlighting a general want of empirical support, Fetchenhauer (2009) posits an inverse relationship between religion and social cohesion. In doing so, the United States is contrasted with Scandinavia, with elevated homicide rates and religiosity co-occurring in the former, but not the latter. In the same critique, Fetchenhauer observes infighting among Christians throughout history, concluding that the "willingness to strongly distinguish their ingroup from any outgroup presumably evolved totally independently from religion." Problems abound in this analysis of multilevel selectionist explanation of religion of the kind that have plagued multilevel selection more generally. Once again, criticism extends beyond understanding. There are three outstanding problems with Fetchenhauer's line of criticism outlined before closing this section on religion. First and most basically, from religion in the context of multilevel selection, Fetchenhauer elides to religiosity, with the latter concept being a gross measure of religious adherence, with no distinction made among religious groups. This conflation of concepts invites descriptive and predictive error, especially when applied to nations that are both religious and sectarian. For example, the United States cannot be treated as a religious monolith, as if there were not a host of religions now present in its polyglot present and a host of Christian sects present from its inception. As such, an increase in violence is precisely what one would expect from this religious diversity. The proper measure then is not national violence, but separate measures of violence calculated within and between groups. Recall, religion is not an indiscriminate violence suppressing mechanism, but a cultural adaptation that, among other functions, serves to differentiate groups within larger populations. Second, one must understand that groups are differentiated by lineal descent, ethnicity, and related biocultural markers, relegating religion to one of many mechanisms of group identification. Considering some of these aforementioned markers of group identity, the United States is then far less united than Scandinavia, even as one is a single country and the other a geographic region with multiple countries. In short, religion must be analyzed alongside other contributory factors when explaining conflict. Third, Fetchenhauer cites medieval Christian infighting as evidence against multilevel selectionist explanations of religion, which ignores the tiered framework along which group formation is nested. Yes, Christian nations fought with one another because, for instance, English Christians had interests at variance with French Christians, but English and French Christians crusaded against the Saracens, who were both religiously and ethnically different. The larger point is that groups form and fracture along multiple fault lines, which sometimes pull in opposite directions.

6 Multilevel Section Expanded

The theory of *multilevel selection*, to which the present book is dedicated, reformulates the question of whether any given evolutionary phenomenon is due exclusively either to individual selection or to group selection. Multilevel selection is now seen as not only theoretically possible but indeed more probable than previously thought. As multilevel selection theoretically predicts simultaneous selection at multiple levels of biological organization, the question then becomes how both group and individual selection can operate simultaneously in any given instance of trait evolution.

As described in the preface to this volume, multilevel selection has indeed been shown to operate at various hierarchically nested levels of biological organization, such as: (1) clonal selection; (2) neuronal selection; (3) respondent and operant/instrumental selection; (4) cultural selection; and (5) biocultural group selection. All these examples are expanded upon there. The point that we are trying to make at present is that the polemics of the past appeared to emphasize the operation of selection at these different levels as *alternative hypotheses*, which were presumably considered to be mutually exclusive, rather than *concurrent processes*. This misstatement of the problem might have been due to the positing of different levels of selection as alternative explanations of the same phenomena, rather than as complementary descriptions of related phenomena that were often concurrent. This was presumably done in the service of parsimony, in an attempt to eliminate the need to posit one or the other level of selection, but this consideration is no longer relevant once evidence for the concurrent operation of selection at multiple levels was established.[6]

The situation is analogous to that described by Darwin (1871) relative to the relation between natural selection and sexual selection. In the famous case of the peacock's tail, Darwin was initially concerned that the extravagant and colorful plumage of the peacock would attract the unwanted attention of potential predators, and thus impose a fitness cost to the peacock in terms of natural selection. To solve this intellectual conundrum, Darwin proposed that this fitness cost was offset by the fitness benefit entailed in concurrently attracting the more desired attention of potential mates, based on the presumed preference by the peahen for the said plumage; and thus plumage is favored by the opposing force of sexual selection, a then novel concept that Darwin introduced at this time and for this very purpose. The new question in Darwin's mind was therefore not whether the peacock's tail was shaped exclusively by natural selection or exclusively by sexual selection. The realization was that it was instead being shaped simultaneously by both and that the length of a peacock's tail was in many ways a compromise between the relative strengths of the two opposing forces.

This mental model of opposing forces was indeed the way that Darwin (1871) characterized the evolution of altruistic traits in humans, with between-group competition favoring them and with ingroup competition disfavoring them. Such opposing forces could vary in relative magnitude between environments, so that it was not possible to presume that one force would generally triumph over the other in all cases. Furthermore, it was not clear that different evolutionary selection pressures would always be in opposition to one another. In the case of sexual selection, Hamilton and Zuk (1982) proposed a *good genes* sexual selection theory in which sexual selection could act in concert with natural selection when female preferences were consistent with traits favoring male survival. One such trait was parasite resistance, which at once enhanced lush and colorful plumage in the males of many bird species and contributed to the viability of the female's offspring sired by resistant males. By analogy, there is no reason to presuppose that individual selection and group selection must always act in opposition to one another. It is relatively easy to imagine scenarios where certain traits could enhance both the survival and the reproduction of an individual and that of the social group to which the individual belongs.

As emphasized by Sober and Wilson (1998), the overall covariance between a phenotypic trait and fitness, comprising group-level and individual-level covariances, should be analyzed on a case-by-case basis. Whereas, in some instances, group selection and individual selection vectors can act in opposite directions, in other cases, these vectors operate in parallel trajectories. Consider the evolution of warriorship in small-scale human societies. Warriors in tribal groups can attain individual fitness benefits by building a reputation for formidability and bravery (Chagnon, 1988) while at the same time providing a service to the community such as decreasing the risk of being victimized by rival groups or bringing goods and captives to the group. Groups with a higher average of "warriorship" can outcompete groups with lower averages. Alternatively, if the warrior does not obtain any fitness gain, but the group still acquires a collective benefit, the trait can persist in the population if the force of group selection outweighs the influence of individual selection.

7 Conclusions

Thus far, across the previous two chapters, we have treated the intellectual history of multilevel selection as an abstract concept, integrating instructive examples, such as that of viruses, cancer, slime molds, and eusocial insects. In this third chapter, we have been preoccupied with reviewing intellectual history, describing animal comparisons, modeling, mathematical proofs, and experimental paradigms which establish the likelihood of human multilevel selection. All of this had to be accomplished; it is in some sense prerequisite to further theorizing. This third chapter turned to humans while still confining itself to the review of human multilevel selection as it has been previously treated by other authors. As we have seen, humans inherit complex cultures alongside genetic information, with their interaction understood through the framework of gene-culture coevolution, of which cultural group selection theory is partially a consequence. After reviewing morality and religion as examples of cultural group selection, our focused review of group selection was incorporated into a broader view of multilevel selection. Thus equipped, one can productively turn to Parts II and III of this book armed with some cognizance of what came before, providing context and contrast, along with an understanding of key theoretical concepts.

Notes

1. As Wilson states:

> Against this background, our distinctiveness as a species can be summarized in a single sentence: We are evolution's latest major transition. Alone among primate species, we crossed the threshold from groups of organisms to groups as organisms. Other primate species cooperate to a degree, sometimes to an impressive degree. But disruptive within group competition for mates and resources is still a strong evolutionary force. Even the cooperation that does take place within primate groups often consists of coalitions competing against other coalitions within the same group. Our ancestors managed to suppress disruptive forms of within group competition, making benign forms of within group selection and between group selection the primary evolutionary forces. (p. 49)

Wilson continues in this vein, comparing humans to eusocial insects, taking the biomass of each as a metric of success in a manner that recalls E. O. Wilson's *The Social Conquest of the Earth*:

> All of the hallmarks of a major evolutionary transition are present in the human case. It was a rare event, happening only once among primates, and the combination of a species that is both functionally organized at the group level and highly intelligent at the individual level is doubly rare, as we shall see. It had momentous consequences. Just as eusocial insects constitute over half of the insect biomass on earth, we and our domesticated animals represent a large fraction of the vertebrate biomass on earth, for better or for worse. (pp. 49–50)

2. As quoted in Jefferson (1984), from his *Notes on the State of Virginia*.

3. These quotes are taken from the 2013 hardback edition's dust cover.
4. Relative to the ability of religion to expand group size, Purzycki et al. (2018) studied ritual and religious commitment's binding effect expressed within human groups using an *experimental economic game* paradigm. Spanning African, Southeast Asian, Siberian, and South American samples, this cross-cultural study supports religion's ability to expand circles of exchange beyond regional loyalties. While partiality to neighboring persons presented as the default condition, participants, through their experimental allocations of resources, showed partiality also extending toward coreligionists, to the extent that gods were moralistic, omniscient, omnipotent, and purportedly capable of supernatural punishment.
5. http://www.tampabay.com/news/perspective/eo-wilson-on-ants-and-god-and-us/903761
6. For a more detailed critique of the limitations of the parsimony argument, see Sober and Wilson (1998).

References

Beja-Pereira, A., Luikart, G., England, P. R., Bradley, D. G., Jann, O. C., Bertorelle, G., et al. (2003). Gene-culture coevolution between cattle milk protein genes and human lactase genes. *Nature Genetics, 35*(4), 311.

Boas, F. (1896). The limitations of the comparative method of anthropology. *Science, 4*(103), 901–908.

Boyd, R., & Richerson, P. J. (1976). A simple dual inheritance model of the conflict between social and biological evolution. *Zygon, 11*(3), 254–262.

Boyd, R., & Richerson, P. J. (1983). The cultural transmission of acquired variation: Effects on genetic fitness. *Journal of Theoretical Biology, 100*, 567–596.

Boyd, R., & Richerson, P. J. (1985). *Culture and the evolutionary process*. Chicago: Chicago University Press.

Boyd, R., & Richerson, P. J. (1992). Punishment allows the evolution of cooperation (or anything else) in sizable groups. *Ethology and Sociobiology, 13*(3), 171–195.

Boyd, R., & Richerson, P. J. (2005). *The origins and evolution of cultures*. New York: Oxford University Press.

Boyd, R., Richerson, P. J., & Henrich, J. (2011a). Rapid cultural adaptation can facilitate the evolution of large-scale cooperation. *Behavioral Ecology and Sociobiology, 65*, 431–444.

Boyd, R., Richerson, P. J., & Henrich, J. (2011b). The cultural niche: Why social learning is essential for human adaptation. *Proceedings of the National Academy of Sciences of the United States of America, 108*, 10918–10925.

Chagnon, N. A. (1988). Life histories, blood revenge, and warfare in a tribal population. *Science, 239*(4843), 985–992.

Cochran, G., & Harpending, H. (2009). *The 10,000 year explosion: How civilization accelerated human evolution*. New York: Basic Books.

Creanza, N., Kolodny, O., & Feldman, M. W. (2017). Cultural evolutionary theory: How culture evolves and why it matters. *Proceedings of the National Academy of Sciences, 114*(30), 7782–7789.

Darwin, C. (1871). *The descent of man and selection in relation to sex*. London: John Murray.

Dawkins, R. (1976). *The selfish gene*. Oxford, UK: Oxford University Press.

Dawkins, R. (1981). Selfish genes and selfish memes. In *The mind's I: Fantasies and reflections on self and soul* (pp. 124–144). New York: Bantam Books.

Dawkins, R. (1982). *The extended phenotype* (Vol. 8). Oxford, UK: Oxford University Press.

Dawkins, R. (1999). The selfish meme. *Time, 153*(15), 52–53.

Dawkins, R. (2016). *The god delusion*. New York: Random House.

Durham, W. H. (1979). Toward a coevolutionary theory of human biology and culture. In N. A. Chagnon & W. Irons (Eds.), *Evolutionary biology and human social behavior* (pp. 39–59). North Scituate, MA: Duxbury Press.

Fetchenhauer, D. (2009). Evolutionary perspectives on religion–what they can and what they cannot explain (yet). In E. Voland & W. Schiefenhövel (Eds.), *The biological evolution of religious mind and behavior* (pp. 275–291). Waldhilsbach, Germany: Springer-Verlag.

Gerbault, P., Liebert, A., Itan, Y., Powell, A., Currat, M., Burger, J., et al. (2011). Evolution of lactase persistence: An example of human niche construction. *Philosophical Transactions of the Royal Society B: Biological Sciences, 366*(1566), 863–877.

Gerbault, P., Moret, C., Currat, M., & Sanchez-Mazas, A. (2009). Impact of selection and demography on the diffusion of lactase persistence. *PLoS One, 4*(7), e6369.

Gintis, H. (2011). Gene–culture coevolution and the nature of human sociality. *Philosophical Transactions of the Royal Society B: Biological Sciences, 366*(1566), 878–888.

Haidt, J. (2001). The emotional dog and its rational tail: A social intuitionist approach to moral judgment. *Psychological Review, 108*(4), 814.

Haidt, J. (2003). The moral emotions. *Handbook of Affective Sciences, 11*, 852–870.

Haidt, J. (2007). The new synthesis in moral psychology. *Science, 316*(5827), 998–1002.

Haidt, J. (2013a). *The righteous mind: Why good people are divided by politics and religion*. New York: Vintage.

Haidt, J. (2013b). Moral psychology for the twenty-first century. *Journal of Moral Education, 42*(3), 281–297.

Hamilton, W. D., & Zuk, M. (1982). Heritable true fitness and bright birds: A role for parasites? *Science, 218*, 384–387.

Hare, B., Wobber, V., & Wrangham, R. (2012). The self-domestication hypothesis: Evolution of bonobo psychology is due to selection against aggression. *Animal Behaviour, 83*(3), 573–585.

Harris, M. (1968). *The rise of anthropological theory*. New York: Harper Collins Publishers.

Henrich, J., Boyd, R., & Richerson, P. J. (2008). Five misunderstandings about cultural evolution. *Human Nature, 19*(2), 119–137.

Henrich, J., & McElreath, R. (2007). Dual-inheritance theory: The evolution of human cultural capacities and cultural evolution. In L. Barrett & R. Dunbar (Eds.), *Oxford handbook of evolutionary psychology* (pp. 555–570). New York: Oxford University Press.

Hertler, S. C. (2017). Closing on a note of conciliation: On the attempt to reconcile science and religion at the American Museum of Natural History's Hall of Human Origins. *Museums & Social Issues, 12*, 1–10.

Hollox, E. (2004). Genetics of lactase persistence-fresh lessons in the history of milk drinking. *European Journal of Human Genetics, 13*, 267–269.

Houkes, W. (2012). Population thinking and natural selection in dual-inheritance theory. *Biology and Philosophy, 27*(3), 401–417.

Hunt, G. R. (1996). Manufacture and use of hook-tools by New Caledonian crows. *Nature, 379*(6562), 249.

Itan, Y., Powell, A., Beaumont, M. A., Burger, J., & Thomas, M. G. (2009). The origins of lactase persistence in Europe. *PLoS Computational Biology, 5*(8), e1000491.

Kaminski, J. P. (2006). *The quotable Jefferson*. Princeton, NJ: Princeton University Press.

Krützen, M., Mann, J., Heithaus, M. R., Connor, R. C., Bejder, L., & Sherwin, W. B. (2005). Cultural transmission of tool use in bottlenose dolphins. *Proceedings of the National Academy of Sciences, 102*(25), 8939–8943.

Laland, K. N., & Hoppitt, W. (2003). Do animals have culture? *Evolutionary Anthropology: Issues, News, and Reviews, 12*(3), 150–159.

Langness, L. L. (1974). *The study of culture* (revised ed.). Novato, CA: Chandler & Sharp Publishers.

Leach, H., Groves, C., O'Connor, T., Pearson, O., & Zeder, M. (2003). Human domestication reconsidered. *Current Anthropology, 44*, 349e368.

Ludeke, S., Johnson, W., & Bouchard Jr., T. J. (2013). "Obedience to traditional authority:" A heritable factor underlying authoritarianism, conservatism and religiousness. *Personality and Individual Differences, 55*(4), 375–380.

Lumsden, C. J., & Wilson, E. O. (1981). *Genes, mind, and culture: The coevolutionary process*. Cambridge, MA: Harvard University Press.

Lumsden, C. J., & Wilson, E. O. (1985). The relation between biological and cultural evolution. *Journal of Social and Biological Structures, 8*(4), 343–359.

Mann, J., & Patterson, E. M. (2013). Tool use by aquatic animals. *Philosophical Transactions of the Royal Society B: Biological Sciences, 368*(1630), 20120424.

Morgan, L. H. (1877). *Ancient society, or researches in the lines of human progress, from savagery through barbarism to civilization*. New York: Henry Holt.

Nakamichi, M., Kato, E., Kojima, Y., & Itoigawa, N. (1998). Carrying and washing of grass roots by free-ranging Japanese macaques at Katsuyama. *Folia Primatologica, 69*(1), 35–40.

Norenzayan, A. (2010). Why we believe: Religion as a human universal. In *Human morality and sociality: Evolutionary and comparative perspectives* (pp. 58–71). New York: Palgrave Macmillan.

Norenzayan, A. (2013). *Big gods: How religion transformed cooperation and conflict*. Princeton, NJ: Princeton University Press.

Norenzayan, A., & Gervais, W. M. (2012). The cultural evolution of religion. In E. Slingerland & M. Collard (Eds.), *Creating consilience: Integrating science and the humanities* (pp. 243–265). New York: Oxford University Press.

Norenzayan, A., & Hansen, I. G. (2006). Belief in supernatural agents in the face of death. *Personality and Social Psychology Bulletin, 32*(2), 174–187.

Norenzayan, A., Shariff, A. F., Gervais, W. M., Willard, A. K., McNamara, R. A., Slingerland, E., et al. (2016). The cultural evolution of prosocial religions. *Behavioral and Brain Sciences, 39*, e1.

Odling-Smee, F. J., & Laland, K. N. (2000). Niche Construction and Gene-Culture Coevolution: An Evolutionary Basis for the Human. In T. N. S. Tonneau F. (Ed.), *Perspectives in Ethology (Vol. Sciences Perspectives in Ethology, 13)*. Boston, MA: Springer.

Odling-Smee, F. J., Laland, K. N., & Feldman, M. W. (2000). Niche construction and gene-culture coevolution: An evolutionary basis for the human sciences. In *Perspectives in ethology* (pp. 89–111). Boston: Springer.

Peterson, A., Soberón, J., Pearson, R. G., Anderson, R. P., Martínez-Meyer, E., Nakamura, M., et al. (2011). *Ecological niches and geographic distributions: Monographs in population biology* (Vol. 49). Princeton, NJ: Princeton University Press.

Purzycki, B. G., Henrich, J., Apicella, C., Atkinson, Q. D., Baimel, A., Cohen, E., et al. (2018). The evolution of religion and morality: A synthesis of ethnographic and experimental evidence from eight societies. *Religion, Brain & Behavior, 8*(2), 101–132.

Richerson, P. J., & Boyd, R. (1978). A dual inheritance model of the human evolutionary process. I. Basic postulates and a simple model. *Journal of Social and Biological Structures, 1*, 127–154.

Richerson, P. J., & Boyd, R. (1984). Natural selection and culture. *Bioscience, 34*, 430–434.

Richerson, P. J., & Boyd, R. (2001). The evolution of subjective commitment to groups: A tribal instincts hypothesis. In R. M. Nesse (Ed.), *Evolution and the capacity for commitment* (pp. 186–220). New York: Russell Sage Foundation.

Richerson, P. J., & Boyd, R. (2005). *Not by genes alone: How culture transformed human evolution*. Chicago: Chicago University Press.

Schoener, T. W. (2009). Ecological niche. In S. A. Levin (Ed.), *The Princeton guide to ecology* (pp. 3–13). Princeton, NJ: Princeton University Press.

Smith, J. M. (1990). Models of a dual inheritance system. *Journal of Theoretical Biology, 143*(1), 41–53.

Sober, E., & Wilson, D. S. (1998). *Unto others: The evolution and psychology of unselfish behavior*. Cambridge, MA: Harvard University Press.

Sterelny, K. (2006). Memes revisited. *The British Journal for the Philosophy of Science, 57*(1), 145–165.

Vaas, R. (2009). Gods, gains, and genes on the natural origin of religiosity by means of bio-cultural selection. In E. Voland & W. Schiefenhövel (Eds.), *The biological evolution of religious mind and behavior* (pp. 25–49). Waldhilsbach, Germany: Springer-Verlag.

Van Lawick-Goodall, J. (1970). Tool-using in primates and other vertebrates. In D. S. Lehrman (Ed.), *Advances in the study of behavior* (Vol. 3, pp. 195–249). New York: Academic Press.

Van Schaik, C. P., Deaner, R. O., & Merrill, M. Y. (1999). The conditions for tool use in primates: Implications for the evolution of material culture. *Journal of Human Evolution, 36*(6), 719–741.

Weiss, K. M., & Buchanan, A. V. (2009). *The mermaid's tale: Four billion years of cooperation in the making of living things*. Cambridge, MA: Harvard University Press.

Wilson, D. (2010). *Darwin's cathedral: Evolution, religion, and the nature of society*. Chicago: University of Chicago Press.

Wilson, D. S. (1980). *The natural selection of populations and communities*. Menlo Park, CA: Benjamin/Cummings.

Wilson, D. S. (1983). The group selection controversy: History and current status. *Annual Review of Ecology and Systematics, 14*(1), 159–187.

Wilson, D. S. (2002). *Darwin's cathedral: Evolution, religion, and the nature of society*. Chicago: Chicago University Press.

Wilson, D. S. (2008). Beyond demonic memes: Why Richard Dawkins is wrong about religion. *Skeptic, 13*(4), 42–51.

Wilson, D. S. (2015). *Does altruism exist? Culture, genes and the welfare of others*. New Haven, CT: Yale University Press.

Wilson, E. O. (2004). *On human nature: Twenty-fifth anniversary edition*. Cambridge, MA: Harvard University.

Wrangham, R. W., Jones, J. H., Laden, G., Pilbeam, D., Conklin-Brittain, N., Brace, C. L., et al. (1999). The raw and the stolen: Cooking and the ecology of human origins. *Current Anthropology, 40*(5), 567–594.

4

Aggregation: From Ethnic and Regional Competition to Group Selection at the Level of States and Nations

Steven C. Hertler, Aurelio José Figueredo, and Mateo Peñaherrera-Aguirre

That intense patriotism which is peculiar to the members of societies congregated within a narrow space was, in such circumstances, strongly developed. London was, to the Londoner, what Athens was to the Athenian of the age of Pericles, what Florence was to the Florentine of the fifteenth century. The citizen was proud of the grandeur of his city, punctilious about her claims to respect, ambitious of her offices, and zealous for her franchises.
—Thomas Babington Macaulay, *The History of England, from the Accession of James II—Volume 1* (2008, Chapter 3, para. 81).

1 Introduction

This chapter and the three that follow collectively comprise Part II of our book, respectively, entitled "Aggregation, Growth, Decline, and Collapse." Thus, these four chapters collectively comprise a theme familiar to declinists and students of cyclical history. As will be reiterated in later chapters, each of these processes must be studied in abstract after the manner of traditional scholarship, as has been pursued by Gibbon, Vico, Toynbee, Ibn Khaldun, and Spengler. However, it will also be important to update

S. C. Hertler et al., *Multilevel Selection*, https://doi.org/10.1007/978-3-030-49520-6_4

some of the theories proposed by these past historians with subsequent discoveries. Furthermore, as this is a book on multilevel selection, we separately and systematically consider the influence of group-selective pressures as they contribute to aggregation and state formation, the decline of mature civilizations, and their eventual collapse.

More specifically with respect to this fourth chapter, we strike out into near-virgin territory, only lightly tread by a select set of publications, such as *Darwin's Cathedral* and *Unto Others*, reviewing world histories and sociological studies of civilization to describe and explain the growth of group size. The Red Queen[1] (Brockhurst et al., 2014) positive feedback loop continued, undulating higher and then lower, but with the trend line steadily pointing upward toward agglomeration and increasing levels of sociopolitical complexity. Competition and cooperation, one compelling through the threat of conquest, and the other persuading with the promise of mutual gain, ineluctably subsumed smaller into larger groups, which cohered through a combination of relatedness and rivalry. Necessity and opportunity together dragged and cajoled lower levels of social organization to give way to higher levels of social organization. This description of social evolution should resonate with students of history, as it traces the arc of civilization populating the historical record wherein competition and cooperation actively spurred social aggregation. No other candidate mechanisms are sufficient, even as some are necessary. For instance, predation pressures and physical ecology, alongside life history traits of developmental speed and food availability, are among the many factors influencing group size. Yet, none of these factors likely propelled human groups beyond tribal limits. Some such factors might be understood to erect upper bounds on group size.[2] For example, many hunter-gatherer bands necessarily migrated along great ranges, or followed herds to support their numbers, modest though they were. Then are factors enabling without compelling social aggregation. For example, though initially yielding a niggardly bounty, first horticulture and then agriculture eventually supplied enough calories to dense conurbations to allow sedentary cities, capital accrual, and land ownership. The point being, however, that this *Neolithic Revolution*, provisioning people with the bounty of improved land management, enabled dense aggregations of stable, sedentary peoples, without necessitating state-sized societies. The imposition of competition and the benefits of cooperation, on the other hand, actively impelled aggregation.

2 Primate Despotism and the Evolution of Human Supralocal Societies

In *Life History Evolution*, Hertler, Figueredo, Peñaherrera-Aguirre, Fernandes, and Woodley of Menie (2018) presented an expanded *Evolutionary Ecological Systems Theory*, with some claim to a degree of cross-cultural validity, which featured the following biologically and anthropologically informed hierarchy of the purportedly natural and species-typical human levels of social organization: (1) self; (2) biological parents and siblings; (3) sexual/romantic partners/mates; (4) genetic offspring; (5) extended consanguineous/genetic kin; (6) extended conjugal/affinal kin; (7) friendship/alliance kith networks; (8) village/community-level organizations; (9) clan/tribal-level organizations; (10) ethnocultural/national-level organizations; and (11) supranational/imperial-level organizations. We recall these hierarchical subdivisions presently in examining how humans extended alliance formations beyond the natal troop or kinship band to larger conglomerations of more distantly related individuals. Progressing across the timeline of recorded history, we can see populations grow and cluster: first into parochial and locally controlled structures, which were thereafter absorbed into larger sociopolitical units such as the national and supranational structures specified above. However, within this present section, we begin with a review of the aggregation process along its initial phases, as it pertains to some of the lower levels of organization enumerated above.

To properly understand the evolution of human sociopolitical complexity, it is essential to determine any variation in intragroup competition between extant hominoid species and small-scale human societies. Plavcan (2002), for example, classified primate species according to the intensity[3] and frequency[4] of male intrasexual agonistic interactions, coding each variable as either high or low. The procedure considered each possible combination of frequency and intensity of conflict generating a four-level ranking system: level 1, low intensity and low frequency; level 2, low intensity and high frequency; level 3, high intensity and low frequency; and, level 4, high intensity and high frequency. The comprehensive nature of this approach allowed Plavcan to examine the coevolution of several socioecological and morphological indicators with competition levels. Although a detailed review of the various correlates is beyond the scope of this chapter (Plavcan, 2002; Plavcan, van Schaik, & Kappeler, 1995), for the present purposes, it is worth mentioning that measures of sexual dimorphism (e.g., body mass or canine size) correlated quite well with Plavcan's levels. Moreover, these results encouraged the author to estimate the level of competition of extant humans, based on measures of morphological sex differences. Plavcan's analyses (2012a) concluded that human sexual dimorphism was higher relative to that of gibbons (level 1), but lower than chimpanzees (level 2) and gorillas (level 3). Due to the flexibility of this approach, Plavcan extended his inquiry to extinct hominid species. Unfortunately, variations in estimation methods and the incomplete nature of the fossil record hindered any clear extrapolation of the levels of intrasexual competition in these taxa (Plavcan, 2002, 2012b). Hence, even though future tests are needed to determine whether the last common ancestor between *Pan* and *Homo* (i.e., *concestor*) inhabited societies featuring high intragroup aggression and dominance hierarchies akin to contemporary chimpanzees (Chapais, 2009), it is reasonable to conclude that ancestral human societies had lower levels of intragroup competition than other ape species.

Despite having the aforementioned lower levels of intragroup competition relative to other species, human groups progressed along an evolutionary trajectory toward the further suppression of intragroup competition, as can be tracked through the anthropological literature. Anthropological theories of sociopolitical evolution have addressed the transition from despotic societies with intragroup coalitional competition, to egalitarian groups vehemently opposing authoritarian individuals. Bernard Chapais (2009) theorized that the evolutionary changes in sociality, mating, and kin discrimination occurred in three distinct phases. In the first phase, the local group included multiple males and females. Intergroup relations were often based on avoidance and hostility. Mating occurred as promiscuous interactions, with kin-group outbreeding and female dispersal. Male philopatry and paternal uncertainty influenced several social dynamics. For example, descent classifications were absent, relationships between brothers and sisters were weak and inconsistent, and ties among primary agnates were restricted to brothers. During the second phase, pair bonding replaced sexual promiscuity, with polygyny and social monogamy coexisting in some groups. Citing the comparative literature, the author argued that stable-breeding bonds rather than evolving in response to the costs associated with parental care (the *provisioning hypothesis*) arose from either mate-guarding behavior (*temporal courtships*) or as a counterstrategy employed by females to decrease the risk of suffering from sexual coercion or infanticide. Furthermore, contrary to some theories endorsing a distinction between polygyny and social monogamy, Chapais (2009) viewed the evolution of the latter strategy as an outcome of the growing costs associated with preserving a polygynous system. Pair bonds increased paternity certainty, laying the foundations for latent patrilineal descent classifications. Brothers and sisters exhibited stronger relationships but these continued to be temporary. Primary agnates established robust bonds of brotherhood, including ties with their fathers. Male philopatry led to patrilocality. Although these ties transformed the local group into a *multifamily* one, supra-groups remained absent. During this phase, avoidance and hostility persisted as the primary expressions of intergroup interactions.

Chapais (2009) described the third phase as follows. Although the multifamily still comprised the local group, individuals' abilities to recognize dispersed relatives enabled the formation of between-group connections and facilitated the appearance of primitive tribes. Hence relations between groups were also altered, including within-tribe cooperation and between-tribe aggression. Residence patterns within the tribe diversified extending beyond patrilocality. Due to variation in residence, descent classifications also experienced modifications, with most societies following either patrilineal or matrilineal organizations. The relations between primary agnates preserved the level of strength observed in the previous phase. Connections between brothers and sisters endured even after dispersal.

3 Egalitarian and Stratified Systems

As has been seen in the previous section, diffuse violence became suppressed within groups and concentrated between groups. Here, in this section, we see the social evolution of stratification altering alongside aggregation. In addition to Chapais' (2009) model addressing the evolution of supralocal structures, anthropologists, such as Christopher Boehm (1999), argued that human egalitarian societies managed to invert the underlying dominance structure and, in the process, generated an anti-hierarchical system countering ancestral forms of hierarchical organization. Boehm hypothesized that this ancestral "political revolution" resulted from adopting an array of antiauthoritarian sanctions (*leveling mechanisms*), some still observed in contemporary egalitarian societies. Public opinion, for instance, has been known to regulate the behavior of individuals in the group (Boehm, 1999). Individuals suffering from reputation loss forfeit benefits, such as access to food, mates, coalitions, and alliances. Similarly, instead of accepting the will of upstarts, group members could instead reject the self-proclaimed *leader* and abandon the group (Boehm, 1999). Even though authoritarians could face sanctions, such as public ridicule, group desertion, or expulsion, it is not uncommon for the community to establish coalitions organized for the sole purpose of eliminating the transgressor (Boehm, 1999). Executions and assassinations, however, could have unintended consequences such as initiating blood feuds and revenge cycles within the group (Kelly, 2000). Hence, to avoid lethal intergroup feuding, it is not unusual for the killers to consult with the relatives of the victim before carrying out assassinations.

The presence of an array of social mechanisms restricting the ascendency of despots and promoting an egalitarian ethos (Boehm, 1999, 2012) is nevertheless arrested and thereafter reversed as one progresses from small- to medium- and large-scale societies. The subsequent origins of inequality associated with later phases of growth seem associated with the accrual and transmission of wealth. Indeed, economic analyses across small-scale societies support the distinction between wealth transmission and inequality in small-scale and larger-scale organizations. For instance, Mulder et al. (2009) calculated the degree to which three types of wealth, *material*, *embodied*, and *relational*, varied across twenty-one small-scale societies. The authors operationalized material wealth as ownership of domesticates, land, and household items. Alternatively, body mass, everyday abilities, and indicators of reproductive success indicated embodied wealth. Lastly, social connections, including political coalitions and alliances, as well as food networks, represented social wealth. Although it is often assumed that material wealth transmission depends on bequests or transfers, the authors also analyzed other dynamics. These included positive assortment, such as between mates and among allies, as well as positive feedback loops that increased the rate of wealth accretion. Building on the various classifications of wealth, and the means through which individuals pass their endowment to others, Mulder and colleagues predicted differences in the preponderance of each type of wealth depending on the society's subsistence economy, such as hunting and gathering, pastoralism, horticulture, and agriculture. As expected, analyses detected that, relative to pastoralist and agricultural societies, hunter-gatherers and horticulturalists exhibit lower wealth transmission. No significant difference existed in either wealth inequality or wealth heritability between hunter-gatherers and horticulturalists. Mulder et al. (2009) viewed the observed variation in institutions, norms, and technology designed for resource acquisition, as fundamental contributors to the amount of wealth inequality and transmission.

Flannery and Marcus (2012) offered a complementary perspective regarding the evolution of inequality. The authors suspected that even though population size and growth, climatic variation, and the adoption of agriculture partially contributed to the origins of inequality, status competition acted as the principal catalyst of social stratification. Flannery and Marcus' ethnographic and archeological compilation identified that this process begins with the presence of prestige systems, such as those related to variation in attributes, as in warriorship, trade skills, and public demonstrations of generosity (e.g., allocation of resources to collective rituals). Interestingly, though generosity and gift giving are often portrayed as prosocial manifestations, these behaviors not only foster social bonding but also generate social debt and thereby social power. Furthermore, according to the authors, prestige competition extends beyond direct displays of generosity between rivals, but these exhibitions may also involve the rest of the lineage, clan, or community (e.g., hosting onerous feasts). Individuals unable to reciprocate may instead opt to serve the donor temporarily. It is not infrequent, however, for this provisional state to change permanently, creating patrons and servants (Flannery & Marcus, 2012). Based on their ethnographic review, Flannery and Marcus (2012) recognized that debt could be the outcome of "exorbitant bride-price, loans to aspiring Big Men, excessive war reparations, or the desperate cries of impoverished kinsmen" (p. 537). Hence, class stratification and hereditary authority positions marked the origins of chiefdoms.[5] Phylogenetic comparative analyses agree with Flannery and Marcus' assessment. For instance, Currie and Mace (2011) analyzed the coevolution of the political organization (classified as *acephalous* or *chiefdoms/states*) and hereditary social stratification (*present* or *absent*) in Austronesian societies. The phylogenetic comparative models demonstrated that acephalous organizations exhibiting class stratification have a high likelihood of developing into chiefdoms/states. Small-scale societies could remain politically acephalous by abandoning hereditary class stratification. Similarly, chiefdoms/states without class stratification display a high likelihood of becoming politically acephalous.

Along with directional changes in intragroup violence and inequality, aggregation brings stratification. Relative to stratification and state evolution, Claessen and Skalník (1978) classified early states into three categories: *inchoate*, *typical*, and *transitional*. The authors suggested that inchoate early states rose from local connections at the level of families or communities with rulers and commoners interacting directly. In this stage, taxation systems remained rudimentary, and social distinctions arose based on reciprocal exchanges. Inhabitants did not specialize in any particular craft or discipline. The authors hypothesized that the origins of typical early states, in contrast to inchoate states, depended on the modification of the individuals' system of allegiances. Hence, inhabitants developed ties toward larger geographical and political territory above and beyond preexisting ties extant within kin groups or communities. In typical early states, individuals also competed for nominations to occupy governmental offices. These positions offset the influence of hereditary authority roles. Hence, the administration was no longer monopolized by a kin group. Despite these differences, reciprocity persisted as the primary mechanism influencing social dynamics between social classes (Claessen & Skalník, 1978). Still, in early transitional states, kinship became less influential, with the appointment of administrative officials replacing the former nepotistic system. Economic dynamics also changed due to the presence of private property, market economies, and confrontations between social classes (Claessen & Skalník, 1978).

Evolutionary perspectives on political complexity have been historically criticized for proposing theories based on universal (*nomothetic*) laws. As discussed in Chap. 3, cultural anthropologists, such as those of the Boasian School, endorsed *historical particularism* (e.g., Boas, 1896), wherein cultures ought to be studied on a case-by-case (*ideographic*) basis. This presumably avoided the impression that some cultures were more "evolved" than others, a hypothesis which the Boasians attributed to ethnocentric bias. The notion of sociopolitical complexity still generates vehement reactions from researchers arguing against the assessment of intersocietal variation based on the level of stratification or specialization. Contra the Boasian perspective, aggregation co-occurs with progression along an underlying dimension of sociopolitical complexity, as seen in the anthropological literature just reviewed as well as in the work of Peter Turchin and colleagues. Turchin et al. (2018) examined the covariation of nine metrics of sociopolitical indicators (e.g., the level of hierarchy, polity population size, capital population size, polity territory, the level of infrastructure, and the presence of a monetary system, among others) computed from 51 measures in a sample of 414 societies across 10,000 years. From there, Turchin extracted a single principal component, revealing a unique underlying sociopolitical complexity dimension. Disaggregating the data and analyzing these temporal trends across different geographic areas, the authors demonstrated that polities tend to increase their level of sociopolitical complexity over time. Moreover, regional differences were detected in principal component scores. For example, Turchin and colleagues estimated that, in addition to the temporal lag between Eurasian and American societies, the latter also exhibited lower principal component values at the time of the Columbian exchange. We now move on to consider this progression toward stratification and complexity as it played out along the later phases of aggregation.

4 The Nomad and the Cultivator

The origins of agriculture, discussed above as a precondition to civilizational growth, are related intimately to the subsequent development of warfare, discussed below as an active driver of aggregation. Agriculture not only provided the ability of persons to aggregate without depleting the local food supply and starving, but allowed them to accumulate wealth and resources in the form of provisions, cultivated land, herds, structures, tools, and, eventually, currency. As in nature we find the theft of food (Brockmann & Barnard, 1979), nesting materials (Hunter & Dwyer, 1997), and nesting sites (Field, 1992), so we find that the aforelisted products of early agricultural civilizations invited conspecific competition, as initially framed within the world histories of William Durant, Ellsworth Huntington, William McNeill, and Arnold Toynbee.

The *Neolithic Revolution* was long in developing. Early farmers were notoriously malnourished, getting by only after inuring themselves to hard physical labor, routine, and monotony. This new way of life, predicated on a future-oriented mode of thought and deferral of gratification, eventually yielded surpluses to those that persisted in its course, as intuitively reconstructed in Durant's magisterial *Story of Civilization*:

> A nomad stock, like the Bedouins of Arabia, may be exceptionally intelligent and vigorous, it may display high qualities of character like courage, generosity and nobility; but without that simple sine qua non of culture, a continuity of food, its intelligence will be lavished on the perils of the hunt and the tricks of trade, and nothing will remain for the laces and frills, the curtsies and amenities, the arts and comforts, of civilization. The first form of culture is agriculture. It is when man settles down to till the soil and lay up provisions for the uncertain future that he finds time and reason to be civilized. Within that little circle of security-a reliable supply of water and food-he builds his huts, his temples and his schools; he invents productive tools, and domesticates the dog, the ass, the pig, at last himself. He learns to work with regularity and order, maintains a longer tenure of life, and transmits more completely than before the mental and moral heritage of his race.

Whereas Durant (1950) contrasts the nomad and the cultivator, positioning them in the part of the ant and the grasshopper in Aesop's fable, others attempted to reconstruct the intimate coevolutionary relationship operative between these groups. Ever alert to instances of nonrandom selection, Ellsworth Huntington (1927)[6] provided the following theoretical rendering of prehistory relevant to group formation within the Neolithic Revolution:

> When a race settles down to agriculture a great selective process takes place. Practically all nomads hate farming. But some hate it worse than others. The ones who hate it most are likely to be the most active, adventurous spirits, whereas those who take to it most readily are the ones who are predisposed to a sedentary life, and who are least averse to steady work, a thing which nomads and hunters usually find extremely irksome and often impossible. Thus, if the pressure of population and the presence of a favoring environment cause part of a race to adopt a sedentary life, and part to remain as hunters or cattle-raisers, there is likely to be a clear-cut line of fission. As time goes on the line will presumably become still clearer, for among the children of those who have ken to farming, there is a strong tendency for some to go back to the nomadic life. (p. 109)

What Huntington infers with respect to prehistory, McNeill (1974) documents with respect to history. Excepting only Japan and Western Europe, McNeill (1983, p. 12) interprets the whole of Africa, Europe, and Asia's early political history as the interplay between "nomad conquest and agricultural revolt."[7] McNeill elaborates thus:

> By A.D. 450 or A.D. 500, therefore, the spreading moldboard plow on the one hand and increasingly seaworthy ships on the other provided a new technical basis for the eventual flowering of high culture in northwestern Europe. But before that result could be achieved, drastic social differentiation had to occur. The free barbarian husbandman had to become a peasant paying dues and services to his social superiors. The principal impulse pushing in this direction was the need for more effective defense against raiding parties, whether coming over land on horseback, like the Huns and many another steppe nomad people, or by sea and river like the Vikings. (p. 68)

Arnold Toynbee's *A Study of History* (1951) also emphasizes raiding by nomads, casting them as *intraspecific parasites*, compelling settled agriculturalists to "live as domesticates, as do the sheep and goat, under the rule of the herding nomad" (volume III; p. 24).

Durant, like Huntington, McNeill, and Toynbee, was a historian with decidedly cosmopolitan interests in prehistory and world history; yet he was not privy to modern archeological data, recent recoveries of ancient wrecks, and physical remains of ancient societies referred to as *nomadic* in his history. Readily available documentary evidence, it must be said in defense of Durant, and indeed of all of these eminent authors, was implicitly biased toward settled agriculturalists, for their societies, unlike those of nomadic pastoralists, developed writing and record keeping, thereby contributing disproportionately, indeed nearly exclusively, to traditional historiography. Conflict and interactions between such cultures were then documented from the perspective of the settled agriculturalist that did not fully know, or care to capture, the social complexities of their rivals.

In addition, we believe that some of this confusion derives from a conflation of so-called "nomadic" hunter-gatherers with "nomadic" pastoralists. First of all, few of these groups are actually truly *nomadic*, in that they wander aimlessly through their environments, but instead *transhumant*, in that they travel cyclically throughout their home ranges as local resources are depleted and regenerated. Although both groups can be contrasted with many horticultural and agricultural societies, which are generally *sedentary*, they are quite different from each other and not equally "primitive" in the scale of complexity. The hunter-gatherers are, of course, the ancestral type, whereas the pastoralists are a derived form and may be quite culturally and technologically "advanced" in comparison. Durant's model of "settled agriculturalists" versus "nomadic raiders" is thus seriously flawed as an evolutionary progression, as it conflates transhumant pastoralism with the ancestral hunter-gatherer lifestyle. It suggests that *transhumant* (again, rarely *nomadic* in the proper sense of the word) pastoralism is somehow more primitive than horticulture, a position without evidentiary support. It is instead likely that pastoralism evolved after horticulture, as a distinct lifestyle fulfilling a distinct ecological niche, and continued on to develop along its own trajectory, often achieving comparable or transcendent levels of social complexity. Although raiding occurred opportunistically, the relationship of herders to farmers was more often characterized by trade than conflict or depredation. Further, the growth of the first empires (e.g., Akkad under Sargon I) was not based on the expansion of settled agriculturalists at the expense of the alleged nomads, but instead on the conquest of one civilized nation (Sumer) by another (Akkad).

Although most of the historians that we have cited seem to suggest that the pastoralists evolved directly from the hunter-gatherers, with settled farmers presumably evolving last, the preponderance of evidence does not support this view. If we examine the archeologically determined dates for the domestication of plants and animals, the data suggest the opposite pattern. The earliest dates of plant domestication precede the earliest dates of animal domestication by millennia. For example, by 9000 BC, humans had domesticated fig trees, emmer wheat, foxtail millet, flax, and peas; by 8500 BC, we had domesticated einkorn wheat, barley, and chickpeas; and, by 8000 BC, we had domesticated bottle gourds, rice, potatoes, beans, and squash (Hirst, 2019a). In contrast, the distribution of the earliest dates of animal domestication is shifted systematically toward later dates. For example, it was not until about 8500 BC that sheep were domesticated; it was not until about 8000 BC that goats were domesticated; it was not until about 7000 BC that pigs and cattle were domesticated; and it was not until about 6000 BC that chickens were domesticated (Hirst, 2019b). Dogs had been domesticated first by hunter-gatherers significantly earlier, perhaps by 30,000 BC, and should therefore not be considered as part of the Neolithic Revolution; cats had been semidomesticated by 8500 BC, but this process was secondary to the storage of grain products and functioned mostly for the control of rodent pests, rather than directly as food sources. Animals for transportation, portage, and traction were among the last to be domesticated, with llamas and alpacas domesticated by 4500 BC, donkeys by 4000 BC, horses and Bactrian camels by 3500 BC, and dromedary camels by 3000 BC (Hirst, 2019b). The suggestion that animal husbandry is therefore a more primitive or ancestral condition than plant cultivation is thus not consistent with the currently available evidence; though, again, the larger point is that both raised environmental carrying capacities considerably, enabling demographic growth and aggregation.

Instead, it is quite likely that animal husbandry originally evolved as a *complement* to plant cultivation, rather than as an *alternative*, the same way that many remaining horticulturalist societies in the Amazon supplement their horticultural activities with hunting in the surrounding forests. Raising animals provided richer sources of protein than cultivating plants and also made possible the use of land for pasture that was not suitable for cultivation. Land that was suitable for farming was usually devoted to horticulture and later to true agriculture, whereas land that was less fertile, less well-watered, or more mountainous was relegated to pasture. For most of history, it is therefore likely that the ecological relations between herders and farmers were generally characterized by mutualistic exchange, or trade, rather than by predatory conflict, or "raiding." This is not to say that conflict did not occasionally occur, as between any human societies, but that it was an exception to the general pattern of symbiotic cooperation. In some groups, like many of the Bantu peoples of Africa, the farming is generally carried out by the women, and the herding is generally carried out by the men *within the same society*. It is quite possible that farming by females was a natural outgrowth of gathering of plant foods by females and that herding by men was a natural outgrowth of hunting of animals by men, such that the sexual division of labor was reflected and somewhat conserved in later developments, such as the sex-biased evolution of animal traction in plow agriculture.

This interpretation is also consistent with the geographical origins attributed to many domesticates. Most of the Eurasian domestic animals originated in roughly the same regions of Southwest Asia as the majority of the Eurasian domestic plants, indicating a probable coevolutionary process with either the same or similar societies living ancestrally in close proximity. Only as the subsistence technology of herding became more advanced and specialized, diverging from ancestral patterns, did those cultures spread across new habitats like the Eurasian steppes, which were significantly less suitable for sedentary agriculture. This latter development eventually led to the evolution of the classical "nomadic pastoralists," such as the highly sophisticated Turko-Mongol tribes and empires of the fourteenth century AD, that so impressed the cited historians with their infamous depredations upon the "civilized" world.

Another similar conflation by many historians is one between the subsistence economies of *horticulture* and *agriculture*. Both are types of cultivation, but horticulture is the more ancestral condition in which small "kitchen" gardens are cultivated, typically by women, for the nearly exclusive consumption of the immediate family or kin group, and is accomplished by "hoe farming." Agriculture is the more derived condition in which larger tracts of land (*ager* meaning "field" in Latin) are cultivated, typically by men, intensifying food production for wider distribution or trade either within or between broader communities. This historic transition was accomplished by "plow farming" with the use of domesticated animals (such as oxen or asses) for traction, an innovation that was introduced as far back as 3500 BC in Sumer (Kramer, 1956). Many historians have subsumed both under the generic term agriculture, incorrectly referring to the "Neolithic Revolution" as the "Agricultural Revolution" in spite of the former having started with simple horticulture. The shift from horticulture to true agriculture was a monumental one in its consequences, selecting for more pronounced concepts of property and patrilineal inheritance in relation to the more intensively cultivated resources, requiring at once higher levels of resource defense and demanding higher levels of female sexual fidelity and consequently male paternity confidence in response to the greater quantity of male parental effort now required (Hrdy, 1999, 2009). These socially consequential shifts were driven primarily by the fact that the use of animal traction for plowing required significantly greater muscular strength than hoe farming and so favored the deployment of men over women in the agricultural workforce (Alesina, Giuliano, & Nunn, 2011, 2013; Boserup, 1970; Burton & White, 1984; Murdock & Provost, 1973).

Although it is still debated whether agriculture acted as the main catalyst for the evolution of early states, comparative analyses identified coevolutionary patterns between subsistence economies and the presence of social classes. Frank Marlowe (2000), for example, used the Standard Cross-Cultural Sample (i.e., SCCS comprising 186 societies) to determine the degree to which population density and social stratification, operationalized as wealth variation, differed in terms of sustenance practices. According to Marlowe the database classified the various economies into four categories: (1) foraging; (2) horticulture; (3) pastoralism; and (4) agriculture. The SCCS also categorized societies into four levels of wealth stratification: (1) none; (2) low to moderate; (3) high; and (4) very high. Marlowe's analyses revealed that social stratification varied depending on the type of subsistence mode. The author attributed the low degree of wealth variation in foragers to their nomadic lifestyle as well as their lack of control over resources. Horticulturalists' use of simple technologies allowed them to increase the extraction of resources leading to the limited accumulation of wealth. Pastoralists displayed greater wealth variation by controlling and defending their livestock. Finally, according to the author, agriculturalists had very high levels of social stratification, a feature in part due to a sedentary lifestyle, along with the use of adequate technologies (such as plows), enabling some individuals to control resources and accumulate wealth. In addition, Marlowe's comparative examination also revealed that population density followed a polynomial pattern, in contrast to the observed linear trend between subsistence mode and wealth differences. Hence, even though pastoralists had a higher level of wealth skewness relative to horticulturalists, the former lived at a lower population density. Consequently, the study demonstrated that pastoralist societies retained considerable levels of wealth monopolization even under low population densities.

A fuller appreciation of these complexities qualifies the binary contrasts drawn by Durant, Huntington, and Toynbee between the nomad and the cultivator.[8] While the dichotomous descriptions of *nomad* and *cultivator* are simultaneously inaccurate with respect to the character of these historical groups, as well as with respect to their alleged complexity, these different subsistence economies did repeatedly compete with one another over territory throughout the course of world history, to the end of selecting for larger and more sophisticated aggregations of peoples. We will pick up the thread of competition as a driver of aggregation in the subsequent section, but we herein pursued Durant et al.'s narrative, corrected and clarified, for its power to show how advances in agricultural technologies, irrespective of strategy, allowed further aggregation that was necessary for subsequent state formation.

5 Did War Make the State?

The narrative continues as one transitions from exclusive emphasis on nomadism to a generalized view of war as an active driver of aggregation. Toynbee wrote extensively about nomadism, but his work contains relevant writings on *challenge and response* arising from various *stimuli* discussed within pages 100–208 in volume II of *A Study of History*. He details climatic challenges such as sea crossings and impoverished soils but also includes war and conflict or what he refers to as the *stimulus of blows*. War is a stimulus to action operating across time to rouse the people, stoke their pride, and stir their unity. As per Toynbee, many of the great mergers and monarchies of the European past were organized, like sedimentary rock, by the stimulus of blows: blows from external groups, including those inflicted by non-European peoples, pressing at the borders. Some of the great aggregations of principalities, duchies, municipalities, regions, and states came of such pressure.[9] To take one of many proffered examples, Toynbee reviews the struggle for possession of early modern Iberia, finding regions in the vanguard of the resistance were ascendant, whereas those sheltered from threat declined in influence and power. He uses the example of Aragon, located in the north east region of Iberia. Aragon had been extremely dynamic and influential, contributing disproportionately to law and art and excelling in other endeavors during the middle ages when the Moorish invasion beat against their gates. However, the Moors were eventually pushed southward, land was reclaimed, and eventually there developed a reliable buffer between Aragon and the Moors. With pressure relaxing, Aragon waned and more southerly regions that either remained in, or were newly thrust into the van, waxed.

The tenor of Toynbee's writings on *the stimulus of blows* was later elaborated. The *war made the state* thesis, associated with the American sociologist Charles Tilly, coauthor of *The Formation of National States in Western Europe* (1975), is more generalized, not specifying nomads or any particular group but similarly emphasizing the aggregating impulses induced by conflict (North, Wallis, & Weingast, 2009). Exchanging nomadism specifically for war generally allows any group to pressure any other group into tighter, more enduring aggregation.[10] Tilly's thesis, known as the *bello-centric* or *warfare-paradigm*, finds the outlines of the state called forth of necessity in the form of military administration, organization, financing, credit, banking, and martial law (Kaspersen & Strandsbjerg, 2017). This bello-centric theory, deemed contributory rather than either necessary or sufficient to state formation, does reflect an apparently real process by which existing institutions and bonds are conditioned into mature state structures (Spruyt, 2017). Consistent with its original formation, Tilly's bello-centric theory is on firmest empirical terrain when describing the "consolidation" and "bureaucratization" of the modern European state, even as primogeniture and related alternatives have been suggested (Gorski & Sharma, 2017). In speaking of Tilly, Azar Gat (2006, p. 358), understanding states as most fundamentally concentrations of "force elevated to a commanding position over society," finds war "predominant in the formation of larger states, which welded together distinct and different communities, and, indeed, separate societies, ethnicities, cultures, and polities." For instance, the Hunnic invasions drove many disunited Germans into the arms of Empire (Gat, 2006). Thereafter, with the end of the *Holy Roman Empire* in 1806, and seemingly in answer to being devastated by a united, nationalistic France under Napoleon, the various provinces of Germany were brought into a federal union.

Again, though designed for modern states and related specifically to war, the bello-centric theory has some pretensions to generalizability relative to early phases of consolidation. Akin to the defensive clusters induced by the aforementioned predatory small-mouthed ciliates (Chap. 2), banditry, lawlessness, and raiding created selective pressures to which embryonic settled societies responded, evolutionarily and culturally, via aggregation. There is safety in numbers! There is also safety behind walls. There are then human walls called armies that communes could raise on the surplus of their labor. This process of aggregation for protection transitioned into ever more formal arrangements, the signature of which is evident in architectural features of the castle, the keep, and the walled city. From small aggregations of equals providing mutual aid, there came formal payment of protection costs to those specializing in warfare (McNeill, 1992).[11] The feudal arrangement is the exemplar of such protection costs that existed in the space between mutual aid and national defense. Feudalism is a laboratory grade model of multilevel selection, with groups of various sizes nested within networks of allegiances, acting to counter threats of individualism from within and conquest from without. The vassal undergoes the *commendation ceremony* swearing *homage* and *fealty* to a lord above, while assuming responsibility for those peasants in tenantry on his newly granted lands. The feudal lord himself was only an intermediate stage in that nested hierarchy of interdependence, the sanctity of which was maintained by various oaths, ceremonies, and the religious worldview embodied in the *great chain of being*.

Any problems with the bello-centric theory come from its being incomplete rather than incorrect. Aggregation comes not only from *push* factors, such as war, but *pull* factors, such as trade, as is the central thesis of Robert Wright's (2001) *Nonzero*. Consistent with statements made in the opening of the fourth section of this fourth chapter, through ups and downs, and notwithstanding setbacks and alternations of leaders and leading groups, Wright understands complexity to augment through time, which is manifest in the complexity of human groups. Yet, the rise of group complexity is alternatively explained. Wright acknowledges war as a stimulus to alliance formation, reviewing supporting theory and evidence; for example, the work of Evans-Pritchard and Robert Carneiro, as well as the accounts of the Nuer of Sudan, and the Boers of South Africa. Nevertheless, consistent with the thesis of his book, Wright emphasizes *pull factors* impelling aggregation; pull factors in the form of benefits accruing to groups cooperating in non-zero-sum interactions. Cooperation within and between groups, respectively, taps the power of divided labor and far-flung trade networks (Wright, 2001). Adam Smith's *invisible hand* and Ricardo's *Law of Comparative Advantage*, like the evolutionary literature on altruism and cooperation, are all called to mind by such pull factors, which impel aggregation while war compels it. Complicating these contrasting forces, one also should recognize that war was sometimes waged by aggressive states. From this view, the threat of war can bind smaller groups into larger groups. When war is waged from the drive for imperial expansion, making war remains a binding agent, but in a manner inconsistent with the bello-centric theory. One should also note that some aggressive states waged war to gain the benefits of cooperation, a slightly confusing mix of motives that blurs the line between pull factors and push factors. Territories were subsumed within empires, not to extirpate their populations or even always with the end of expunging their leaders, but to integrate the productive powers of their populace.

6 *Asabiyyah*

The foregoing sections of this fourth chapter addressed the initial phases of growth from early tribal-sized societies, rapidly progressing to feudal nodes and emerging nation-states. The benefits accruing to cooperation, and the necessity of cooperating from the threat of war, were stressed in this initial phase of aggregation. But we now ask, beyond this broad pressure promoting aggregation, what are the features that are thereafter important to the growth of a national group, the perpetuation of the stable state, the superior cohesiveness of some aggregates vis-à-vis others, and, finally, the expansion into empire? In other words, in these various ways, we are asking *why some groups expand to the marginalization, displacement, or absorption of others*. We do justice to the reader in humbly approaching an answer, while paying due deference to its inexorable intricacy and its manifold nature. Chance historical events, ecological variation, technology, and cultural complexity, as well as a group's mean values for *life history speed*[12] and cognitive ability, are among the variegated variables demanding representation when attempting to answer the question above. Nevertheless, we submit that group selection is among the brightest stars in this constellation of explanatory variables. Thus, we can say, all else being equal, highly group-selected societies expand to *the marginalization, displacement, or absorption of others*. With the question so reframed, we now need only operationalize group selection and thereafter explain what indicates that one society has been subjected to more group-selective pressure than another.

The highly group-selected society is endowed with *asabiyyah*, a cohesive tribalism and martial vitality (Enan, 2007; Fromherz, 2011) described in Ibn Khaldun's (1377/1958) classic work, *The Muqaddimah*. Our use of the term *group selected* also shares substantial variance with the European concept of *esprit de corps*. Regarding recent publications, there is overlap with *vigor*, which is a cultural legacy of toughness, grit, and resilience wherein locally well-integrated groups conquer rival groups (Penman, 2015).[13] While systematic surveys might yield further synonymous terms, it is relatively easier to identify overlapping terms characterizing some subset of group selection's larger meaning: *Élan*, *panache*, and *dash* evoke group selection's martial enthusiasm; *comradery*, *loyalty*, and *compatriotism* evoke group selection's fraternal solidarity; and *jingoism*, *nationalism*, and *patriotism* evoke group selection's national cohesion. Despite descriptive overlap, the term *group selection* extends from the evolutionary literature and is consequently not only descriptive but also connotes causality. Thus, when one speaks of a group-selected population, one is simultaneously suggesting a phenomenon *and* its cause; in other words, one is implying intragroup coherence while simultaneously suggesting that it developed as a result of evolutionary pressures. Thus, unlike all other terms, even *asabiyyah*, which is sometimes subsequently used for its descriptive excellence,[14] group selection is not open to charges of mystical vagary because it, alone, specifies a mechanism by which it is generated.[15]

A group evolves *asabiyyah* when subject to a selective regime favoring cooperation, most often compelled by competition from without, as described in foregoing sections. Groups respond genetically *and* culturally to competitive selective pressures through gene-culture coevolution, which then can create positive feedback loops wherein cultural values consistent with *asabiyyah* amount to an anthropogenic selective pressure productive of maintaining or increasing *asabiyyah*. Groups that emerge as stably high on Ibn Khaldun's trait of *asabiyyah*, or in other words those that emerge as highly group selected both in terms of genetic evolution and cultural inheritance, are those exposed to *optimal doses* of cohesion-inducing competition from rival groups over *extended periods of time*. Arnold Toynbee was perhaps the first to capture both of these important variables, namely, the *strength* and *duration* of competitive pressures. Toynbee described a golden mean, relevant to all the *stimuli* listed in his aforementioned writings on *challenge and response*, including those imparted by climate and war. The effect is curvilinear. The saying *too much of a good thing* illustrates the point that stimuli can be tridirectional in their effects, evoking different responses in small, medium, and large doses. For example, a sick person might continue to be sick in the absence of a vitamin or medicine, become well with the proper dose, and thereafter became ill once more if given an overdose of that vitamin or medicine. With respect to martial competition among human groups, an absence leads to an absence of a selective regime productive of *asabiyyah*, while an excess shatters a society, fracturing the group. Moreover, Toynbee also recognized the need for competitive pressures to be temporally sustained, which we observe is necessary for the cultural, and especially the genetic, evolution of *asabiyyah*. For instance, Toynbee (1951; volume II) writes about Western development wrought of sustained Islamic pressure, which

> declared itself on the battle-field of Tours in A.D. 732 continued in force and increased in momentum on this front until, some eight centuries later, its impetus was carrying the Portuguese vanguard of Western Christendom right out of the Iberian Peninsula and onward overseas round Africa to Goa and Malacca and Macao. (p. 204)

There are two contemporary authors, MacFarlane and Turchin, whose writings, respectively, allow us to expand on Toynbee's insights regarding the strength and duration of competitive stimuli. MacFarlane (2003), in his *Savage Wars of Peace*, offers a comparative history of England and Japan, illustrating the special attributes accruing to island nations near the mainland (MacFarlane, 2003). This affords an opportunity to consider how a factor, geography in this example, can moderate competitive pressures. The channel, scarcely twenty miles wide at the *Strait of Dover*, nevertheless buffered England from many continental wars, thwarted invasions as with the *Spanish Armada*, or discouraged their attempt, as with *Operation Sea Lion*. MacFarlane understands the channel as a necessary, though of course not sufficient, condition for England's advanced development of industrialization. As for the development of *asabiyyah*, the channel may well have buffered the English excessively from the selective pressures of rival nation-states, arguably enabling civil wars and revolution, contributing to the early growth of internal political factions and the delayed advent of peak powers, relative to continental powers, such as the *asabiyyah* optima experienced in fifteenth-century Spain and seventeenth-century France. England was certainly not afforded the natural refugium of a remote island nation sequestered in the distant open ocean, but, then again, neither was it subject to the dissolutions and partitions (Bain, 1891; Halecki, 1945; Lord, 1915; Perkins, 1896) to which Poland has been subjected, being within reach of France, bordered by Germany and Russia, and having rivers bisecting and dividing the country rather than flanking it.[16]

Relative to duration, Turchin's (Turchin, 2004, 2006, 2008, 2012; Turchin & Gavrilets, 2009) writings on the sustained conflict imparted by *meta-ethnic frontiers* are productive of *asabiyyah*. The meta-ethnic frontier is the setting for various forms of interaction including conflict. "People originating on fault-line frontiers," Turchin (2006, p. 6) writes, "become characterized by cooperation and a high capacity for collective action, which in turn enables them to build large and powerful territorial states." As such, the Danube region and the *Danubian Limes* qualify as meta-ethnic frontiers, as this was one of the long-held zones between imperial Rome and the Germanic tribes. As at the Danube and other meta-ethnic frontiers, competition was extremely intense, such that the Romans were pushing outward along the meta-ethnic frontier during their phases of territorial expansion, while the Germans were pushing inward in search of imperial spoils. The meta-ethnic frontier, because it separates such culturally distinct peoples, facilitates imperial expansion and agglomeration of culturally similar peoples behind the meta-ethnic frontier. Turchin demonstrates that "world empires arise from regions where civilizations clash" (Turchin, 2006, p. 169). As one of many examples, Turchin provides a comparative history of Ibn Khalid's Arabs and Ermak's Cossacks, noting that both were highly cohesive groups wrought of the hardships inherent in frontier life and thus able to prevail against overwhelming odds.[17] Turchin's writings are particular to multi-ethnic empire formation, though there is no reason why these sustained competitive pressures, operating among nations and even city states, would not function similarly.

7 The Ecology of *Asabiyyah*

More than unidimensional changes in *asabiyyah*, competition, and clashes across meta-ethnic frontiers may occasion biocultural divergence. Ethnic groups frequently differ genetically as well as culturally (Wang et al., 2007), as is expected by the theory of gene-culture coevolution. Further studies are required to determine the degree to which groups may genetically vary due to dynamics such as population viscosity. This process recalls the ecological concept of *character displacement* (Grant, 1972;

Pfennig & Pfennig, 2009; Schluter, 2000), as documented in stickleback fish (Schluter & McPhail, 1992), finches (Grant & Grant, 2006; Schluter, Price, & Grant, 1985), frogs (Gerhardt, 1994; Höbel, & Gerhardt, 2003), salamanders (Adams, 2004), and insects (Tynkkynen, Rantala, & Suhonen, 2004), as well as many plant species (Armbruster, Edwards, & Debevec, 1994; Beans, 2014; Muchhala & Potts, 2007; Murray et al., 1987). Character displacement occurs at bordering ecological ranges between two similar species wherein competition renders each species more different than one another where they interface, compared with those ecological zones without such overlap (Hutchinson, 1957). Related to character displacement, Hutchinson's (1959) *Theory of Limiting Similarity* describes the maximum allowable overlap in ecological niches between two similar species. This principle has been applied to the biohistory of interethnic conflict in a recently published work, *The Ecology of Empire* (Figueredo et al., 2019), as recounted in Chap. 11 of this volume. In this section of the present chapter, we more broadly treat the concept of *asabiyyah* in ecological perspective, examining: (1) whether environmental harshness sustains *asabiyyah*; (2) whether intragroup competition increases as one or more adjacent groups approach its environmental carrying capacity; (3) whether distance from a polity's center affects *asabiyyah* levels; and (4) whether physical geographic features such as mountains and rivers integrate or isolate population to the end of augmenting or attenuating aggregation.

Relative to the question as to whether environmental harshness sustains *asabiyyah*, one can view the specialized forms of so-called nomadic pastoralism and sedentary agriculturalism from an ancestral condition of mixed farming, through the lens of quantitative theoretical ecology. This divergence also marks the shift in symbioses from one of pure mutualistic cooperation and trade to one of resource competition and depredation, at times even devolving into one of intraspecific parasitism (as per Toynbee, 1951). As a result of their great success in overcoming many nations of settled agriculturalists during the rapid expansion of the Islamic Arab empire, Ibn Khaldun (1377/1958) attributed heightened *asabiyyah* to the camel-herding desert Bedouin, ascribing this advantage to various different sources. One was the harsher lifestyle of the desert-living Bedouin, as compared with the more comfortable and luxurious living conditions presumably experienced by the city dwellers, which purportedly sapped their moral and military fortitude. For example, each individual Bedouin was perpetually armed and vigilant for the purpose of self-defense, whereas the settled urbanites delegated their defense to the local militias. Moreover, tribal Bedouins were bound by "blood" (genetic) ties, whereas larger settled populations were more heterogeneous, as distinct lineages became confused and purity was lost; the latter were furthermore less genetically related to their leaders and this eroded their solidarity as a group. In addition, the life of comfort and abundance led by the city dwellers has robbed them of their "vigor" and "violence," as it has with their "dumb" domesticated animals such as cows and donkeys; whereas the "savage" Bedouin have retained the "wildness" of their "desert habits," embodying "bravery," "enterprise," "courage," and *asabiyyah*. Nevertheless, whenever a previously "savage" Arab tribe takes up a civilized life of greater ease, becoming more "friendly," "meek," and "sociable," the same process occurs, analogous to domestication, and this leads to the decline of their group, lineage, or dynasty. After a fashion then, Ibn Khaldun is pitting his concept of *asabiyyah* against inevitable products of aggregation, such as attenuated genetic relatedness, and the delegation of defense to a warrior caste.

Relative to whether intragroup competition increases as one or more adjacent groups approach their environmental carrying capacity, it is relevant to cite Peter Turchin's *Historical Dynamics: Why States Rise and Fall*. Turchin (2003) hypothesized that the ratio of population to resources, in addition to intergroup conflict and ethnic boundaries, also influences the persistence of intragroup prosocial solidarity. The author argued that populations at carrying capacity experience greater intragroup competition. Although polities could respond to predictable cycles of resource scarcity, Turchin emphasized that the groups' inability to adequately address these unpredictable and severe shortages increases the levels of intragroup social antagonism. Under these conditions collective action is considerably compromised. Furthermore, societies experiencing a deficit in the extraction, production, and distribution of resources not only are more susceptible to internal crises, but, for Turchin, also halt any military campaigns abroad and often leave the group's territorial boundaries defenseless. Military and political reversals both increase the likelihood of suffering from, as well as succumbing to, external threats such as foreign invasions. In contrast to the former scenario, meta-ethnic frontier theory predicts that societies living below the ecology's carrying capacity not only exhibit sociopolitical stability but also manage to generate food surpluses necessary for sustaining collective actions. This prosperity also allows the group to successfully defend its borders and initiate or continue their territorial expansion.

With respect to whether distance from a polity's center affects *asabiyyah levels*, one can consider the work of Turchin once more. Turchin (2003) also recognized that the level of *asabiyyah* varies depending on the physical distance from the polity's center. At the same time, Turchin recognized that traditional group selection theory and the meta-ethnic frontier theory differed in the predicted direction of the effect between distance from the center and level of social solidarity. According to the author, traditional group selection theory predicts that institutions enforcing prosocial behaviors, located closer to the polity's center, would foster greater cooperation. Alternatively, the meta-ethnic frontier theory argues that since the center also presents higher degrees of social and political inequality, social solidarity should remain relatively low, increasing as one gets closer to the polity's boundaries. Turchin suspected that lower political and social inequalities along with the existence of meta-ethnic fault lines generate the adequate sociopolitical context for prosocial solidarity.

With respect to whether physical geographic features such as mountains and rivers integrate or isolate populations to the end of augmenting or attenuating aggregation, one should recognize that one or both outcomes are possible depending on both current circumstances and historical factors. Turchin (2003) suspected that geographical features, such as mountains, provided groups with natural defensible positions. This natural defense, however, increased the level of isolation between groups and decreased the necessity for establishing coalitions and alliances with neighbor groups in response to common threats (Turchin, 2003). For the author, early states also faced considerable difficulty conquering or integrating a mountainous group to a larger polity. This is partially ascribed to the tactical and strategic difficulties associated with maneuvering in rough terrain but also because of the mountainous group's limited interest in accepting any form of unification. Turchin viewed rivers, in contrast to mountains, as geographical routes for the spread of culture and trade, increasing the likelihood of group integration.

8 Conclusions

The birth of a nation comes of cooperation. Very often, ingroup cooperation is compelled by outgroup competition. Cooperative aggregates formed and ratcheted up in terms of complexity and size through *Red Queen* effects. This is not a concise story of cultivators growing in complexity through the pressure of nomadic raiding, but a more generalized process of competitive warfare between groups practicing any subsistence economy. Networks of associations developed into feudal arrangements as competition compelled aggregation to the end of state formation. From the surplus of agriculture, a warrior class, and later leviathan-like governments monopolizing power, directed the energies of groups against one another. Those highly group-selected societies that expanded to the marginalization, displacement, or absorption of rival groups experienced pull and push factors compelling cooperation in measured proportions over sustained periods. These conditions, in effect, created selective regimes, productive of *asabiyyah*, which then became anthropogenically self-reinforcing, at least so long as prevailing conditions were sustained. As with the seventh-century Arabs and the eighteenth-century Cossacks, group-selected populations grew, prevailed, conquered, and expanded by virtue of elevated *asabiyyah* beyond what might be expected from other parameters relevant to the between-group competition, such as mean cognitive ability, life history speeds, absolute numbers, technological sophistication, and economic resources. Subsequent sections comprising Chap. 5 will proceed to examine how a society that evolved to be highly group selected deploys cultural mechanisms to exaggerate and perpetuate psychological, behavioral, and functional group-selected qualities.

Notes

1. The Red Queen effect refers to an evolutionary "arms race," a metaphor wherein although competing individuals invest in activities directly related to the conflict, neither party gain any direct advantage over its corresponding rival (Gat, 2006; Ridley, 2003; Robson, 2005). This concept was inspired by Lewis Carrol's *Through the Looking-Glass*. In this fiction, individuals competing in a game had to run faster just to remain in the same place.
2. For example, food availability causes many Southeast Asian primates to exist in bonded pairs or in solitude, with larger aggregations only appearing during masting events. Like the great ungulate herds on the African plains, mass aggregations of animals so often migrate because, if stationary, they would devour all local foodstuffs and thereafter starve.
3. Plavcan (2002) distinguished between high-intensity species featuring dominance hierarchies and intra-male intolerance and low-intensity taxa characterized for the presence of tolerant interactions between males.
4. Plavcan (2002) also defined frequency of conflict based on the number of breeding males in the group. The author classified multi-male societies as high frequency and single-male systems as low frequency.
5. Following Service's description (1975), Claessen and Skalník (1978, p. 22) referred to chiefdoms as "socio-political organizations with a centralized government, hereditary hierarchical status arrangements with an aristocratic ethos but no formal, legal apparatus of forceful repression, and without the capacity to prevent fission."
6. Huntington (1927) provides this additional insight relative to later stages of the contest between nomads and agriculturalists:

   > Almost all nomads are hard to rule and are often at enmity with their government. One reason is that when they have made a raid or otherwise incurred the displeasure of the officials, they can easily slip away into the desert and disappear. Moreover, they live in such small groups that it is impossible for the government to maintain officials among them. So they have their own patriarchal form of self-government, and bitterly resent any attempt to force anything else upon them. (p. 116)

7. A reading of this thesis suggests that early husbandmen were first hunted like wild animals and then kept like domesticates. Cultivators first clustered in rudimentary rings of defense and then behind walls in response to nomadic raiders who preyed upon their labor, only later exchanging this crippling uncertainty for regularized tribute to a professional warrior class to which they outsourced their defense. This warrior class sometimes derived from within the group and sometimes from a subset of the nomadic raiders themselves. With respect to the latter circumstance, we see something akin to the evolution of virulence wherein parasites evolve in such a way that bioenergetic claims on their hosts are balanced against host survivability.
8. In describing the adaptiveness of traits, Nettle (2011) states that "all carry some degree of risk, of accident, disease, social conflict or resource depletion." The long alleles of the D4DR gene, Nettle asserts, are found in higher proportion among currently or historically nomadic human populations. Being that the D4DR gene is associated with "personality traits and behaviors related to extraversion."
9. Tainter (1988) reviewed the political literature addressing the origins of the state. The author's summary classified the various theories into four main types of explanations. *Managerial* approaches emphasized the rise of hierarchical institutions necessary to preserve adequate levels of socio-political integration, such as resource extraction and distribution, when societies increase in population size or face potential disruption. *Internal conflict* theories described how stratified organizations evolved to guarantee the elite's monopolization of resources at the expense of the rest of the population. Theories of *external conflict* proposed that confrontations between polities demanded the development of institutions necessary for territorial defense, military campaigns, and administration of conquered territories. Finally, *synthetic* perspectives viewed the evolution of the state as the outcome of interacting forces, such as the adoption of a different system of economy and production influencing social structures.
10. Sober and Wilson (1998, p. 147) understand that individuals, like groups, must be cooperative units. There are *rules* and *regulations* that must be followed if multicelled life is to become an *adaptive unit*. Sober and Wilson call our attention to the language of within-organism cooperation, using words such as *outlaw genes*, *sheriff genes*, and *parliaments of genes*; these mimic the language of group conflict and cooperation.

In short, *the evolution of group-level functional organization cannot be explained on the basis of natural selection operating within groups. On the contrary, natural selection operating within groups tends to undermine group-level functional organization*. This statement holds not only for traits that appear overtly altruistic, such as the suicidal sting of a honey-bee worker, but also for the low-cost coordination of behaviors for the good of the group, such as deciding upon the best nest cavity. "The statement is so important that E. O. Wilson called it 'the central theoretical problem of sociobiology' in his 1975 book *Sociobiology*" (Wilson, 2015, p. 21).

D. S. Wilson also provides a similar comment, illustrating the interplay of selfishness and altruism as it relates to multicellular organisms, and groups of organisms: "E. O. Wilson and I provided this one-foot summary of sociobiology in a 2007 article: "Selfishness beats altruism within groups. Altruistic groups beat selfish groups. Everything else is commentary" (Wilson, 2015, p. 23).

11. Without secure property and the ability to reap what is sown, the whole endeavor is undermined. Illness, predation, and parasitism were among the many factors threatening reliable returns on investment. Perhaps a modest group size, enough to sustain perpetual fires, manufacture weapons, and post sentries, was sufficient to stave off predation. Parasitism was greatly reduced for those populations moving to temperate climates; and density-dependent diseases might remain rare with modest group sizes. *Conspecific competition*, competition between the members of the same species, was not so easily tamed. Perhaps then it was conspecifics that pressed early agriculturists to ever denser living arrangements. These would not be fellow agriculturists, for at the time, land, even fertile and fecund land, was relatively plentiful, making it easier to find than to fight for. No, agriculturists were threatened instead by those who had not yet transitioned to an agricultural existence.
12. Life history speed refers to the placement on a continuous distribution, from fast-selected to slow-selected people or populations, which is reviewed at length in Hertler et al. (2018).
13. Consider, for instance, that "A high level of V can help tribes and nations in their struggle for survival." More to the point, Penman (2015, p. 128) writes that high-V peoples are "warlike, confident, and well organized at the local level and frequently conquer or drive out neighboring groups." Competitive advantage, maintenance of birth rates, growth, and hierarchy are understood to result from high vigor. Further direct reading of

Penman's *Biohistory* establishes that highly vigorous societies are highly group-selected societies. For example, Penman cites nineteenth-century English public schools as vigor inducing through the use of "sport, cold showers, physical punishment, and institutionalized bullying through the fagging system" (Penman, 2015, p. 130). The Duke of Wellington is said to have observed, Penman reminds us, that "the battle of Waterloo was won on the playing fields of Eton."

14. Though we will sometimes use asabiyyah for its particular descriptive overlap, it herein denotes the group-selected society.
15. Turchin addresses this concern on page 323 of *War and Peace and War*. Concepts like fighting spirit or asabiyyah, he notes, are apt to appear inchoate, making measurement and scientific descriptive rigor important.
16. Notwithstanding Hannibal and his elephants, the Alps effectively sealed off the only exposed side of the Italian peninsula, regulating the threat of conquest, as was accomplished by the Aegean with respect to the Greek city states and dikes and dams with respect to Holland. Speculating further, geographically delimited areas might also stage the operation of martial competition progressing toward aggregation, allowing, for instance, the Romans to subsume and vanquish the Sabines, Samnites, Etruscans, Lucanians, Messapians, Bruttians, and Frentani before conquering and subjugating Sicilia, Macedonia, Syria, and Gallia.
17. Of note, Turchin mentions multilevel selection theory on page 7 of *War and Peace and War*, showing that multilevel selection theory, as opposed to rational choice theory, is consistent with his claims. However, he does not apparently go on to mention multilevel selection theory in that work.

References

Adams, D. C. (2004). Character displacement via aggressive interference in Appalachian salamanders. *Ecology, 85*(10), 2664–2670.

Alesina, A., Giuliano, P., & Nunn, N. (2011). Fertility and the plough. *American Economic Review, 101*(3), 499–503.

Alesina, A., Giuliano, P., & Nunn, N. (2013). On the origins of gender roles: Women and the plough. *Quarterly Journal of Economics, 128*(2), 469–530.

Armbruster, W. S., Edwards, M. E., & Debevec, E. M. (1994). Floral character displacement generates assemblage structure of Western Australian triggerplants (*Stylidium*). *Ecology, 75*(2), 315–329.

Bain, R. N. (1891). The second partition of Poland (1793). *The English Historical Review, 6*(22), 331–340.

Beans, C. M. (2014). The case for character displacement in plants. *Ecology and Evolution, 4*(6), 862–875.

Boas, F. (1896). The limitations of the comparative method of anthropology. *Science, 4*(103), 901–908.

Boehm, C. (1999). *Hierarchy in the forest: The evolution of egalitarian behavior*. Cambridge, MA: Harvard University Press.

Boehm, C. (2012). *Moral origins: The evolution of virtue, altruism, and shame*. New York: Basic Books.

Boserup, E. (1970). *Women's role in economic development*. London: Allen & Unwin.

Brockhurst, M. A., Chapman, T., King, K. C., Mank, J. E., Paterson, S., & Hurst, G. D. (2014). Running with the Red Queen: The role of biotic conflicts in evolution. *Proceedings of the Royal Society B: Biological Sciences, 281*(1797), 20141382.

Brockmann, H. J., & Barnard, C. J. (1979). Kleptoparasitism in birds. *Animal Behaviour, 27*, 487–514.

Burton, M. L., & White, D. R. (1984). Sexual division of labor in agriculture. *American Anthropologist, 86*(3), 568–583.

Chapais, B. (2009). *Primeval kinship: How pair-bonding gave birth to human society*. Cambridge, MA: Harvard University Press.

Claessen, H. J. M., & Skalník, P. (1978). The early state: Theories and hypotheses. In H. J. M. Claessen & P. Skalník (Eds.), *The early state* (pp. 3–30). The Hague, The Netherlands: Mouton Publishers.

Currie, T. E., & Mace, R. (2011). Mode and tempo in the evolution of socio-political organization: Reconciling 'Darwinian' and 'Spencerian' evolutionary approaches in anthropology. *Philosophical Transactions of the Royal Society B: Biological Sciences, 366*(1567), 1108–1117.

Durant, W. (1950). *The story of civilization*. New York: Simon & Schuster.

Enan, M. A. (2007). *Ibn Khaldūn: His life and works*. New York: The Other Press.

Field, J. (1992). Intraspecific parasitism as an alternative reproductive tactic in nest-building wasps and bees. *Biological Reviews, 67*(1), 79–126.

Figueredo, A. J., Peñaherrera-Aguirre, M., Fernandes, H. B. F., Lomayesva, S. L., Woodley of Menie, M. A., Hertler, S. C., et al. (2019). The ecology of empire: The dynamics of strategic differentiation-integration in two competing Western European biocultural groups. *Journal of Politics and the Life Sciences, 38*(2), 210–225.

Flannery, K., & Marcus, J. (2012). *The creation of inequality: How our prehistoric ancestors set the stage for monarchy, slavery, and empire*. Cambridge, MA: Harvard University Press.

Fromherz, A. J. (2011). *Ibn Khaldun*. Edinburgh, UK: Edinburgh University Press.

Gat, A. (2006). *War in human civilization*. New York: Oxford University Press.

Gat, A. (2009). So why do people fight? Evolutionary theory and the causes of war. *European Journal of International Relations, 15*(4), 571–599.

Gerhardt, H. C. (1994). Reproductive character displacement of female mate choice in the grey treefrog, *Hyla chrysoscelis*. *Animal Behaviour, 47*(4), 959–969.

Gorski, P., & Sharma, V. (2017). Beyond the Tilly thesis "family values" and state formation in Latin Christendom. In L. B. Kaspersen & J. Strandsbjerg (Eds.), *Does war make states?* (pp. 98–124). New York: Cambridge University Press.

Grant, P. R. (1972). Convergent and divergent character displacement. *Biological Journal of the Linnean Society, 4*(1), 39–68.

Grant, P. R., & Grant, B. R. (2006). Evolution of character displacement in Darwin's finches. *Science, 313*(5784), 224–226.

Halecki, O. (1945). The sixth partition of Poland. *The Review of Politics, 7*(2), 142–155.

Hertler, S. C., Figueredo, A. J., Peñaherrera-Aguirre, M., Fernandes, H. B. F., & Woodley of Menie, M. A. (2018). *Life history evolution: A biological meta-theory for the social sciences*. New York: Palgrave Macmillan.

Hirst, K. K. (2019a, August 22). *Plant domestication*. Retrieved from https://www.thoughtco.com/plant-domestication-table-dates-places-170638

Hirst, K. K. (2019b, August 22). *Animal domestication – Table of dates and places*. Retrieved from https://www.thoughtco.com/animal-domestication-table-dates-places-170675

Höbel, G., & Gerhardt, H. C. (2003). Reproductive character displacement in the acoustic communication system of green tree frogs (*Hyla cinerea*). *Evolution, 57*(4), 894–904.

Hrdy, S. (1999). *Mother nature: A history of mothers, infants and natural selection*. New York: Pantheon.

Hrdy, S. (2009). *Mothers and others: The evolutionary origins of mutual understanding*. Cambridge, MA: Harvard University Press.

Hunter, C. P., & Dwyer, P. D. (1997). The value of objects to satin bowerbirds *Ptilonorhynchus violaceus*. *Emu-Austral Ornithology, 97*(3), 200–206.

Huntington, E. (1927). *The character of races as influenced by physical environment, natural selection and historical development*. New York: Charles Scribner's Sons.

Hutchinson, G. E. (1957). *A treatise on limnology*. Hoboken, NJ: Wiley.

Hutchinson, G. E. (1959). Homage to Santa Rosalia, or why are there so many kinds of animals? *The American Naturalist, 93*, 145–159.

Ibn Khaldūn, A. R. (1377/1958). *The Muqaddimah: An introduction to history*. Translated from the Arabic by Franz Rosenthal. 3 vols. New York: Princeton.

Kaspersen, L. B., & Strandsbjerg, J. (Eds.). (2017). *Does war make states?: Investigations of Charles Tilly's historical sociology*. New York: Cambridge University Press.

Kelly, R. C. (2000). *Warless societies and the origin of war*. Ann Arbor, MI: University of Michigan Press.

Kramer, S. N. (1956). *History begins at Sumer: Thirty-nine firsts in man's recorded history*. Philadelphia: University of Pennsylvania Press.

Lord, R. H. (1915). *The second partition of Poland: A study in diplomatic history* (Vol. 23). Cambridge, MA: Harvard University Press.

Macaulay, T. B. (2008). *The history of England, from the accession of James II—Volume 1*. Project Gutenberg. Retrieved from http://www.gutenberg.org/ebooks/1468

MacFarlane, A. (2003). *The savage wars of peace: England, Japan and the Malthusian Trap*. New York: Palgrave Macmillan.

Marlowe, F. (2000). Paternal investment and the human mating system. *Behavioural Processes, 51*(1-3), 45–61.

McNeill, W. H. (1974). *The shape of European history*. New York: Oxford University Press.

McNeill, W. H. (1983). *The great frontier: Freedom and hierarchy in modern times*. Princeton, NJ: Princeton University Press.

McNeill, W. H. (1992). *The global condition: Conquerors, catastrophes, and community*. Princeton, NJ: Princeton University Press.

Muchhala, N., & Potts, M. D. (2007). Character displacement among bat-pollinated flowers of the genus *Burmeistera*: Analysis of mechanism, process and pattern. *Proceedings of the Royal Society B: Biological Sciences, 274*(1626), 2731–2737.

Mulder, M. B., Bowles, S., Hertz, T., Bell, A., Beise, J., Clark, G., et al. (2009). Intergenerational wealth transmission and the dynamics of inequality in small-scale societies. *Science, 326*(5953), 682–688.

Murdock, G. P., & Provost, C. (1973). Factors in the division of labor by sex. *Ethnology, 12*(4), 203–225.

Murray, K. G., Feinsinger, P., Busby, W. H., Linhart, Y. B., Beach, J. H., & Kinsman, S. (1987). Evaluation of character displacement among plants in two tropical pollination guilds. *Ecology, 68*(5), 1283–1293.

Nettle, D. (2011). Evolutionary perspectives on the five-factor model of personality. In D. M. Buss & P. H. Hawley (Eds.), *The evolution of personality and individual differences* (pp. 5–28). New York: Oxford University Press.

North, D. C., Wallis, J. J., & Weingast, B. R. (2009). *Violence and social orders: A conceptual framework for interpreting recorded human history*. New York: Cambridge University Press.

Penman, J. (2015). *Biohistory*. Newcastle upon Tyne, UK: Cambridge Scholars Publishing.

Perkins, J. B. (1896). The partition of Poland. *The American Historical Review, 2*(1), 76–92.

Pfennig, K., & Pfennig, D. (2009). Character displacement: Ecological and reproductive responses to a common evolutionary problem. *The Quarterly Review of Biology, 84*(3), 253–276.

Plavcan, J. M. (2002). Reconstructing social behavior from dimorphism in the fossil record. In J. M. Plavcan, R. F. Kay, W. L. Jungers, & C. P. van Schaik (Eds.), *Reconstructing behavior in the primate fossil record* (pp. 297–338). New York: Springer Science & Business Media.

Plavcan, J. M. (2012a). Sexual size dimorphism, canine dimorphism, and male-male competition in primates. *Human Nature, 23*(1), 45–67.

Plavcan, J. M. (2012b). Body size, size variation, and sexual size dimorphism in early Homo. *Current Anthropology, 53*(S6), S409–S423.

Plavcan, J. M., van Schaik, C. P., & Kappeler, P. M. (1995). Competition, coalitions and canine size in primates. *Journal of Human Evolution, 28*(3), 245–276.

Ridley, M. (2003). *The Red Queen: Sex and the evolution of human nature*. New York: Harper Collins.

Robson, A. J. (2005). Complex evolutionary systems and the Red Queen. *The Economic Journal, 115*(504), F211–F224.

Schluter, D. (2000). Ecological character displacement in adaptive radiation. *The American Naturalist, 156*(S4), S4–S16.

Schluter, D., & McPhail, J. D. (1992). Ecological character displacement and speciation in sticklebacks. *The American Naturalist, 140*(1), 85–108.

Schluter, D., Price, T. D., & Grant, P. R. (1985). Ecological character displacement in Darwin's finches. *Science, 227*(4690), 1056–1059.

Service, E. R. (1975). *Origins of the state and civilization*. New York: Norton.

Sober, E., & Wilson, D. S. (1998). *Unto others: The evolution and psychology of unselfish behavior*. Cambridge, MA: Harvard University Press.

Spruyt, H. (2017). War and state formation: Amending the bellicist theory of state making. In L. B. Kaspersen & J. Strandsbjerg (Eds.), *Does war make states?: Investigations of Charles Tilly's historical sociology* (pp. 73–97). New York: Cambridge University Press. http://www.emersonkent.com/map_archive/central_europe_1477.htm

Tainter, J. (1988). *The collapse of complex societies.* Cambridge, UK: Cambridge University Press.

Tilly, C., & Ardent, G. (1975). *The formation of national states in Western Europe* (Vol. 8). Princeton, NJ: Princeton University Press.

Toynbee, A. J. (1951). *A study of history.* New York: Oxford University Press.

Turchin, P. (2003). *Historical dynamics: Why states rise and fall* (Vol. 26). Princeton, NJ: Princeton University Press.

Turchin, P. (2004). Dynamic maps of evolution of the state system and metaethnic frontiers in Europe during the two millennia. http://www.eeb.uconn.edu/people/turchin/Clio.htm

Turchin, P. (2006). *War and peace and war: The rise and fall of empires.* New York: Plume.

Turchin, P. (2008). A theory for the formation of large agrarian empires. *Santa Fe Institute working paper, 05*-024.

Turchin, P. (2012). Failed states and nation-building: a cultural evolutionary perspective. *Cliodynamics: The Journal of Quantitative History and Cultural Evolution, 3*(1), 158–171.

Turchin, P., Currie, T. E., Whitehouse, H., François, P., Feeney, K., Mullins, D., et al. (2018). Quantitative historical analysis uncovers a single dimension of complexity that structures global variation in human social organization. *Proceedings of the National Academy of Sciences, 115*(2), E144–E151.

Turchin, P., & Gavrilets, S. (2009). Evolution of complex hierarchical societies. *Social Evolution & History, 8*(2), 167–198.

Tynkkynen, K., Rantala, M. J., & Suhonen, J. (2004). Interspecific aggression and character displacement in the damselfly *Calopteryx splendens. Journal of Evolutionary Biology, 17*(4), 759–767.

Wang, S., Lewis Jr., C. M., Jakobsson, M., Ramachandran, S., Ray, N., Bedoya, G., et al. (2007). Genetic variation and population structure in Native Americans. *PLoS Genetics, 3*(11), e185.

Wilson, D. S. (2015). *Does altruism exist? Culture, genes and the welfare of others.* New Haven, CT: Yale University Press.

Wright, R. (2001). *Nonzero: The logic of human destiny.* New York: Vintage.

5

Growth, Maintenance, Control, and Competition

Steven C. Hertler, Aurelio José Figueredo, and Mateo Peñaherrera-Aguirre

1 Introduction

Where Chap. 4 reviewed the formation of sociopolitically complex civilizations, this chapter reviews their growth and maintenance. Mature states invariably come to encompass expanding territories and consequently absorb populations distinct in dialect and language, ethnicity and race, and culture and religion. As discussed herein, maintaining integrity at a particular level of group size comes from managing both sources of threat: managing one's own population while defending against rival groups. Populations must be bound by some combination of custom, sanctions, religion, and legal infrastructure. To the extent that this can be accomplished, a state must radiate control stably through time, as indicated by the Roman Empire and contraindicated by the conquests of Alexander the Great. Controls necessary for stable growth can be (1) psychological, as with propaganda; (2) legal, as with incarceration; (3) social, as with

S. C. Hertler et al., *Multilevel Selection*, https://doi.org/10.1007/978-3-030-49520-6_5

banishment; (4) martial, as with conscription; or (5) economic, as with taxation. Some controls woo and win elites, ensuring allegiance through shared interest, title, rank, privilege, estates, orders, and garters. With reliable money as a medium of exchange, states ensnare citizens within a tightening cage of mutual interest, trade, dependencies wrought of divided labor, and the attractive ability to solve collective action problems. All such forms of control foster growth and allow for societal maintenance.

Drawing on authors as varied as Wimmer, Vico, Ellul, Tainter, Ferguson, and Padover, we review (1) myth, (2) propaganda, (3) punishment, (4) societal interdependence via systemic differentiation, and (5) mechanisms by which impediments to growth are removed. Each of these five elements of social control is a device by which historical societies have attempted to meet the centripetal and centrifugal forces intrinsic to stable social growth and is fully consistent with the theories of cultural group selection that we have reviewed. With an eye toward future research, by the end of each section, we attempt to specify whether these historically applied mechanisms were efficacious and, if so, whether they are universal or particular with respect to time, group size, or phase of growth. Each section frames its respective device, a mechanism historically expressed to solve the adaptive problem of societal growth.

Regarding this characterization of societal growth as an adaptive problem, it is important to begin with the caveat that some of the authors whose work we review appear to present pro-nationalist or pro-imperialist biases. Our goal in this chapter is neither to promote nation-building or empire-building nor even to advocate for them as particularly worthy (or unworthy) endeavors in human affairs. As evolutionary scientists, we instead seek to specify the conditions that appear conducive to the establishment, preservation, and expansion of such state and imperial-level sociopolitical organizations from the standpoint of their relative advantage in intergroup competition under the shaping forces of multilevel selection. For this purpose, we review the works of authors that were clearly favorable to these goals and adulatory of their achievements, as such authors represented the past scholars that were seemingly most motivated to identify what societal adaptive strategies were most likely to either lead to success or end in failure. We therefore review their insights for the purpose of examining the efficacy of such strategies descriptively and dispassionately, rather than prescriptively to synthesize a normative manual for empire. We accordingly disavow in advance the various biases evident in these partisan sources, while seeking to employ their pragmatic insights in our evolutionary analysis for the *selection by consequences* of these group-level adaptations, as discussed in our Preface.

2 Of Men and Myths: Heroes, Hero Worship, and National Narratives

Do societies require a collective consciousness in the form of myth and mythic heroes? The writings of Giambattista Vico, an eighteenth-century Italian political philosopher, suggest an affirmative answer (Pompa, 2010).

Epicurus, Hobbes, Machiavelli, Grotius, Selden, and Pufendorf are equally taken to task by Vico for the fault of viewing history from the time of monarchical society, a procedure which emphasizes the *preservation of mankind* (meaning the human species as a whole) to the exclusion of *preservation of nations* (meaning particular biocultural groups). It seems that Vico considers prehistory, or at least the traditions of early civilizations using oral transmission to preserve their culture, a prerequisite to understanding the histories focused on by the aforementioned writers. He specifically argues for the serious consideration of the roles of gods and heroes within those cultural traditions. Organizing myths are "collective phenomena" owing their powers to "collective participation" (Ellul, 1973, pp. 116–117), with roots sunk far down into the subconscious, providing a sense of permanence and place, more felt than thought, and possessing strong motive force. Religion can also be seen as national or supranational myth. Consequently, Vico warns against ignoring religion, understanding it to be the fount of Roman greatness, for instance, but one poorly understood by Polybius, Plutarch, and Machiavelli alike. Religion is considered the source of solemnity in marriage, Patrician patronage of Plebeians, and valor in war, spurring the Romans *to conquer or to die with one's own gods* (Vico, 2002, p. 86).[1]

Cohesive groups require terrestrial as well as celestial fathers, metaphorically speaking. Vico finds several commonalities among "fathers" of populations, families, and cities from which states are thereafter derived:

1. Of imagining deities;
2. Of begetting certain children with certain women through certain divine auspices;
3. Of being, therefore, of heroic or Herculean origin [for the following reasons]:
   (a) Because they possessed the science of the auspices, that is, of divination;
   (b) Because they made sacrifices in their houses;
   (c) Because of their infinite power over their families;
   (d) Because of the strength with which they slew the wild animals, tamed the uncultivated lands, and defended their fields against the impious vagabonds who came to steal their harvests;
   (e) Because of the magnanimity with which they received into their asylums the impious vagabonds who, endangered by the quarrels of Hobbes' violent men in the state of bestial communion, sought refuge in them;
   (f) Because of the height of fame to which their virtue in suppressing the violent and assisting the weak had raised them;
   (g) Because of the sovereign ownership of their fields that they had acquired naturally through such exploits;
   (h) Because, consequently, of their sovereign command of arms, which is always conjoined with sovereign ownership;
   (i) And, finally, because of their sovereign will over the laws, and therefore also punishments, which is conjoined with sovereign command of arms.

In reviewing these features, one finds essential elements of cohesion, identification, altruism, and related aspects of culturally group-selected societies. For the Greeks, Ajax was a colossus, representing their strength, just as Ulysses was a fox, representing their cunning. Roland is submitted as a heroic Gallic composite of valor. Hebrews, Assyrians, Persians,

Egyptians, and Greeks respectively looked to Levites (strong), Chaldeans (sages), magi (diviners), priests, and divinari (diviners). The prince or hero is so often described as holding back hordes, defending a bridge, or turning the tide of war by his single will. The reputations of such heroes are *resplendent*, reflecting their luster across their representative people. Vico understands these as poetic *personae*, emblematic of a house, a coat of arms, or "a kind of genera in which many men are comprehended" (Vico, 2002, p. 205). Continuing this line of reasoning into the Victorian age, Carlyle describes an enduring and inevitable process of transformation through the ages, from the *naïveté* of worshiping man as a divinity to the necessity of at least admiring the heroic individuals among us.

A founding figure, be it Moses or Joseph Smith, communes with a deity. As Chap. 3 illustrates, a deity may serve as a central fault line in group identification and disidentification. Vico's founding father figure descends from on high to beget progeny or is otherwise associated with inaugurating a seminal, exalted, or hybridized godlike lineage, from which comes the heroic founding. The heroic founding imparts powers ranging from divination, to strength, to wisdom. Clemency combines with control to allow the founder to bring others into the fold, suppress dissension, rule with justice, and consequently become sovereign with a monopoly of authority over a unified people. Vico uses the term *father* advisedly for it represents the relationship between founder and followers, which is patterned on father and family in the primitive state. Indeed, such familial language is ubiquitous: Priests and friars are sometimes called fathers and brothers, army units are bands of brothers; this may well amount to a general principle, to wit that kinship is the template of association from which larger aggregations are extrapolated.

Whether in reality or myth, the right of life and death over subjects and the responsibility of maintaining order and liberty among them may be transferred from a personalized founding figure to an impersonal civil order. To illustrate this in action, we turn to the example of George Washington's role in the founding of the United States. Washington's distance from the present at once renders him sufficiently modern for instructive documentation and sufficiently remote to accrue the organic patina of myth and legend. Tall, grave, and martial in bearing, the alpha status of Washington, reminiscent of many a tribal leader, gave power and precedent to a nascent nation of abstract laws. Padover (1955/1989) expertly captures this transition from a tribe to a republic, from a nation of men to a nation of offices:

> It was the sheer personality of Washington that was the decisive element in the three crucial events of early America—the Revolutionary War, the Constitutional Convention, and the first national administration. Hardly anything more than his willpower held together the ragged Revolutionary army in times of darkest despair; a weaker man would have given way to hopelessness as the troops deserted, provisions gave out and funds dwindled to near nothingness. The Commander-in-Chief complained with furious bitterness, but stuck to his guns. Similarly, it was his presence that helped to weld the Constitutional Convention in 1787. Chairman of the Convention, he was a silent member, but the fact that he was there was a guarantee of the importance of the meeting and cemented the disparate viewpoints. Washington's immense prestige was a major factor in the adoption of the bitterly assailed Constitution…His exquisite sense of balance and steadying wisdom reconciled clashing interests and opposing sections and gave the new nation the fundamental shape that it has retained to this day. (p. 8)

Padover's writings reflect Washington's genuine greatness, but national myths and heroic adulation demand preternatural greatness, purging flaws and foibles while exaggerating virtues and competencies. "Few other national heroes have ever been greater targets of assiduous idolatry, hagiography, iconolatry, myth-making, and breathless patriotic oratory." "Young America," Padover writes, was "hungry for a hero," and so from history we transition to hagiography. The hagiography of Parson Mason Weems is attributed to *naïveté* by Esmond Wright (1995), author of *A History of the United States of America*. However, Weems was most probably not a naïve historian, but a wise storyteller. He generated myth, purposefully and consciously. Weems was simply the first and worst among the historical myth makers. Others followed in his path, including Jared Sparks who edited and reformulated Washington's literary style making it more felicitous and flowing; also, there was Washington Irving who cited him, contrary to available evidence, as a faithful Sunday worshiper and a man who married for love alone. Wright is aware that "the process of glorification was quite deliberate…it was even more the work of artists than of writers" (Wright, 1995, p. 164). Indeed, "Washington was made into a graven image for the nation to worship" (Padover, 1955/1989, p. 1). Similarly, at the end of the Civil War, Constantino Brumidi supplied the nation with some much-needed unifying fodder by painting Washington into a Renaissance-like fresco on the rotunda of the United States Capitol Building accompanied by Columbia, Minerva, Mercury, Neptune, Vulcan, and Ceres. Thus, Washington passed from man to myth in writing as well as iconography.

Contemporaries witnessed the beginning of this deification. John Adams raged against it, both from thinly veiled jealousy and because he hated the process by which history was amalgamated with the dross of myth. Another contemporary, Marshal Jean-Baptiste Donatien de Vimeur, comte de Rochambeau, tolerated and acquiesced in the accretion of myth from conscious recognition of its unifying function. Rochambeau commanded France's expeditionary force, dispatched to aid the American bid for independence. He presided over a delicate impasse wherein he repeatedly sought to persuade Washington to assume a southern strategy, while Washington trained tenaciously north, looking for vindication in New York where he was so soundly defeated in the early phases of the American Revolution.[2] The two together would decide exactly how to use French naval power. Washington was the supreme commander of American forces, which were the significant force on land; Rochambeau was the supreme commander of French forces, which were the significant force at sea (Whitlock, 1929). Nevertheless, coordinated action was indispensable.[3] Not only did Washington resist the call to decisive victory at Yorktown in favor of probable defeat at New York, but once the southern campaign was decided upon, he contributed little to its success. According to Ferling, it was the experience of French military engineers with their sapping and mining techniques that was responsible for the siege's quick success. It is true that Washington symbolically struck a spade into the ground and lit the first cannon, but the necessity of his presence at Yorktown seemed not to extend much further. Despite this, the ultimate success of the battle would be, if attributed to anyone, attributed to Washington. As recounted in his memoirs, Rochambeau recognized that the success of the siege was more to his honor than it was to

Washington's. More accurately, it was the honor of the French military engineers who had safely maneuvered the British into inevitable capitulation and the fleet who had made the siege possible in the first place by defeating the British navy and thereby disallowing retreat via the Chesapeake Bay. The older Rochambeau wanted to leave an accurate account for the historical record, but explained why he had written this later and not spoken earlier. He was silent at the time because he and the French generally recognized the need the young America had for a hero like Washington (Ferling, 2010).

One should be aware, however, that in taking the example of the American Revolution, we see just one level at which group selection is operating. Looking at the larger whole, we can see that the American Revolution itself was an internecine conflict, often referred to as a civil war or as a *cousins' war*. Just a generation prior, colonists were fighting alongside Britons against Frenchmen. It was the very success of the Anglo-American alliance that opened the way for the revolution. Having so decisively won, the French ceased to become a threat to the colonists, making the British dispensable. These are the fault lines across which between-group competition cleaves. We should also keep in mind that the American Revolution constitutes an aggregation event from the American perspective but a decline event from the British perspective. We cannot perhaps say that this was the beginning of the end for the British Empire, which only in the nineteenth century achieved its full grandeur with its Victorian jewels in India and Africa. It was nevertheless a prelude to that decline. It can also be considered a budding event wherein, though it became a rival and eventually dominant, the American colonies were analogous to a reproductive propagule of the mother country as depicted in some forms of group selection (MLS2, as introduced in Chap. 2 and detailed in Chap. 8). Such is the fertility of history through the lens of evolution.

In correcting the errors of recent contemporaries, Vico championed to a fault the previously underestimated importance of myth. Notwithstanding Vico's overstatement, his instincts are productive of intuitive claims, supported by the seeming ubiquity of founding myths and mythic heroes, not only in the examples proffered but also in most foundational texts, sacred and secular. Narrative myths may well be most important during early stages of aggregation, even as they may have an enduring binding force thereafter.

3 Propaganda

Experiments in perception suggest that human eyesight has evolved to exaggerate the border between shapes, a deviation from reality which nonetheless aids in making figure-ground discriminations. Similarly, both the impulse to create and credit propaganda relate to a tribalistic aspect of human nature, productive of exaggerating differences between neighboring groups.

Following the post-World War I usage relating to willful misinformation,[4] propaganda's principal goal is arguably to delineate groups, one from another, making propaganda eminently relevant to cultural group selection. In *Propaganda: The Formation of Men's Attitudes*, Ellul (1973, p. 212) includes a major subsection entitled *Propaganda and Grouping*, which begins by noting that "all propaganda has to set off its group from all the other groups." Propagandized groups allow characterizations to stand in for the reality of rival groups, ceasing "altogether to be open to an exchange of reason, arguments, points of view" (Ellul, 1973, p. 213). Writing without knowledge of multilevel selection theory, Ellul (1973) nonetheless recognizes the various levels of aggregation upon which propaganda can act to bind or divide:

> This partitioning takes place on different levels—a unionist partitioning, a religious partitioning, a partitioning of political parties or classes; beyond that, a partitioning of nations, and, at the summit, a partitioning of blocs of nations. (p. 213)

"The Manichean universe of propaganda" (Ellul, 1973, p. 69) routinely reinforces within-group allegiance, while at the same time devaluing those outside the group, a combination which Ellul (1973, p. 213) refers to as a "double foray on the part of propaganda, proving the excellence of one's own group and the evilness of the others." We review four major uses of propaganda: (1) enhancing intragroup cohesion; (2)

demarcating and vilifying rival groups; (3) welding smaller groups into larger wholes; and (4) severing larger wholes into smaller groups. The former two attempt to maintain the integrity of existing groups, whereas the latter two attempt to increase or decrease group size.

Enhancing intragroup cohesion is perhaps the best-known function of propaganda, as is reflected in Ellul's (1973) definition of propaganda, which refers to mass collective action, as denoted by the terms *participation*, *mass*, *unification*, and *organization*:

> Propaganda is a set of methods employed by an organized group that wants to bring about the active or passive participation in its actions of a mass of individuals, psychologically unified through psychological manipulations and incorporated in an organization. (p. 61)

Effective propaganda mobilizes action and opinion by tapping a group's preexisting threads of commonality, its values, myths, and thoughts, as Ellul (1973) explains:

> Propaganda must not only attach itself to what already exists in the individual, but also express the fundamental currents of the society it seeks to influence. Propaganda must be familiar with collective sociological presuppositions, spontaneous myths, and broad ideologies. By this we do not mean political currents or temporary opinions that will change in a few months, but the fundamental psycho-sociological bases on which a whole society rests, the presuppositions and myths not just of individuals or of particular groups but those shared by all individuals in a society, including men of opposite political inclinations and class loyalties. (pp. 38–39)

Containing no actionable message, political or otherwise, *sociological preconditioning* directly enhances cohesion while preparing the populace to be actively propagandized by building those shared myths, suppositions, and broad ideologies discussed above (Auerbach & Castronovo, 2013). Sociological preconditioning is compared to plowing by Ellul; the ground is prepared for seeding by more pointed propagandistic aims. American civic education is understood by Ellul (1973) to be a form of sociological preconditioning, with state-sponsored education generally apt to inculcate societal principles, ideologies, and myths. Thus, education, often understood as a "prophylactic against propaganda," can be co-opted by the state and thereby amount to *pre-propaganda* that distributes disparate facts with the end of preparing a mind to receive state-sponsored narratives (Ellul, 1973, p. vi). Intragroup cohesion can also be enhanced simply by suppressing dissent, as when the Bolsheviks sabotaged and suppressed communications, both means and content, by shutting down some newspapers and co-opting others (Werth, 1999). Long-term investments in sociological preconditioning can then be leveraged by the state during war, transforming "normal feelings of patriotism into a raging nationalism." Binding myths are "sharpened" and invested with "the power of shock and action" (Ellul, 1973, p. 41). To impulses to protect home and hearth, the modern state must add additional motivations to suffer the strain of the trench, the forced march, shot and shell, disease, death, and desperation. Propaganda then supplements self-preservation, inducing a man to make "super-sacrifices" when "pushed to the very limit of his nervous and mental endurance, and in a sort of constant preparation for ultimate sacrifice" (Ellul, 1973, pp. 142–143). To induce a mass of individuals to temporarily transform into something of a superorganism in which the good of the part is subordinated to the good of the whole, modern states propagandize their populations, manufacturing patriotic sentiments, ideological screeds, heuristic glosses, and doctrinal explanations (Ellul, 1973). The individual then becomes a "cell organized into the social unit" in the "anatomy of society, with its interlocking group formations and loyalties" (Bernays, 2005, p. 55).

Propaganda channels the human capacity for enmity and division as much as it does fraternity and solidarity. The demarcation and vilification of rival groups are the complementary inverse of creating internal cohesion and the corollary of maintaining distinct groups. Emotions such as rage and actions such as murder, which are prohibited within the group, are encouraged when applied outside the group in the context of war:

> propaganda will permit what so far was prohibited, such as hatred, which is a dangerous and destructive feeling and fought by society … Propaganda offers him an object of hatred, for all propaganda is aimed at an enemy. And the hatred it offers him is not shameful, evil hatred that he must hide, but a legitimate hatred, which he can justly feel. Moreover, propaganda

> points out enemies that must be slain, transforming crime into a praiseworthy act. Almost every man feels a desire to kill his neighbor, but this is forbidden, and in most cases the individual will refrain from it for fear of the consequences. But propaganda opens the door and allows him to kill the Jews, the bourgeois, the Communists, and so on, and such murder even becomes an achievement. (p. 152)

Whereas group cohesion is bolstered by framing collectives as *bands of brothers*, *founding fathers*, and related terms meant to activate kin-selected psychology, the time-worn method of contrasting self and other is to strip the other of their humanness and humanity. For instance, in an English engraving, freedom and peace can be seen to look across the channel at devils and skeletons representing the universal destruction of revolutionary France. Such representations served as iconographic analogues to literary pleas, such as Edmund Burke's *Reflections on the Revolution in France*, which was meant to forestall Jacobin sympathies on the home front for France's radical revolution (Ben-Israel, 2002; Verhoeven, 2013).

In addition to keeping extant groups distinct, propaganda can weld smaller groups into larger wholes. History has examples of protean diplomatic marvels like Franklin weaving people together at all levels of organization. Franklin's political career was heavily marked by propagandistic and persuasive attempts to augment group size, creating alliances among previously unallied smaller powers to combat rivals: he did this at the local level by organizing community groups against criminal elements; he did this at the colonial level through his *join or die* motif, which was meant to unite the colonies against Franco-Amerindian incursions; and he did this at the national level through his diplomatic efforts to solicit alliances with European powers. Franklin enlisted multiple methods, such as private whisperings, leaks of information, published writings, as well as state struck medals, the latter being part of what Olson (2004) deems *rhetorical iconography*, as when Franklin designed the *Libertas Americana* to solidify and perpetuate the Franco-American alliance during his attempts to solicit ever more aid (Olson, 2004).

Indeed, Franklin's rebellion was actually his one great deviation, though it serves as an example of the ways in which propaganda, opposite the welding aims described above, can be used to sever one group from another. Consequently, there is *integration propaganda*, described as a "complex tool to weld individuals to the collective body of the state" (Castronovo, 2014, p. 90), and *agitation propaganda*, which "unleashes an 'explosive moment' that seems too volatile to suit the purposes of a durable nationalism" (Castronovo, 2014, p. 91). Successful revolutions are begun with agitation propaganda and ended with integration propaganda (Ellul, 1973):

> the transition from one type of propaganda to the other is extremely delicate and difficult. After one has, over the years, excited the masses, flung them into adventures, fed their hopes and their hatreds, opened the gates of action to them, and assured them that all their actions were justified, it is difficult to make them re-enter the ranks, to integrate them into the normal framework of politics and economics. What has been unleashed cannot be brought under control so easily, particularly habits of violence or of taking the law into one's own hands. (p. 77)

Franklin, and other colonial revolutionaries, illustrates both the tension between levels of selection and its reflection in the literature on propaganda as they became American statesmen. Through a group selectionist lens, former colonists fractured themselves off from the larger group of which they were tied by bonds of kinship, language, and history (Phillips, 2000), only to then tug hard at the reins of the revolutionary forces they unleashed, which tended toward perpetuating democratic freedoms and individualistic impulses. Better at tearing down than building up, Thomas Paine's *Common Sense* was such an instance of agitation propaganda, which would not brook the growth of a viable state if not later countered by a different vision. The 1790s witnessed Washington and Hamilton subduing fellow colonists rebelling under the duress and discontent that prompted rebellion against British rule in the 1770s. Similarly, from Samuel Adams in the 1770s to John Adams in the 1790s, the colonials turned Americans reversed course, from spreading propaganda to suppressing it.

Samuel Adams seemed to have conspired with Paul Revere, a goldsmith and engraver, to produce an iconic image of the *Boston Massacre*, part of the propaganda that framed a "motley rabble of saucy boys,

Negroes and mulattoes, Irish teagues and outlandish jacktars"[5] as unalloyed victims of British aggression. The Federalist and Arch-Federalist elites holding the reins of power through the 1790s acted thereafter to suppress propaganda, most notably through the *Alien and Sedition Acts*,[6] which were then attacked by the internal dissidents they were meant to control (Rosenfeld, 1997). The Alien and Sedition Acts were leveled also against emerging political rivals later embodied in the first instantiation of the Republican Party (Miller, 1953).

Our reading of Lippmann, Bernays, Davidson, Castronovo, Kidder, and Ellul suggests that propaganda serves an analogous function to national myth, with both propaganda and myth contrasting the ingroup with the outgroup. Beyond systematically or quantitatively studying propaganda's relationship to group cohesion, future research can perhaps support or negate our supposition that propaganda, as compared with myth: (1) emerges in the late phases of aggregation; (2) uniquely vilifies outgroups; (3) comes of conscious top-down creation; (4) confines to complex sociopolitical societies; and (5) often represents elite manipulation of non-elites, be it to exploit them or to promote the survival of the state.

4 Compelling Compliance: Punishment from the Bottom-Up and the Top-Down

Cooperation and conformity within groups are ensured through punishment (Chen & Szolnoki, 2018; Deakin, Taylor, & Kupchik, 2018; Gottschalk, 2006) meted out from the top-down and the bottom-up. Even as no large society neglects either mechanism, freer, decentralized governments are more reliant on locally administered punishment from the bottom-up, whereas tightly caged monarchical and authoritarian regimes are more reliant on centrally administered punishment from the top-down.

Readers of Alexandre Dumas' *Count of Monte Cristo*, in identifying with protagonist *Edmond Dantès* as he plots a long-deferred revenge against three unjust men, will be learning something about their intrinsic disposition toward moralistic punishment. Punishment from the bottom-up comes from values of fairness, justice, righteous anger, and outrage, which are derived of innate emotion (Haidt & Joseph, 2007), even as the parameters past which these emotions are triggered, like the behavioral response to norm violations, are culturally specific (Haidt, 2003). *Contempt*, *anger*, and *disgust* (Hutcherson & Gross, 2011) are among the moral emotions elicited in a social context that have corresponding *action tendencies* redounding to the regulation of others within the group. These action tendencies predict the punishment of rule breakers, norm violators, cheats, and dissidents even at personal expense (Ostrom, 2000), as Hoffman and Goldsmith (2003) describe at length:

> Punishment by the group addresses a central problem: the free rider. The individual who does not participate in the hunt or who otherwise shirks group responsibility can become a social parasite, using resources obtained by the sweat of others' brows and consequently sowing discord among the rest of the group. The individual who shirks his social duties annoys and angers us. We feel motivated to punish because the miscreant's behavior has violated some intrinsic sense of fairness that is latent in each of us, and which helps protect our self-interest in social exchanges.

Punishment confers *competitive superiority* within societies facing direct competition from rival societies, "creating circumstances that are highly favorable for the evolution of accompanying group-functional behaviour" (Boyd, Gintis, & Bowles, 2010; Sääksvuori, Mappes, & Puurtinen, 2011).

Murder is among the multitudinous mechanisms of punishment. For instance, there is some suggestion that group members developing psychopathic personality structures, notorious for exploitation, self-serving cunning, and feckless dealings during iterative interactions, are killed within Inuit tribes (Hoebel, 1964). Intragroup lethal forces are known among the Hadza, and !Kung, albeit after a manner that does not always rigorously delineate murder and capital punishment (Knauft et al., 1987). And of course, lethal injection, electrocution, and hanging are mainstays of ultimate punishment in some modern societies. Short of murder, and often after first restricting access to vital resources (Gat, 2010), comes ostracism (Liddell & Krusch, 2014), a penultimate solution used by groups against noncooperative individual members. The *Pathan Hill* tribes, for instance, ostracize those group members putting their individual freedoms above "the necessity of tribal unity" (Mahdi, 1986). An ostracized group member may fall prey to predators, rival groups, exposure, and starvation but also may become subject to murder by agnates absolved from the traditional sanctions against intragroup killings by the victim's newfound otherness (Boehm, 1985). Indeed, ostracism is understood as a cross-cultural method of punishing burdensome or nonconforming individuals meant to induce behavioral change to the ultimate end of protecting group integrity (Wesselmann, Williams, & Wirth, 2014). Having precedents among social animal species, ostracism marks the historical record across a range of social complexity (Williams, 2009).[7] Ostracism, as it operated in extended kinship bands and tribal societies, should nevertheless be contrasted with ostracism as practiced in ancient Greek poleis, which entailed exodus from Attica, within ten days and for a period of ten years, as decided by formal vote (Forsdyke, 2009; Mattingly, 1991). This should then be differentiated from similar cultural institutions such as *banishment* as practiced by the Romans, which was more severe in that it imposed an indefinite exile along with the loss of status and property.[8] Excommunication is then the religious counterpart to political exile.

Methods of social control multiplied within Medieval Europe, with their use extending into early modernity. Norm violators were apt to be pilloried, having their head and limbs caged between wooden boards in public spaces, simultaneously punishing and humiliating; such was the fate of Daniel Defoe, convicted of seditious libel in 1703, fined and pilloried on three separate occasions (Richetti, 2015). The *pranger* and *stocks* served similar functions. The *rack* and the *strappado*, or *corda*, stretched their victims into agonizing contortions. Perhaps most interesting from a multilevel selectionist perspective are the cropping of ears (Hatfield, 1990), the slitting of noses (Kollmann, 2006), and the branding of skin (Jones, 1987)—all ways in which deviants and dissidents were marked, allowing them to remain within the group, but burdened by a lasting representation of their transgression. Montgomery and Kilroy, two British privates convicted of manslaughter for their part in the Boston

Massacre, were branded with the letter *M* on their thumbs, marking their conviction of manslaughter and disallowing future clemency using the benefit of clergy by way of an indelible cutaneous sign forever recalling their deed.

These many methods of social control appearing in medieval and early modern societies are intermediate methods of social control, short of outright ostracism, or capital punishment. These and other punishments were sanctioned by increasingly powerful and legitimate rulers and ruling bodies laboring to maintain order in the face of demographic growth. With demographic growth and the anonymity of urban living, providing punishment from the bottom-up becomes a less effectual form of social control, for it was at once harder to detect and recall dissident acts, track reputation, and ensure efficacy. *Diffusion of responsibility* operates to diminish the motivation of the altruistic punisher among large groups, wherein an ever increasingly diffuse societal benefit is enjoyed by the group at large. Meting out punishment then becomes a collective action problem of unsolvable proportions without layering top-down controls on preexisting bottom-up controls, as Boyd and Richerson (2005) imply:

> as group size increases, the average frequency of cooperative strategies typically declines to a quite low level…groups in which cooperation occurs over the long run, can remain at substantial levels even when groups are large. One must keep in mind, however, that this conclusion presupposes that individual punishers can afford to punish every noncooperator in the group. (p. 176)

Boyd and Richerson assure us that cooperation among large groups is possible if only punishers "can afford to punish every noncooperator in the group." Lone altruists cannot afford to do this. Therefore, we collectively bear the burden of enforcing cooperation by funding punishers. Modern societies levy a small tax borne by individuals, then concentrate its power within agencies that mete out justice and punishment. We pay prosecutors, judges, and police officers salaries so that they can afford to punish. Concentrated power, delegation of authority, and monetary capital are required if punishment is to continue efficacious with the growth of group size. Punishing nonconforming behavior within a large society becomes a massive undertaking beyond the capacity of individuals at the local level. It must be sustained, consistent, and omnipresent. Altruistic punishment, inspired by moral emotion, is quite the opposite; it is inconsistent, particular, passionate, and diffuse. Similar trends are observable in generalized forms of altruism, such as alms giving, which, as societies grew, were undertaken by church organizations, private benevolent associations, and finally bureaucratized state apparatuses.

The French *gendarmerie* leveraged developed military authority, pointing it inward toward the social control of the national population, and so France was comparatively law abiding. Rudimentary police forces likewise arose in thirteenth-century Italian city states (Roberts, 2019). In contrast, the English, ever jealous of their liberties, slowly rationalized internal controls. Aside from Oliver Cromwell's miscarried plans for a districted military police force in the 1650s, professional rigor was instituted by 1753, but with continued reliance on nonprofessional constables, a medieval institution of amateur rank. It was only in 1829 that the *Metropolitan Police Force* was established (Lyman, 1964). England belatedly followed the European trend of police force professionalization, progressing from civic, to municipal, to state policing (Denys, 2010), which then extended to international policing as early as the nineteenth century (Deflem, 1996). Fully federalized and centralized agencies, using data-driven tracking technologies, lavishly funded and having de facto global reach, were the twentieth-century's logical extension of early international policing. *Roman*, *canon*, *common*, *customary*, and *feudal* law (Karras, Kaye, & Matter, 2013) likewise evolved to meet the demands of social control arising from societal growth. Comparative to mores, norms, and customs, law was less parochial and capricious, as seen from the Norman conquest in English common law, which "was 'common' to the entire kingdom of England, superseding purely local laws and customs" (Neal & Cameron, 2016).

In sum, whether administered from the bottom-up or top-down, punishment suppresses selfish individualists threatening to undermine cohesion and collective action. Evidence abounds. We feel the moral emotions within ourselves and recognize them in others; we sense the consequences of violating sacrosanct norms and are told the punishments for intragroup harm and unfairness. Buttressing the aforementioned historical instances, laboratory findings on so-called *ultimatum games* furnish examples of subjects for going the offer of an unfair share of some resource for the pleasure of punishing a stingy partner (De Quervain, Fischbacher, Treyer, & Schellhammer, 2004; Gowdy, 2008; Hardy-Vallée, 2007). Such experiments show that, despite personal costs, some are motivated to engage in altruistic punishment (Boyd, Gintis, Bowles, & Richerson, 2003; Fowler, 2005), a phenomenon inexplicable through the lens of *kin selection, signaling theory*, or *reciprocal altruism* (Fehr & Gächter, 2002). Contrariwise, far from having trouble explaining its existence, multilevel selection in humans is believed to require altruistic punishment. Recall that critics of multilevel selection often state that group selection is possible, but remains exceedingly improbable because of the relative strength of individual selection. The salience and efficacy of altruistic punishment, actuated by moral emotions, mechanistically explain the ubiquity of multilevel selection in small-scale societies; thereafter, cultural institutions evolved in concert with expanding group size and ensured the scaling of culturally group-selected cooperation. Together, punishment from the bottom-up and the top-down imposed a strong selective pressure amounting to a process of anthropogenic selection or self-domestication wherein some percentage of selfish individualists were culled each successive generation, culminating in highly group-selected societies.

5 Societal Interdependence via Systemic Differentiation

Here we discuss the augmentation of interdependence, economically, socially, and militarily. Within the declinist literature, societies are said to *senesce*, being compared to the aging process afflicting most life forms. We submit that this analogy is equally applicable to growth. Bacteria and protozoa represent a wide variety of diminutive organisms, the size and volume of which allow for direct exchange with the external environment. Aquatic salamanders, as well as certain species of frogs and fish, are among a small segment of vertebrates relying fully or partially on cutaneous gas exchange for respiration, with increasing needs being met through flat morphologies, capillary formation, and expanded surface area using specialized skin folds (Feder & Burggren, 1985). Augmenting volume renders this impracticable and then impossible. With the evolution of large vertebrates, one sees complex branching lungs with gas-exchanging alveoli, along with circulatory systems that, in humans, total 60,000 miles end to end (Cabin & Henry, 1992), intestinal length in pinnipeds stretching hundreds of yards, and giant squid axons measured in feet (Young, 1977). By way of analogy, as societies grow, simple, parochial structures must give way to differentiated systems. While mindful that this is merely a suggestive analogy, we review: (1) military expansion, which is comparable to immune system evolution; (2) economic diversification, which is comparable to cellular differentiation; and (3) infrastructure, which is comparable to innervation.

If space afforded, we could dedicate a full chapter to considering military growth from a multilevel selectionist perspective. It is interesting to think about the kin-selected metaphors for military units, as in a *band of brothers*. Military drill could be dilated on for its ability to form a rabble into a functional mass capable of coordinated action, with the most conspicuous transformation from one to the other deriving from Baron von Steuben's[9] drill of the green troops of the American Revolutionary Army wintering in Valley Forge. Yet, we confine ourselves here to a few observations as guided by our aforementioned analogy. Immune systems are physiologically costly and gained only through bioenergetic trade-offs with growth and other important properties (Kubinak, Nelson, Ruff, & Potts, 2012). The same is true of armies, which require immense capital to field and maintain. It is said that the army in late imperial Rome had *eaten up* the fruit of thrift, which would not come again in such abundance for centuries (Tainter, 1988, p. 71). More concretely, one can view the share of central government expenditures dedicated to military defense, which ranged between 35 and 80% in Germany from 1875 to 1913 (Castillo, Lowell, Tellis, Munoz, & Zycher, 2001). This is an extreme example from one of the most embattled states of the long nineteenth century, though these levels have been approached in the recent past during phases of active war.

Immune systems and armies impose costs, but they also impose risks, compounding the issue. Rheumatoid arthritis (Oelzner et al., 1998), lupus (Huisman et al., 2001), type I diabetes (Hyppönen, Läärä,

Reunanen, Järvelin, & Virtanen, 2001), Graves' disease (Yamashita et al., 2001), psoriasis (Staberg, Oxholm, Klemp, & Christiansen, 1986), Crohn's disease, and multiple sclerosis are but a few well-known representatives of a large class of autoimmune disorders wherein a system designed to manage parasitical invasion, attacks the host it was evolved to protect. The same risks pertain to armies designed to meet foreign threats, which can become the engines of rebellion and revolution. Allegiance of course was always at issue among armed forces where functional cooperation is so crucial, and so we see attempts to suppress individually selected selfishness through courts martial and summary execution. These are paired with oaths of allegiance and basic training, the use of promotion, epaulettes, stars, decorations, and other military baubles manufactured to induce group cohesion and commitment. Continued growth exacerbated the problem. The warrior king in the mold of Charles XII of Sweden, ruling from the saddle instead of the throne, had to watch for court intrigue but worried less about military coups. As states expanded, specialization ensued, with attributes beyond martial valor becoming increasingly important for state administration, while manifold demands on monarchs increasingly precluded direct intervention in distant adventures. This resulted in a division of labor between military and civil authority, another complexity necessitated by growth. For a society dividing labor along these lines, its continued growth, even its stable existence, sometimes hinged upon subordinating the military to the civil arm of government. This was done ably by Justinian and Belisarius of the Eastern Roman Empire, resulting in the reconquest of a portion of the Western Roman Empire. History furnishes examples of quite the opposite kind. Recall the juvenile Peter the Great of Russia witnessing the Streltsy Rebellion of 1698. Then there was the 1806 revolt of the Janissaries in the Ottoman Empiree; also, one can observe the effective civil power the late imperial Roman Army assumed after having been thoroughly interpolated with Germanic peoples. Indeed, subordination of military to civil power is a hallmark of stability within an emerging state or empire, as exemplified by Washington, who, when the American Revolutionary War was won, presented himself before Congress to formally tender his resignation and surrender his sword. This action purportedly inspired King George III to remark, "If he does that, he will be the greatest man in the world." This earned Washington comparisons to Cincinnatus and Cato the Younger (Whitney, 2007), reminding us that Washington acted in accord with acute consciousness of the Roman tradition, which strictly subordinated military power to civil authority, at least as in the great days of the Republic. Having won battles and restored order, this is what Napoleon Bonaparte failed to do for the French Republic.

Moving on, one can see parallels between economic diversification (Cameron, 1993) and cellular differentiation, both of which are necessitated by augmented size. Complex, multicellular bodies are federations of cells combining into tissues, tissues to organs, and organs to systems. Rather than having the powers of procreation, motility, digestion, and defense within a single cell, large, multicellular life has evolved gametes, limbs, intestines, and scales, all differentiated parts. Like an eye without a brain or a foot without a leg, a single neuron or nephron serves no evolutionarily relevant function—it can neither survive nor procreate. The same is true of many specialized roles and functions within complex societies. The number of people dedicated to farming and husbandry has steadily fallen within the developed world. Surplus manpower, also gained through tapping fossil fuels, can then be invested in research, technological development, engineering, and a myriad other specializations. Persons then dedicate years in higher education to training and acquiring specialized knowledge, while correspondingly forgoing opportunities to acquire generalized knowledge relating to raising crops, managing herds, shelter construction, hunting, and all such activities occupying the time of persons living within less-differentiated societies.[10] As discussed by Adam Smith, differentiated production processes could yield much greater efficiency and output for society, though at the cost of denying a diverse skill set to individuals. This process continued with the rationalization of assembly line production, as practiced to such good effect by Henry Ford. The responsibility of the laborer is reduced to a single specialized skill, acquired at the expense of broad mechanical learning, and having no application outside the overarching production process. The roles of individual persons within such an advanced industrial society are analogous to those of the specialized castes seen among eusocial insects within their superorganismic colonies.

Rationalized, liquid money has been crucial to the advent of these circumstances. Money can lubricate the friction that would otherwise prevent fluent exchange among individual societal entities. Money can then be amassed as capital, which can be deployed to achieve otherwise impossible feats of collectivization, such as fielding the armies discussed above and building the infrastructure discussed below. Thereafter, it was the joint stock company that allowed further growth, representing "a response to the growing needs of a developing economy for a more flexible and efficient way of organising business activity" (Johnson, 2010, p. 108). Another important innovation was the *sinking fund*, a form of standing government debt capable of, at once, opening the purses of elite investors and aligning their interests with those of the state. Hamilton's Assumption Bill, passed into law over the suspicions of Jefferson and Madison, assumed state-accrued Revolutionary War debt within a sinking fund capitalized by elite investors who consequently found their interests allied with that of the federal government (Schachner, 1946). The modern financial market generally and the joint stock company and sinking fund specifically promote successful intergroup competition because it takes a cacophony of unrelated individuals, aligns their interests, and concentrates their power in the form of capital, as can be seen in the ascendency of Britain over France in the latter part of the eighteenth century. France retained higher land mass and population size, but still lost out to Britain in the race for empire, as can be seen in the outcome of the *Seven Years' War*. Though France under Jean-Baptiste Colbert progressed along this trajectory, Britain led the way with the *British East India Company* and other joint stock companies inaugurating a modernized economy, which allowed Britain to militarily punch beyond its weight a century later.

Lastly, one can see parallels between infrastructure and innervation. As eyes, ears, and the distal tips of fingers, to confer any functional advantage, must be wired with brains, so individuals, districts, and distant frontiers must be connected with capitals. Infrastructure enables penetration, allowing centralized rulers and agencies to administer regions directly, decreasing their reliance on local power brokers (Wimmer, 2013). The Roman example is again illustrative. The *Appian Way* is emblematic of approximately 100,000 kilometers of roads, which facilitated the movement of legions, wheeled transportation, communications, commerce, tax collection, and migration, collectively forging "an imagined body politic that transcended the purely local, creating in the process an empire which bore all the hallmarks of an early and sustained globalization" (Hitchner, 2012). Navigable rivers similarly facilitate transport and trade within regions and empires, less often serving as boundaries than is customarily thought (Turchin, 2006). Rome was also reliant on water transport, pound for pound more economical than overland. Indeed, maps of Rome clearly show the Empire to be organized around the Mediterranean, connecting Europe, Southwest Asia, and North Africa into a coherent organization. David Livingstone's hopes of a profitable and prosperous colonization scheme were dashed when Africa's Zambezi River proved utterly unnavigable due to shallows and falls (Ferguson, 2008). As seen in the great age of canal building, however, successful penetration sometimes required the alteration of natural waterways or their wholesale creation. New York earned the cognomen, the *Empire State*, after its creation of the Erie Canal (Shaw, 1966), stretching from Lake Erie to the Atlantic Ocean via the Hudson River. As Neal and Cameron (2016) explain, this project was one among a sustained effort to tie the original thirteen colonies to those territories and new states further west:

> Another advantage of the size of the United States was its potential for a large domestic market, virtually free of artificial trade barriers. But to realize that potential required a vast transportation network. At the beginning of the nineteenth century the sparse population was scattered along the Atlantic seacoast; communication was maintained by coastal shipping supplemented by a few post roads. Rivers provided the only practical access to the interior, and that was severely limited by falls and rapids. To remedy this deficiency the states and municipalities, in cooperation with private interests (the federal government was scarcely involved), engaged in an extensive program of "internal improvements," meaning primarily the construction of turnpikes and canals. By 1830 more than 11,000 miles of turnpikes had been built, mainly in southern New England and the mid-Atlantic states. Canal construction got seriously underway after 1815 and reached a peak in the 1820s and 1830s. By 1844 more than 3,000 miles of canals had been constructed and more than 4,000 by 1860. (p. 257)

Public works projects like the Erie Canal inaugurated more than an explosion of commerce and industrialization: They assuaged fears that the continental interior to the west of the Appalachians would be dissevered from America's Eastern Seaboard (Bernstein, 2005). Eclipsing canal building, rail eventually connected the commercial activities of the original thirteen colonies with Western territories and newly minted states. Rail revolutionized Europe, having both economic and military implications, especially as it related to colonial possessions. Rail construction facilitated penetration, control, and resource extraction within colonized regions as seen in Japanese Korea (Duus, 1998), British West Africa (Dumett, 1975), and Germany's East African possessions (Henderson, 2012). Indeed, Wimmer (2018) demonstrated a correlation between rail length and per capita voluntary associations, linguistic homogeneity, and political representation. Moving information was just as important as moving people and goods. Consider the importance of the *Postmaster General*, which was instated prior to American independence and which held the status of a cabinet position for more than a century (Gallagher, 2016). Then came the telegraph, which was in use in British India by the 1850s and helped suppress later mutiny, with one mutineer identifying the telegraph cable as *that accursed string that strangles me* (Ferguson, 2008, p. 141). By 1880, nearly 100,000 miles of telegraph cable stretched over sea and land alike, connecting Britain with her colonies, past and present. Collectively inaugurating the Victorian revolution in global communications (Ferguson, 2008, p. 142), all in a concentrated space of industrialization, the telegraph, when combined with the steamship and rail system, presided over a shrinking world that was easier to integrate and control. Thus, we find the image of Cecil Rhodes as depicted in the *Rhodes Colossus*: astride the continent, draped in a telegraph line, with one foot in North Africa and the other in South Africa. All electronic messaging simply improved the speed, ease, and efficiency of communications throughout the telecommunications revolution. Never again would generals fight on in the field long after their civil authorities had sued for peace, as happened in the Battle of New Orleans.

In concluding this review of military, economic, and social interdependence, evolutionary comparisons seem ever more apt. As with the structures derived of *convergent evolution* wherein selective pressures evoke similar adaptive solutions across time, space, and phylogeny, these cultural innovations have considerable manifest variation, even as they conserve fundamental latent similarities. Thus, even as their evolutionary histories and underlying mechanics are differently developed, the bat's and bird's wings enable flight in the same way that the semaphore and telegraph enable communication. As with the adaptive solutions of evolution, cultural evolution will produce more or less capable adaptive variations which will be culled through a selective process as groups compete with one another. Finally, just as the human brain and the giraffe neck meet with upper bound limitations, so too will interdependencies experience a point of diminishing returns as they tend toward the baroque in the service of unwieldy growth.

6 Removing Impediments to Growth: Ethnic Closure, Chosen Peoples, and Mortal Men

This final section considers the final phase of growth, that of empire. Confining our review to the example of the British Empire, we consider whether it is necessary to ethnically and religiously absorb subjects to progress from nation-state to empire and whether it is also necessary to socialize subjects via transcendent institutions radiating out from the center.

With island holdings from the South Pacific to Canada, added to India and the British Isles themselves, amounting to a 1909 peak territory of 12.7 million square miles (Ferguson, 2008), the British created the largest empire in history, on which the sun truly never set. Collectively, the foregoing facts deserve explanation. Why was the British Empire, and the Roman Empire before, so astonishingly successful? Any answer might occupy an entire volume or even ten, though, before even outlining an answer to this question, we should explain and operationalize *success*. Commonly endorsed social desiderata such as *democratization*, *liberalism*, or *benevolence* might be understood as inherently laudable signs of success. Yet, if so defined, the British Empire was only successful with respect to those subjects sharing in a full measure of citizenship. Colonial subjects, in contrast, often experienced semipermeable social hierarchies and inequitable, insalubrious, or iniquitous vocational roles and, in many other ways, were subject to exploitation. We approach the subject from a more staid, biological, and quantifiable view, understanding success as akin to power and stability; and the ability to subordinate a mass into a functional whole, to keep internal order, to act beyond one's numerical influence in competition with other societies as borders are defended and national integrity is upheld. This includes stability through time in a country's institutions, cultural continuity, and the like. This operationalization then follows from an understanding of group selection and is grounded in the historical competition taking place between groups. In short, by success, we mean successful aggregation. Needless to say, it has nothing whatsoever to do with ethics of goodness, human flourishing, or social desirability, all value-laden terms applicable to applied political science but inapplicable to understanding group aggregation.

Most obviously, without industrialization and advances in infrastructure technology, the vastness of the British Empire could not have been. More than this, Victorian Britons had an unusual mixture of temperament, at once entrepreneurial and pious, which they shared with some of their Puritanical forebears. Thus, at least a moiety of Britons seemed to believe that the act of colonization was not unambiguously exploitative, even as objective metrics demonstrate the realities of resource extraction, abuses of power, and lethal skirmishes. The whitewashed perception of colonization is seen in Rudyard Kipling's assumption of the *white man's burden*, which represents a sense of paternalistic responsibility. The missionary zeal expressed in Lachlan Macquarie's government of Australia and David Livingstone's proselytizing efforts in sub-Saharan Africa are among the many examples of mixed motives governing British colonization (Ferguson, 2008). In these efforts, one can read the spirit of a group-selected society, confident in its righteous might, which was exporting its people and culture to distant lands. Overall, innumerable examples of slaughter, confiscation, and oppression belie these professions and protestations; however, in some persons and places, their expression may well have been genuine, a potential fact relevant insofar as it served to soften resistance to colonization.

Like Rome in the vigor of its youth, Victorian Britain was demographically fecund, shipping her sons forth to subdue, administer, and populate distant possessions. Like Rome, Britain also exported order. Good governance was ensured by meritocratic advancement on tests, as had so long bolstered Confucian China. Yet, the essence of good government relates to the stance taken toward conquered peoples, where empires are concerned. Though it did not show the same wisdom with respect to sub-Saharan Africa, and notwithstanding the recalcitrant pique exhibited during the war for American independence, Britons could show respect, or make a show of respect, for a peoples' cultural heritage that palliated the pain of colonization, as they did with India. Furthermore, Britons showed themselves capable of conscious progress along these lines. In the wake of the *Sepoy Mutiny*, Queen Victoria issued an 1858 proclamation, assuring the peoples, princes, and chiefs of India that the crown had assumed the reins of governance from the *East India Company* and would henceforth ensure freedom of worship and equal protection under law (Ferguson, 2008). Britain had realigned its policies to keep India in the fold. This allowed a profitable British-Indian Empire to endure through the nineteenth century and even elicited an outpouring of support for the "mother country" as she was pressed by the German World War I offensive. More than this, in the half century before World War I, Britain had enlisted Indian troops in more than a dozen campaigns (Ferguson, 2008). During World War II, five million troops were raised throughout the empire, matching those raised in the United Kingdom itself. All this was gained on the cheap. Looking at the year 1898, we see military expenditure at 2.5% of gross national product and more Indian soldiers in the field than those of ethnically British stock (Ferguson, 2008). This is not to depict the British Empire as a positive good, it is only to say that the British strategy of colonization, like the Roman, was less onerous than it might have been. Where the British were best able to keep the Empire together, the yoke was light, the benefits obvious, and the reprisals for resistance extreme. Under such a regime, cost-benefit analyses of sticks and carrots routinely militated against resistance. Such concrete factors could be debated, and also multiplied, but we forbear. The strength of its institutions, extensive military power, and the ability to enlist local elites within a durable bureaucracy were among many centripetal forces counterbalancing the centrifugal forces created by exploitative arrangements, the incapacity to mount viable rebellion, and the inability to present a united front against the colonizing outgroup.

Andreas Wimmer's models and historical analyses are supportive of several conclusions above. First, weakly voluntary associations are enabled by the provision of public goods and some manner of political participation. Both Britain and Rome provided this. Additionally, both Britain and Rome transitioned from nation-state to empire by avoiding rigid extremes of *ethnic closure*. Wimmer (2013) contrasts how ethnic nationalism restricts imperial growth:

> Imperial elites thus had incentives to conquer other states and to permanently incorporate their territories into their domain. Nation-states, however, cannot legitimately rule over vast numbers of ethnic others, given that they are built on national self-rule as their legitimizing principle. (p. 27)

Ethnic exclusivity can create strong cohesion but creates upper bounds to the inclusion rate, thus, limiting growth and group size. The British, like the Romans, were able to incorporate peripheries into the core, sewing on appendages to the body politic, because many British possessions ultimately identified with the mother country on some level. This is true of America before and after the revolution; it was also true of India, Australia, and Canada. Britons exported Britishness, a commodity not to be lightly spurned. This was a cultural ethos pregnant with durable institutions, an inspired religion, and patriotic sentiment capable of extrapolation and importation. The value of these features is highlighted by contrast: opposite the stable system of governance radiating across the British Empire was the autocracy of Napoleonic France, with its cult of the emperor; opposite the insularity of Judaism, there is the universalist propensity of Christianity; and opposite the ethnic nationalism of twentieth-century Germany, there was the patriotism of Britishness. Take pause at this subtle point. We would judge that Napoleonic France, Jews, and nationalist socialist Germany were all comparatively more group selected. The point, however, in this chapter on growth, is that their transience in the first instance and insularity in the latter two, while serving as pillars of group-selected cohesion, created a closed loop difficult for outsiders to penetrate.

Colonized peoples would never be Britons, but some were allowed to be British. This afforded the right balance between inclusion and exclusion, which Moffett (2019) finds to underlie enduring political arrangements. Colonizing peoples do not then feel diluted, and colonized peoples do not feel utterly absorbed. In the language of multilevel selection theory, one would say that this allows colonized peoples to layer a level of superordinate group affiliation over and above preexisting subordinate group affiliations. Additionally, it is clear that colonized elites must be co-opted rather than defied if the colonizing country is to avoid the repressive costs of a police state, with subjects ready to rebel when the opportunity presents. Wimmer explains that, with a multi-ethnic region or state, and certainly within an empire, peace is fostered by inclusionary power sharing, tying elites to other segments of society. These many factors discussed by Wimmer, and applicable to the case of the British Empire, are generalized by Gat (2006) in a passage that merits extended quotation:

> Furthermore, whatever other mechanisms—economic, social, or religious—contributed to the formation of state authority in relatively small and close-knit communities, military power and war were predominant in the formation of larger states, which welded together distinct and different communities, and, indeed, separate societies, ethnicities, cultures, and polities. In such expansions, the state was all the more an instrument of power, ruling through conquest, subjugation, and coercion, at least until other bonds of cohesion evolved. For, in due course, spreading state power had a unifying effect on its realm, as contact and integration increased through the binding effect of the state's apparatus, state's religion and language, improved communications, cultural diffusion, elite integration, population movement, larger-scale economy, and military service. The expansion of the state thus had the effect of gradually diminishing tribal and local boundaries within the same ethnos, and of reducing the differences between separate ethnies in multi-ethnic states and empires, subsuming them within supra-ethnic identities, even to the point of creating new, transformed, and larger ethnic identities. (p. 358)

As trade-offs, bioenergetic and otherwise, are ubiquitous in biology, we surmise that trade-offs will become evident when systematizing these inquiries into empire building. Extrapolating for now from the British model, there appears to be a trade-off between *group size* and *group cohesion*. While the British Empire avoided ethnic closure, insularity, and cults of personality, all of which facilitate imperialism, it could temporarily cobble together subordinate group allegiances, but not efface them. Bankrupt and hobbled after the world wars, Britannia could no longer project power over its subject peoples. What took three centuries to create took three decades to dismantle (Ferguson, 2008). Britain itself did not disappear, but retracted to its isles, becoming a nation-state, more internally cohesive, if less grand.

7 Conclusions

Feudal networks eventually formed the local nodes upon which many nation-states were mapped, with founding myths and national heroes serving as a sort of psychological substrate of the nation-state. Aggregated against an enemy, sewn together by myths and legends, led by heroes, a mass of people can form into a nation, or bud from an existing nation, with the aid of propaganda and punishment. At the behest of moral emotions, nonconformists are punished from the bottom-up, and thereafter, with the growth of the group, from the top-down. Dissent and defection are suppressed through punishment, or otherwise its source is eradicated through exile or execution (e.g., Wilson, 2002). With these engines of group cohesion in place, further growth is facilitated by and requires economic, social, and martial interdependence via systemic differentiation. As has been illustrated with the British model, the final phases of growth into enduring empires proceed from the removal of impediments to growth. A cultural ethos must be preferred to strict ethnic closure; ecumenical proselytization must be preferred to an insular body of chosen people; and enduring institutions must be preferred to mortal men. Even when all these barriers have been transcended, as happened in the British Empire and to some extent the Roman Empire, the strength of the whole derives from its mass more than its density; that is to say, even in the most successful empires, there will be hierarchically nested groups based on closer ethnic ties and regional commonalities, which can be the object of reversion and disaggregation when the projection of empire weakens. Lower levels of organization, such as ethno-states, can be more highly group selected and thereby committed, zealous, and competitive; but these, in turn, will have to compete with the vast weight of empire.

Notes

1. Religion only competes with language in its ability to propagate across national boundaries, capturing large swaths of diverse persons across populations as illustrated by the definition of Catholic as inclusive, or through ecumenical efforts aimed at reconciling Christian sects, or by virtue of the great traditions of missionary proselytization taking place from biblical times to the European colonization of Africa and the Americas. Religion may well have mapped more directly onto lower levels of social organization, such as tribes, regions, and states, but the great religions have long transcended these confines. Analogous to supranational organizations, federated unions, and vast empires, religions play a crucial role in human evolution, allowing conflict and competition to play out at higher levels of organization, as seen in the crusades waged between Christians and Muslims.
2. Rochambeau understood the difficulties that New York presented. New York sometimes swelled with loyal opposition, but, more than this, it was heavily fortified by British forces, a process that had commenced after Washington's defeat and had proceeded conscientiously for five years.
3. This particular phase in the annals of the Franco-American alliance can be productively studied from a multilevel selection lens, though we focus here on Rochambeau's later reflections and their significance specifically to men and myths.
4. Presently, we distinguish between rhetoric and propaganda; the former persuades, the latter persuades by deception. Eloquence, argument, and marshaled facts change men's minds, and this can surely shade into sophistry; but propaganda crosses a line into misinformation. Of note, this usage is traced back to World War I, where the term first firmly acquired its association with "the transmission of fraudulent information."
5. This is a quote of John Adams, speaking about the Americans that confronted British soldiers during the Boston Massacre.

6. We should note that we have focused only on the sedition portion of the Alien and Sedition Acts. This legislation's focus on limiting alien dissidents is likewise significant for group selection.
7. Ostracism is sometimes referred to as *shunning* among modern insular, genetically homogeneous groups, such as the Amish (Gruter & Masters, 1986).
8. https://www.britannica.com/topic/ostracism
9. A purported ranking member in the Prussian Army.
10. The process is gradual, with gentlemen scientists of the nineteenth century like Darwin exposed to much more generalized knowledge than a modern geneticist, while an eighteenth-century statesman like John Adams still managed a farm.

References

Auerbach, J., & Castronovo, R. (2013). *The Oxford handbook of propaganda studies*. New York: Oxford University Press.

Ben-Israel, H. (2002). *English historians on the French Revolution*. New York: Cambridge University Press.

Bernays, E. (2005). *Propaganda*. New York: IG Publishing.

Bernstein, P. L. (2005). *Wedding of the waters: The Erie Canal and the making of a great nation*. New York: WW Norton & Company.

Boehm, C. (1985). Execution within the clan as an extreme form of ostracism. *Information (International Social Science Council), 24*(2), 309–321.

Boyd, R., Gintis, H., & Bowles, S. (2010). Coordinated punishment of defectors sustains cooperation and can proliferate when rare. *Science, 328*(5978), 617–620.

Boyd, R., Gintis, H., Bowles, S., & Richerson, P. J. (2003). The evolution of altruistic punishment. *Proceedings of the National Academy of Sciences, 100*(6), 3531–3535.

Boyd, R., & Richerson, P. J. (2005). *The origin and evolution of cultures*. New York: Oxford University Press.

Cabin, H. S. (1992). The heart and circulation. In B. L. Zaret, M. Moser, & L. S. Cohen (Eds.), *The heart book* (pp. 3–10). New Haven, CT: Yale University School of Medicine.

Cameron, R. E. (1993). *A concise economic history of the world: From Paleolithic times to the present*. New York: Oxford University Press.

Castillo, J., Lowell, J., Tellis, A. J., Munoz, J., & Zycher, B. (2001). *Military expenditures and economic growth* (No. RAND-MR-1112-a). Santa Monica, CA: Rand.

Castronovo, R. (2014). *Propaganda 1776: Secrets, leaks, and revolutionary communications in early America*. New York: Oxford University Press.

Chen, X., & Szolnoki, A. (2018). Punishment and inspection for governing the commons in a feedback- evolving game. *PLoS Computational Biology, 14*(7), e1006347.

De Quervain, D. J., Fischbacher, U., Treyer, V., & Schellhammer, M. (2004). The neural basis of altruistic punishment. *Science, 305*(5688), 1254.

Deakin, J., Taylor, E., & Kupchik, A. (2018). *The Palgrave international handbook of school discipline, surveillance, and social control.* New York: Palgrave Macmillan.

Deflem, M. (1996). International policing in nineteenth-century Europe: The police union of German states, 1851–1866. *International Criminal Justice Review, 6*(1), 36–57.

Denys, C. (2010). The development of police forces in urban Europe in the eighteenth century. *Journal of Urban History, 36*(3), 332–344.

Dumett, R. E. (1975). Joseph Chamberlain, imperial finance and railway policy in British West Africa in the late nineteenth century. *The English Historical Review, 90*(355), 287–321.

Duus, P. (1998). *The abacus and the sword: The Japanese penetration of Korea, 1895–1910* (Vol. 4). Oakland, CA: University of California Press.

Ellul, J. (1973). *Propaganda: The formation of men's attitudes.* New York: Vintage books.

Feder, M. E., & Burggren, W. W. (1985). Cutaneous gas exchange in vertebrates: Design, patterns, control and implications. *Biological Reviews, 60*(1), 1–45.

Fehr, E., & Gächter, S. (2002). Altruistic punishment in humans. *Nature, 415*(6868), 137.

Ferguson, N. (2008). *Empire: The rise and demise of the British world order and the lessons for global power*. New York: Basic Books.

Ferling, J. (2010). *The ascent of George Washington: The hidden political genius of an American icon*. New York: Bloomsbury.

Forsdyke, S. (2009). *Exile, ostracism, and democracy: The politics of expulsion in ancient Greece*. Princeton, NJ: Princeton University Press.

Fowler, J. H. (2005). Altruistic punishment and the origin of cooperation. *Proceedings of the National Academy of Sciences, 102*(19), 7047–7049.

Gallagher, W. (2016). *How the post office created America: A history.* London: Penguin.

Gat, A. (2006). *War in human civilization.* New York: Oxford University Press.

Gat, A. (2010). Why war? Motivations for fighting in the human state of nature. In P. M. Kappeler & J. B. Silk (Eds.), *Mind the gap* (pp. 197–220). Berlin, Germany/Heidelberg, Germany: Springer.

Gottschalk, M. (2006). *The prison and the gallows: The politics of mass incarceration in America.* New York: Cambridge University Press.

Gowdy, J. M. (2008). Behavioral economics and climate change policy. *Journal of Economic Behavior & Organization, 68*(3–4), 632–644.

Gruter, M., & Masters, R. D. (1986). *Ostracism as a social and biological phenomenon: An introduction.* New York: Elsevier.

Haidt, J. (2003). The moral emotions. In *Handbook of affective sciences* (Vol. 11, pp. 852–870). Oxford: Oxford University Press.

Haidt, J., & Joseph, C. (2007). The moral mind: How five sets of innate intuitions guide the development of many culture-specific virtues, and perhaps even modules. *The Innate Mind, 3,* 367–391.

Hardy-Vallée, B. (2007). Decision-making: A neuroeconomic perspective. *Philosophy Compass, 2*(6), 939–953.

Hatfield, S. A. (1990). Criminal punishment in America: From the colonial to the modern era. *USAF Academic Journal of Legal Studies, 1,* 139.

Henderson, W. O. (2012). *Studies in German colonial history.* New York: Routledge.

Hitchner, R. B. (2012). Roads, integration, connectivity, and economic performance in the Roman Empire. *Highways, Byways, and Road Systems in the Pre-modern World, 9,* 222–234.

Hoebel, E. A. (1964). *The law of primitive man: A study in comparative legal dynamics.* Cambridge, MA: Harvard University Press.

Hoffman, M. B., & Goldsmith, T. H. (2003). The biological roots of punishment. *Ohio State Journal of Criminal Law, 1,* 627.

Huisman, A. M., White, K. P., Algra, A., Harth, M., Vieth, R. D., Jacobs, J. W., et al. (2001). Vitamin D levels in women with systemic lupus erythematosus and fibromyalgia. *The Journal of Rheumatology, 28*(11), 2535–2539.

Hutcherson, C. A., & Gross, J. J. (2011). The moral emotions: A social–functionalist account of anger, disgust, and contempt. *Journal of Personality and Social Psychology, 100*(4), 719.

Hyppönen, E., Läärä, E., Reunanen, A., Järvelin, M. R., & Virtanen, S. M. (2001). Intake of vitamin D and risk of type 1 diabetes: A birth-cohort study. *The Lancet, 358*(9292), 1500–1503.

Johnson, P. (2010). *Making the market: Victorian origins of corporate capitalism*. New York: Cambridge University Press.

Jones, C. P. (1987). Stigma: Tattooing and branding in Graeco-Roman antiquity. *The Journal of Roman Studies, 77*, 139–155.

Karras, R. M., Kaye, J., & Matter, E. A. (Eds.). (2013). *Law and the illicit in medieval Europe*. Philadelphia: University of Pennsylvania Press.

Knauft, B. M., Daly, M., Wilson, M., Donald, L., Morren Jr., G. E., Otterbein, K. F., et al. (1987). Reconsidering violence in simple human societies: Homicide among the Gebusi of New Guinea. *Current Anthropology, 28*(4), 457–500.

Kollmann, N. S. (2006). Marking the body in early modern judicial punishment. *Harvard Ukrainian Studies, 28*(1/4), 557–565.

Kubinak, J. L., Nelson, A. C., Ruff, J. S., & Potts, W. K. (2012). Trade-offs limiting MHC heterozygosity. In G. E. Demas & R. J. Nelson (Eds.), *Ecoimunology* (pp. 225–258). New York: Oxford University Press.

Liddell, T. M., & Krusch, J. K. (2014). Ostracism and fines in a public goods game with accidental contributions: The importance of punishment type. *Judgment & Decision Making, 9*(6), 523–547.

Lyman, J. L. (1964). The metropolitan police act of 1829. *Journal of Criminal Law and Criminology, 55*, 141–155.

Mahdi, N. Q. (1986). Pukhtunwali: Ostracism and honor among the Pathan Hill tribes. *Ethology and Sociobiology, 7*(3–4), 295–304.

Mattingly, H. B. (1991). The practice of ostracism at Athens. *Antichthon, 25*, 1–26.

Miller, J. C. (1953). *Crisis in freedom. The alien and sedition acts*. Boston: Little, Brown and Company.

Moffett, M. W. (2019). *The human swarm: How our societies arise, thrive, and fall*. New York: Basic Books.

Neal, L., & Cameron, R. (2016). *A concise economic history of the world, from Paleolithic times to the present*. New York: Oxford University Press.

Oelzner, P., Müller, A., Deschner, F., Hüller, M., Abendroth, K., Hein, G., et al. (1998). Relationship between disease activity and serum levels of vitamin D metabolites and PTH in rheumatoid arthritis. *Calcified Tissue International, 62*(3), 193–198.

Olson, L. C. (2004). *Benjamin Franklin's vision of American community: A study in rhetorical iconography*. Columbia, SC: University of South Carolina Press.

Ostrom, E. (2000). Collective action and the evolution of social norms. *Journal of Economic Perspectives, 14*(3), 137–158.

Padover, S. K. (1955/1989). *The Washington papers: Basic selections from the public and private writings of George Washington*. Norwalk, CT: The Easton Press.
Phillips, K. (2000). *The Cousins' Wars: Religion, politics, civil warfare, and the triumph of Anglo-America*. New York: Basic Books.
Pompa, L. (2010). *Vico: The first new science*. New York: Cambridge University Press.
Richetti, J. (2015). *The life of Daniel Defoe: A critical biography*. Hoboken, NJ: John Wiley & Sons.
Roberts, G. (2019). *Police power in the Italian communes, 1228–1326*. Amsterdam: Amsterdam University Press.
Rosenfeld, R. N. (1997). *American Aurora: A democratic-republican returns; the suppressed history of our nation's beginnings and the heroic newspaper that tried to report it*. New York: St. Martin's Press.
Sääksvuori, L., Mappes, T., & Puurtinen, M. (2011). Costly punishment prevails in intergroup conflict. *Proceedings of the Royal Society B: Biological Sciences, 278*(1723), 3428–3436.
Schachner, N. (1946). *Alexander Hamilton*. New York: D. Appleton-Century Company, Inc.
Shaw, R. E. (1966). *Erie water west: A history of the Erie Canal, 1792–1854*. Lexington, KY: University Press of Kentucky.
Staberg, B., Oxholm, A., Klemp, P., & Christiansen, C. (1986). Abnormal vitamin D metabolism in patients with psoriasis. *Acta Dermato-Venereologica, 67*(1), 65–68.
Tainter, J. A. (1988). *The collapse of complex societies*. New York: Cambridge University Press.
Turchin, P. (2006). *War and peace and war: The rise and fall of empires*. New York: Plume.
Verhoeven, W. (2013). *Americomania and the French revolution debate in Britain, 1789–1802*. New York: Cambridge University Press.
Vico, G. (2002). *The first new science*. New York: Cambridge University Press.
Werth, N. (1999). A state against its people: Violence, oppression and terror in the Soviet Union. In S. Courtois, N. Werth, J. Panné, A. Paczkowski, K. Bartošek, & J. Margolin (Eds.), *The black book of communism: Crimes, terror, repression* (pp. 1–31). Cambridge, MA: Harvard University Press.
Wesselmann, E. D., Williams, K. D., & Wirth, J. H. (2014). Ostracizing group members who can (or cannot) control being burdensome. *Human Ethology Bulletin, 29*(2), 82–103.
Whitlock, B. (1929). *La Fayette*. New York: D. Appleton & Company.

Whitney, G. (2007). *George Washington, the greatest man*. Grand Valley State University Scholar Works, Ask Gleaves, paper 18.
Williams, K. D. (2009). Ostracism: A temporal need-threat model. *Advances in Experimental Social Psychology, 41*, 275–314.
Wilson, D. S. (2002). *Darwin's cathedral: Evolution, religion, and the nature of society*. Chicago: Chicago University Press.
Wimmer, A. (2013). *Waves of war: Nationalism, state formation, and ethnic exclusion in the modern world*. New York: Cambridge University Press.
Wimmer, A. (2018). *Nation building: Why some countries come together while others fall apart*. Princeton, NJ: Princeton University Press.
Wright, E. (1995). *A history of the United States of America. 1. The search for liberty: From origins to independence*. Oxford, UK: Blackwell.
Yamashita, H., Noguchi, S., Takatsu, K., Koike, E., Murakami, T., Watanabe, S., et al. (2001). High prevalence of vitamin D deficiency in Japanese female patients with Graves' disease. *Endocrine Journal, 48*(1), 63–69.
Young, J. Z. (1977). The biology of cephalopods: *The Proceedings of a Symposium Held at the Zoological Society of London on 10 and 11 April 1975* (No. 38). Academic Press: Ecumenical Empire.

6

Decline

Steven C. Hertler, Aurelio José Figueredo, and Mateo Peñaherrera-Aguirre

Customs are first barbaric, then severe, next noble, later refined, and finally dissolute and corrupt.
—Vico (Cited, page 83, in Cambridge University Press's 2002 edition of Giambattista Vico's *The First New Science*).

1 Introduction

Where Chap. 5 reviewed the growth and maintenance of states, this chapter reviews their deterioration and decline. At the outset, however, it is important to precisely specify what, if anything, is declining. Most obviously, we can speak of this process after the fashion of the great declinists: Montesquieu, Gibbon, Spengler, Toynbee, and McNeill. Nevertheless, we must ask whether the process of *civilizational* decline described, while relevant, is distinguishable or not from the results of declining group-selective pressures. The relative strength of group-selective pressures experienced may vary among disparate societies while fluctuating within societies over evolutionary time.[1] Thus, we can speak directly of increasing or decreasing group-selective pressures which drive the evolution of group-selected traits, such as an utter absence of internal warfare and commercial competition. We can also speak of decline among

S. C. Hertler et al., *Multilevel Selection*, https://doi.org/10.1007/978-3-030-49520-6_6

a society's group-selected traits, such as loyalty and altruism. Lastly, we can speak of societal decline as traditionally discussed in the declinist literature, involving martial and economic power. While conceptually distinct, these processes are causally related. When group-selective pressures decline, it lowers mean levels of group-selected traits within societies, leaving them vulnerable to generalized civilizational decline. All three forms of decline concern us, and all will be addressed. Nevertheless, it is necessary to specify which particular group characteristics or social dimensions are enhanced by increasing group selection and eroded by decreasing group selection. To begin with, however, we review insufficient responses to competition followed by elite betrayal and factionalism, ending with a review and analysis of declining group-selected traits within the population at large.

2 The Scylla of Stagnation and the Charybdis of Progress

Decline can result from either failing to change[2] or changing too rapidly. In the first case, a society can be overtaken by rivals and, in the second, risk its stability. Continuous, moderate changes made in response to rival groups allow societies to walk the fine line between these two perils, just as species evolve incrementally in response to environmental challenges and changes.

Red Queen effects have operated throughout history, wherein competition from rival states required ever-advancing military and economic efficiency. "The right to continued existence," North and Thomas (1973) assert, depended on increasingly efficient government extractions of revenue. States were imperiled by war with one another as always, though more frequently in modernity this threat was fiscal as much as military, with the interrelation between the two captured by North and Thomas (1973):

> The magnitude of the increasing cost was staggering. A year of warfare represented at least a fourfold increase in costs of government – and most years were characterized by war, not peace. Monarchs were continuously

beset by immense indebtedness and forced to desperate expedients; the specter of bankruptcy was a recurring threat and for many states a reality. The fact of the matter is that princes were not free – they were bound to an unending runaway fiscal crisis. (p. 95)

When one state began taking loans against the future to augment war-making capacity in the present, other nations were forced to either follow suit or succumb. These loans could "tide a king through a war but then he faced the awesome task of repayment" (North & Thomas, 1973, p. 96). Decline often followed from the failure to secure loans or otherwise from the inability to repay them. Confiscation was a tempting option, as was reneging on the loans. Both methods undermined the stability and credibility of the state. In the case of confiscation, future property rights were imperiled, which curtailed investment and economic productivity. In the case of reneging, this only augmented future interest rates at which later loans could be secured, necessary to offset the increased risk to the lender. Rulers were also tempted to transgress customs, prerogatives, and rights of lesser elites in attempting to solve fiscal problems in the face of Red Queen competition; in this, they were in danger of crossing the line, on the other side of which lay revolt. It was necessary to increase revenues by more efficient extraction and by securing new revenue sources or otherwise face fiscal insolvency, which would, in turn, differentiate those states which would decline from those that would survive, grow, or expand.[3]

The declining Spanish Empire illustrates these Red Queen effects precisely. Citing expansion as "the price of survival," Maltby (2009, p. 15) compiles threats from the west, north, and south, respectively from Portuguese commercial hegemony, French military conquest, and Muslim rule. A succession of events, including the marriage of Isabella I to Ferdinand II, the Reconquista, and the accession of Charles I, brought Aragon, Castile, and the Kingdom of Grenada under the umbrella of a unified Spain (Maltby, 2009), which, with the aid of South American silver, expanded rapidly to an empire. Reminiscent of modern-day oil-rich *OPEC* countries, the abundance of silver inhibited agricultural, social, and economic reforms necessitated in France and England among other contemporary Northern European states (Parsons, 2010).[4] In the short term, the Spanish monarchy could lord over its own nobility rather than making concessions to them and expand imperially at the expense of other nations rather than making trade agreements with them. With reserves of silver spent, the Spanish Empire found itself disadvantaged, overextended, and vulnerable (Flynn, 1982). By 1562, Spain assumed loans on which the interest absorbed more than 25% of its annual budget. The remedy proved worse than the disease, for Spain battled indebtedness across several cycles wherein "lengthening maturities, reducing interest rates, [and] raising the price of gold" eventually ended in bankruptcy. Decline was expressed in the form of external signs and internal symptoms. For instance, in an effort to fund wars of the Empire, the Spanish Crown so rapaciously "mulcted" its populace that nobles began to look upon their titles as *positive liabilities*, while merchants were subject to arbitrary confiscation and taxation to the detriment of credit and commerce (Elliott, 1961; North & Thomas, 1973). In a way, the influx of silver proved a stimulus to northern countries, inducing them to adopt the agricultural, social, and economic reforms neglected by Spain. The decline of Spain most especially relates to the rise of the Netherlands, the inhabitants of which united and modernized in the face of Spanish domination.

Rapid change can be as dangerous as stagnation. We see the danger in Burke's writings on the French Revolution, which looks with revulsion upon the unbridled impulse that unseated order, custom, law, and religion in revolutionary France (De Bruyn, 2004; Hirst, 1935). Burke was an *organicist*, a philosophical position inviting comparisons between societies and organisms, holding the survivability of a system dependent upon the connected functionality of its constituent parts. It was from this position that, in his, *Reflections on the Revolution in France*, Burke remarks, "All circumstances taken together, the French Revolution is the most astonishing that has hitherto happened in the world" (1790/1992, p. 11). Conferring upon themselves an unauthorized extension of delegated authority, while dispensing with custom, form, law, and religion, in a matter of months France's *National Assembly* tore down centuries of accreted refinements that had been superadded to an existing governmental structure over many years. As evident in the full title of this work, *Reflections on the Revolution in France and on the Proceedings in Certain Societies in London Relative to that Event*, Burke feared the spread of Jacobinism across the channel, the uprooting of the aristocracy that bound English society (Brody, 2000), and the destruction of the "organic nature and historical legitimacy of the constitution" as well as the "sanctity of the Crown" (Philip, 2014, p. 129). As was the case with the Spanish Empire *vis-à-vis* the Netherlands, the decline of France through the Revolution and Napoleonic Wars served to stimulate British development, as described by Philip (2014):

> Largely as a result of the war with France, the early nineteenth-century British state became a nation state, no longer just a fiscal-military apparatus; the people were no longer an agglomeration, but a mobilized unity with national symbols and a national identity. Although France did not cause all this, its revolutionary and internationalist path, together with the level of mobilization required by the war, forced the pace of existing trends. (pp. 129–130)

To survive the warmaking capacity of France's *levée en masse*, while simultaneously suppressing Irish revolts and Jacobin cadres and clubs within London, England cautiously changed by reluctantly taking slow steps toward the liberal democratic institutions that had so suddenly replaced France's *Ancien Régime* (Philip, 2014). Avoiding decline by walking the line between stagnation and progress can also be seen in Peter the Great's reforms of custom and administration, aimed at keeping Russia competitive with Western rivals but evoking rumblings short of insurrection from noble and religious elites (Massie, 2012). Likewise, one can look to the slow-liberalizing concessions from the top-down in nineteenth-century Germanic polities, which vented revolutionary steam, which might have otherwise ushered in decline and collapse (Evans & Von Strandmann, 2002; Robertson, 1952).

3 Disunity: From Patrician Parasitism to Faction, Rebellion, and Revolution

Decline follows disunity. Disunity manifests in manifold manners. Like cancerous cells, rebellious subgroups can promote their interest at the expense of the collective interest. It can be perpetrated by demagogic and degenerate elites; yet, disunity often comes from the bottom-up in the form of military coups, political factions, regional separatists, or religious schisms. We take up each in turn.

When ruling elites subject ethnic or religious factions to exploitative *exchange relationships* and deny to them some measure of political inclusion (Wimmer, 2018), the disaffection that ensues leaves groups vulnerable to external exploitation and eventual conquest, as happened when the *British East India Company* exploited divisions between Indians and their Muslim overlords within a waning Mughal Empire (Parsons, 2010). Indeed, as Parsons (2010) states:

> The most vulnerable societies were those divided sharply along the lines of class, religion, ethnicity, or some other form of identity. These divisions led to military weakness, hindered organized resistance, and made it easier for conquering powers to recruit local allies. (p. 13)

By leading well or exploiting selfishly, elites either stitch these lower-order group allegiances into a superordinate group identity or otherwise expose the polity to conquest by division. As in the French Revolution where climatic events initiated a succession of meager harvests, elites are now and then unjustly blamed for fully exogenous hardships. However, by engaging in peculation, demagoguery, or any other self-serving vice, elites manufacture endogenous hardships. There may be notable exceptions such as the prosperous American colonists, though, generally speaking, subjects rebel only when the status quo becomes insupportable. Before irresponsible elites drive their populaces to insurrection, they may first come to a point of lackluster disaffection or even allow, acquiesce in, or aid, external conquest, as illustrated by England's *Glorious Revolution* wherein English and Scottish elements enabled William of Orange to effect an unopposed landing from the Netherlands and a bloodless coup to the English Throne.

Still, especially when coming in succession as did the Roman Emperors Nerva, Trajan, Hadrian, Antoninus Pius, and Marcus Aurelius, competent elites can have quite the opposite effect as that which was described above. Yet, even this string of conscientious competence gives way to plebian parasitism, which extends from the bottom-up, sapping the foundation of the state. Elite leadership only forestalls decline because reduction in the strength of group selection evokes downstream changes in altruism, valor, deference, and other fitness-relevant meta-population variables operating evolutionarily and culturally over time scales beyond the reign of a single man or even those of five great men.

Danger also derives from the military. As discussed in Chap. 5, delegating martial authority is necessitated by growth, though not all generals subordinate themselves to civil authority as did Washington. Again, elaborate bureaucracies, deliberative bodies, and powers balanced one against the other promote stability in peace, but their dilatory inability to respond to imminent existential threats motivates the delegation of extraordinary powers. Foreign threats, Skocpol (1979, p. 51) reminds us, come of the ability to "mobilize extraordinary resources from the society and to implement… structural transformations." This can be seen in the Roman office of dictator, the efficacy of which Machiavelli (2010) describes:

> And truly, of all the institutions of Rome, this one deserves to be counted amongst those to which she was most indebted for her greatness and dominion. For without some such an institution Rome would with difficulty have escaped the many extraordinary dangers that befell her; for the customary proceedings of republics are slow, no magistrate or council being permitted to act independently, but being in almost all instances obliged to act in concert one with the other, so that often much time is required to harmonize their several opinions; and tardy measures are most dangerous when the occasion requires prompt action. And therefore, all republics should have some institution similar to the dictatorship. (p. 165)

Machiavelli[5] praised Roman law for having the wisdom to establish legal and controlled means of temporarily vesting dictatorial power in a single person relative to the invariable threat of conquest by an outgroup.

Without in any way detracting from Machiavelli's logic, one will observe that Julius Caesar was a dictator and by his hand the Republic ended. The dictator and the external rivalries that make it necessary illustrate the reality of group selection. Most obviously, rival nations are competing biocultural groups. In turn, from the broader multilevel selectionist perspective, we see the dictator as imposing group unity, with the capacity to harshly suppress individually selected behavior through courts martial, execution, imprisonment, and confiscation, among other means. Nevertheless, as is the point of revisiting the issue here in this sixth chapter, distancing a state from external danger by means of elite fighting forces and standing armies, which were widely suspected through the early modern period, preciptates military coups (Braddick, 2015; Brownson, 1865/2005; Burgess, 1915; Casey, 2007; Cressy, 2006).

As with military defection, political factions can fracture a larger group into constituent parts where the cohesiveness of those constituent parts exceeds the binding force of the larger group. In consequence, much thought has been given to factionalism's ills and remedies. For instance, Montesquieu (1965) thought that a republic had to be small so as to avoid faction and fractionalization, whereas Madison thought that faction could check faction. In either case, faction was recognized as potentially fatal, with only the prescription differing. Large republics, nations, and of course empires, having no hope in Montesquieu's solution, can only employ Madison's. Factions and fractionalization within groups straightforwardly promote fissures and fission but indirectly enable sociopolitical aggregation by piecemeal incorporation into neighboring empires, which often succeed in conquering fragmented groups by exploiting parochial differences (Parsons, 2010). External groups then exploit prevailing divisions and provincial rivalries, as when the Umayyad Caliphate, Spanish *Conquistadores*, and Napoleon, respectively, outmaneuvered the Visigoth Kingdom of Spain, the so-called Inca Empire (*Tawantinsuyu*), and pre-nation-state Italy. Related to this latter point, Wimmer observes that ancient empires easily expanded over lower-level political structures but, in doing so, created local cohesion while providing templates for infrastructure and bureaucracies, later co-opted by emergent nation-states. In Wimmer's view, exposure to empire inoculates populations from future incorporation by virtue of evoking national consciousness.

The tenuous and tumultuous confederation of Great Britain was often exploited by enemies stoking Ireland's longing for independence via diplomacy, aid, or military expedition. Thus, at the instigation of the Spanish during the Anglo-Spanish War (1585–1604) and the French during the 1789 Revolution, we see that *England's difficulty* was used as *Ireland's opportunity*. Though Ireland was not dissevered from England, America was with the aid of England's longtime rival, France. Distance and custom, compounded by a policy of benign neglect and de facto self-rule practiced for more than a century, operated like the cultural analogue to the evolutionary process of *genetic drift*, allowing cultural and political distinctiveness to grow and later form the fault line of division between rulers and revolutionaries. The American revolutionary struggle witnesses the cleaving and welding of group identities across the Atlantic, across colonial lines, and across the allegiances that separated patriots and loyalists. The winners of the Revolution were those that were able to maintain superior cohesion. The British policy was calamitous, as embodied in the actions of the Howe brothers, General William Howe and Admiral Richard Howe. The Howe brothers proceeded to prosecute the war as if one could marry *Pax*, goddess of peace, to *Mars*, god of war. Together, the Howes seemed to lackadaisically pursue the rebels, offering conferences and peace overtures, showing spurts of strategic success, but without the dogged follow-through that was required. Recalling Chap. 4, we see this policy of ineffectual oppression kept the *stimulus of blows* at the optimum arc along the aforementioned curvilinear relationship. In other words, British military presence, being intermediate in strength, could not subdue and suppress. Rather, it was ultimately effectual only in creating colonial cohesion against a shared enemy. This is why Machiavelli (2010) suggests that a threatened state temporize with rivals to the extent that they cannot vanquish those rivals decisively. Temporizing will defer or diminish a threat excessively formidable, whereas ineffectually opposing a threat will only serve to spur an enemy to unity and action, as was the case of Rome when belatedly attacked by a league of neighboring tribes:

> For their league had no other effect than to unite the people of Rome more closely, and to make them more ready for war, and to cause them to adopt new institutions that enabled them in a brief time to increase their power. (p. 163)

Learning from the British, the new American government placed the laurel branch and arrows in separate claws of their national eagle. Moreover, above the squeamishness of staid Republicans, Hamilton was among the influential voices insisting that the newly formed federal government appear as a colossus in the field when faced with rebellious factions, as it did when suppressing the *Whiskey Rebellion* by conscripting a force of federalized militia that was comparable to those armies fielded during the Revolution. Presenting as a colossus in the field accessed the strategy of the fit antelope whose high spring (*stotting*) serves as an *honest signal* of fitness to a would-be predator, cautioning against a costly exchange that would be likely to end unfavorably for the hunter. Nearly unique among the founders, the immigrant Hamilton was able to promote this unyielding policy precisely because he was not divided in his loyalties, as were the other founders who generally identified more closely with their respective states than with the federal government they created. As we have seen, such divided loyalties had been at issue with the Howe brothers before them, who similarly sought to walk the untenable middle road between suppression and appeasement partially from sympathies with the co-ethnics they were supposed to subdue. An admirer of Julius Caesar (Morris, 1973), and possessed by an imperialist temperament, Hamilton hailed from the West Indies and thereafter ranged across the American colonies at the behest of education, war, and politics, leaving him unattached to any local region (Elkins & McKitrick, 1993). It was thus that the fear of foreign threat had no counterweight in his mind, leaving Hamilton to characterize the constitution as "frail and worthless" (Lodge, 1898, p. 262), favoring a *completely sovereign* centralized government to the annihilation of "state distinctions and state operations" (Newman, 2004, p. 48).

As with secular authority, religions employ coercive measures meant to compel adherence. The *Spanish Inquisition* is unjustly cited as the unparalleled exemplar of religious coercion but is truly only one in a long series of campaigns mounted to root out heretical nonconformists and nonconforming ideas in the interest of forging a national unity based on shared religious loyalties. The Protestants, having loosed the tether and slipped the trace of Catholicism, recapitulated the trend of devolution as they degenerated into innumerable sects: *Calvinists*, *Puritans*, *Anabaptists*, *Methodists*, *Shakers*, *Quakers*, and so on. This was justifiable based on the originally declared right of separation from Catholicism, which was an act of conscience and an appeal directly to God that would brook no intermediary. The Bible translated into the vernacular was increasingly preferred. The Puritans, the most fiercely separatist sect, would not get their scripture second hand, filtered through pagan Platonism, the teachings of Augustine or Aquinas, the Church of Rome, the Holy See of Peter, or the Council of Trent or Nicaea. No, they would read the word of God for themselves as it had been translated into English by King James. In doing so, they would think, they would judge, and they would theorize. Among the Puritans then, the process is once more repeated. After having sailed to America in order to establish the fabled "city on a hill," a mythical beacon for reformed Christendom sheltered from the Old World's corruption, the Puritans found themselves acting the part of inquisitors, banishing dissidents, such as Anne Hutchinson, Roger Bacon, and other founders of Rhode Island.

Thus, in both the secular and sacred spheres, decline toward lower levels of aggregation is often precipitated from rebellious internal factions where the *hegemon* can no longer radiate credible authority over its client states. As a body in motion stays in motion except when acted on by an outside force, the process of disaggregation, once started by a faction, must be arrested thereafter or otherwise persists down to lower and lower levels of aggregation.

4 "Mystical" Theories of Decline

When we speak of decline, especially when referring to decline in the meta-population traits evolved by multilevel selection itself, it calls to mind the writings of the great declinists, such as Ibn Khaldun, Montesquieu, Vico, Spengler, and Toynbee. Vico spoke of shirking civic responsibilities in favor of individualistic goals, which brought men back to barbarism. Toynbee (1951) spoke of spirit and spiritualism, being morally routed or experiencing a loss of *élan* or moral courage. Writing in Wagnerian tones of *Gotterdammerung*, Spengler (1991) referred to inexorable societal degeneration (Farrenkopf, 2001). In explaining decline, these authors recur to difficult-to-operationalize intangibles, classed as *mystical* explanations by Tainter (1988), to which he adds Adams's *energetic material*, Dawson's *vital rhythm and balance*, Sorokin's *value systems*, and Griffin's *cultural fatigue*. As part of a larger critique, Tainter savages the mystical genre of declinist writings for relying on biological growth analogies[6] and being imbued with value judgments, though he ultimately classes them as mystical and deems them irremediable for their reference to *intangibles*. "Mystical explanations," Tainter (1988, p. 85) states, "simply fail to identify any isolatable, observable, measurable factor controlling cultural change." Focusing on Spengler and Toynbee as exemplars of the genre, Tainter (1988, p. 84) finds references to biological vitalism, which are "unknowable, unspecifiable, unmeasurable, and unexplainable."

In the eighth chapter of our *Life History Evolution: A Biological Meta-Theory for the Social Sciences*, we share Tainter's criticism in speaking thus of Toynbee:

> Though he rested his hopes of surpassing Spengler via superior causal explanation, it was precisely in the arena of explanation that Toynbee first faltered; for, at the outset, he seemed to reject the lens of Western science, with its fixed laws and reductive methods, in favor of mythical and religious allegory, pivotal leadership, and particular descriptions. (p. 133)

If Tainter were correct, a truly "mystical" genre would be indeed ungrounded, particular, pontifical, vague, murky, superficial, dogmatic, and idiosyncratic. However, we do not believe that all of these concepts fall completely out of the reach of careful psychological measurement. An alternative approach to the operationalization of this traditional wisdom is suggested by multilevel selection theory, which is one of several candidate mechanisms that can stand in the stead of intangible references to *asabiyyah*, *élan*, or *spirit*. Future studies must further operationalize the trait-based products of group selection in terms of measurable biological phenomena that might be able to capture what the so-called mystical authors were attempting to communicate in their metaphorical narratives of decline. Nevertheless, we contend that this declinist theme persists because it is indeed referring to an actual psychological property of human groups, eminently important for the cohesion of a society and its ability to subsist amid a competitive landscape. We insist that this is not semantic legerdemain. We are not simply entering another name into the lists, spuriously arguing for its semantic superiority to *vigor*, *virtue*, *asabiyyah*, or any like variant.[7] To say a society is *group selected*[8] denotes not only a description of what *is*, but an explanation of how it *came to be*. A *group-selected* society is one that has been forged within a selective regime favoring cooperation among members of a group, in competition with members of a rival group.

By reviewing *soft selective regimes* in the remainder of this section, and by providing a historical lexicographic analysis of *asabiyyah* in the subsequent section, this chapter closes with a preliminary attempt to explain how replacing vague, immeasurable terms, with measurable group-selected traits, can rescue what we believe has been mischaracterized as the "mystical" genre of decline.

In beginning to describe soft selective regimes, we first turn to Vico. "The century of Roman virtue," Vico (Pompa, 2010, p. 112) insisted, "lasted until the Carthaginian Wars." Whether measured by territorial extent, military power, or economic activity, the future efflorescence of Rome lay far ahead, well past the end of the Punic Wars of which Vico speaks. Machiavelli finds the same sentiments in Juvenal:

> conquest of foreign countries had caused the Romans to adopt foreign manners and customs, and that, in exchange for their accustomed frugality and other most admirable virtues, gluttony and luxury dwell there, and will avenge the conquered universe. (p. 282)

Toynbee (1951; volume IV) writes in the same vein, thinking Gibbon's epic (1846) oxymoronic:

> The degree of Gibbon's hallucination is betrayed by the very title of his great work. The History of the Decline and Fall of the Roman Empire! The

> author of a history that bears this name is surely beginning his narrative at a point which is very near the end of the actual story; for the Roman Empire itself was a monumental symptom of the far-advanced decline. (p. 61)

Careful reading finds some semblance of specificity across these authors. For example, a group assuming hegemonic status by subjugating all rivals eliminates the martial pressures shaping its founding; in Toynbee's terminology, it eliminates the *stimulus of blows*. There is then the curse of conquest. The taking of slaves, gold, harvests, and booty of all kinds export toil and hardships to the conquered. In speaking of the effects of luxuries, wealth, slaveholding, and hegemonic peace, Vico, Toynbee, Ibn Khaldun, and likeminded declinists are speaking of decadence. In biocultural terms, decadence amounts to a selective regime favoring the more individualistically inclined, self-interested members of the population. Making one further inference, one might say that, *in founding a successful state, founders change the selective regime which created them.* Though much more must be said by way of elaboration, this is the driver of cyclical history: The group selected found a society that thereafter selects against their ilk.

In our aforementioned book, *Life History Evolution: A Biological Meta-Theory for the Social Sciences* (2018), we noted several parallel themes relevant to group selection, nowhere more so than when writing Chap. 8, which treated Toynbee's work. In addition to reviewing the significance of nomadic incursions in Chap. 4 of this book, we here recall our prior discussion of selective regimes, equally relevant to life history evolution and group selection:

> Extending the cooperative venture of small settlements to the national level, however difficult to initiate, proved more difficult to maintain…The selective pressures associated with state formation…slowly relax as the state matures. Entropy ensues. It does so especially when civilizations become hegemonic universal states free from the fitness enhancing group selective pressures that come in the guise of war and competition. Within walls erected against external conquest, no matter if they are the stone ramparts of Constantinople, riverine or montane barriers, or a phalanx of mercenary

> arms, there arises a changed selective regime, increasingly opening niches to Machiavellian leaders, free riders, psychopathic manipulators, thieves, mendicants, adulterers, and dissidents…[who] come to thrive on the increasing anonymity, trust, and abundance, within the walls. (p. 135)

Without presently involving ourselves in any detailed discussion of the complex evolutionary relationships between group selection and life history theory, it suffices to say that some productive combination of group-selected and slow life history-selected elements create stable societies; these then become, in effect, altered selective regimes subject to invasion by individually selected and fast life history-selected elements. The security of a well-secured stable state can produce dissidents and decadence, but it affords cultural and evolutionary progression toward more benign forms of individualism, which can in time prove nearly as inimical to group-selected societal strength. These are values often celebrated in the modern West, classed as rights and liberties. To say nothing of rights and liberties as a positive good from a cultural perspective, their ability to undermine vigorous collective action remains. Emphasizing equity and equality, due process, legal protections, and related rights and liberties is a luxury indulged in by strong states that have substantially reduced external threat. Rival nations can augment powers by dispensing with any excess of rights and liberties and thereby more effectively select in favor of group-selected persons and principles that come to threaten freer, mature societies.

5 Operationalizing *Asabiyyah*

As we have seen in the foregoing section, the many terminological variants employed by authors writing in the so-called mystical genre seem to be describing a decline in the psychological traits found among highly group-selected populations. Following that inference, we can render these many variants tractable by specifying mechanistically how they arise. As we have seen, soft selective regimes, those inviting decadence and sheltering the populace from competition and war, change the basis of economic and reproductive success. Self-sacrifice and bravery, abstemiousness, and disinterested patriotism, those Ciceronian values that can seem so starkly extreme, become less relevant and thus are less often propagated and rewarded. These traits can decline generation after generation, slowly eroding the strength of the state, which attains to its heights of grandeur on the inertia of group-selected founders.

Coming before all the other aforementioned authors within the *mystical* genre, and most precisely describing the trait underlying a group-selected society, was Ibn Khaldun and his concept of *asabiyyah*. *Asabiyyah* is augmented in the tribal or barbarian conquest phase (called "savagery" by Ibn Khaldun, in contradiction to the more precise taxonomy of Morgan, 1877) and thereafter depleted amid effete and effeminate dissipation of high civilization (Irwin, 2018). For these reasons, and also to illustrate the importance of selective regimes and the possibility of measuring their effects, we performed a test of Ibn Khaldun's (1377) theory of the decline of *asabiyyah* as a consequence of increased wealth ("luxury") and ease of living. Ahead of the following details and methodological descriptions, these analyses can be taken as empirically supporting Ibn Khaldun's (1377) hypothesis that declining *asabiyyah* is historically associated with increasing wealth, independent of the effects of time.

Following Sarraf, Woodley of Menie, and Feltham (2019), we performed a historical lexicographic analysis of the use of words associated with each of the five *Moral Foundations* identified by Haidt (2012): (1) care; (2) fairness; (3) loyalty; (4) authority; and (5) sanctity. The diachronic utilization of these specific classes of English language words was evaluated via their relative frequencies of usage through *Google Ngram Viewer* (Michel et al., 2011), an interactive textual corpus encompassing over 5.9 million texts and 500 billion written words from AD 1500 to 2008. The lexicographic data were obtained in the form of frequency counts of each word within its respective language across the 200 years spanning AD 1800–1999.

The lists of words for each of the five Moral Foundations were harvested[9] from keywords used in the online *Moral Foundations Questionnaire*[10] and in Haidt (2012), when describing these theoretical constructs. The words then used as items in each of these lexicographic scales were psychometrically selected on the basis of obtaining adequate part-whole correlations for each word to the corresponding aggregate scale score for each lexicographic scale. We thus empirically selected the best words from the initially larger lists based on their convergent validity with respect to each other, as indicated by the internal consistency of the constructed scales. Our psychometric selection procedure was deemed to be the most straightforward way of identifying the best item-level indicators of these constructs under the presumption that the items would be differentially valid as a function of how well they reflected the central latent constructs, which were the five Moral Foundations.

Unit-weighted common factor scales (Gorsuch, 1983) were estimated as the means of the standardized scores for the lexicographic items on each scale (Figueredo, McKnight, McKnight, & Sidani, 2000). As per Moral Foundations theory, the five scales, as depicted below in Fig. 6.1, were aggregated into two lower-order factors: *binding* and *individualizing*. By reverse-scoring the individualizing factor, we further aggregated these two lower-order factors into a single higher-order factor (*asabiyyah*), based on the preliminary results reported by Sarraf et al. (2019), indicating that these two trends were diverging from each other systematically throughout the twentieth century and might therefore indicate a single tendency for one to increase at the expense of the other.

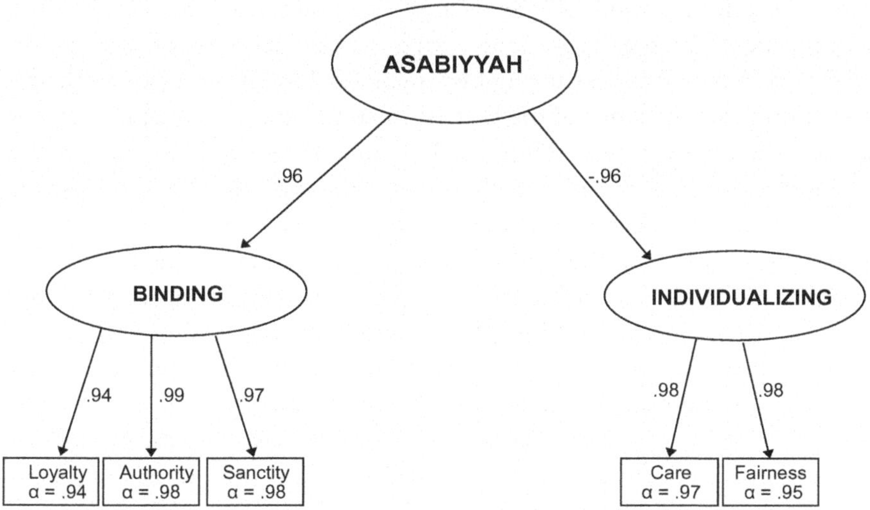

Fig. 6.1 Latent hierarchical structure of lexicographic *asabiyyah*

The unit-weighted factors were then used as manifest variables in longitudinal multilevel models (MLMs). Four nested MLMs were estimated to test the need for increasing parameterization as alternative hypotheses: (1) *MLM1* estimated a single intercept and a single logarithmic slope (unconditional "*asabiyyah*") for all lexicographic factors, lexicographic scales, and lexicographic items (words) over time; (2) *MLM2* estimated a separate intercept and a separate logarithmic slope for each lexicographic factor over time but the same intercepts and logarithmic slopes over time for all scales within each factor and for all words within each scale; (3) *MLM3* estimated a separate intercept and a separate logarithmic slope for each lexicographic scale over time but the same intercepts and logarithmic slopes over time for all words within each scale; and (4) *MLM4* estimated a separate intercept and a separate logarithmic slope over time for each word.

Table 6.1 displays the pertinent nested model comparisons. The systematic AIC and -2RLL comparisons performed among the nested models representing the specific variance components accounted for by each level of the aggregative hierarchy indicated that most of these incremental improvements in model fit were statistically significant ($p<0.05$) but relatively trivial in magnitude. Comparisons of squared multiple correlations among the four nested MLMs yielded essentially the same results. The magnitude of the specific variances explained at each level of aggregation (ΔR^2) were found to be negligibly small (<<1%) in contrast with the common factor variance of the highest unconditional "*asabiyyah*" level of aggregation, which was found to be quite large (>75%). Given the strength of these findings, we chose to retain the model parameters for only the unconditional "*asabiyyah*" level (MLM1), as the extra model parameters added by the lower levels of aggregation (MLM2, MLM3, and MLM4) were virtually irrelevant to an adequate account of the diachronic variances in the Moral Foundations factors, scales, and words. The logarithmic slope of this unitary higher-order "*asabiyyah*" construct over time was negative and statistically significant: $r = -0.96$ (90% *CI:* −0.97, −0.95), $F(1198) = 2546.99$, $p<0.0001$. No significant serially autoregressive effects were found ($ARH1 = 0$) and the incremental variance due to curvilinearity was statistically significant but negligibly small: $r = -0.05$ (90% *CI*: −0.19, 0.09), $F(1197) = 10.99$, $p<0.001$.

Table 6.1 Fit indices for nested multilevel models (MLMs) for lexicographic indicators of moral foundations factors, scales, and psychometrically selected words

Multilevel model	MLM1	MLM2	MLM3	MLM4
	Year	+ Factor + Factor*Year	+ Scale + Scale*Year	+ Word + Word*Year
AIC	7986.7	7982.2	7994.1	8023.1
-2RLL	7870.7	7958.2	7982.1	8015.1
	$\Delta\chi^2$ =	**87.5***	**23.9***	**33.0**
R^2	0.7538	0.75524	0.75628	0.76006
	ΔR^2=	**0.00145***	**0.00104***	**0.00378***
NDF	1	3	9	55
	ΔNDF=	2	6	46

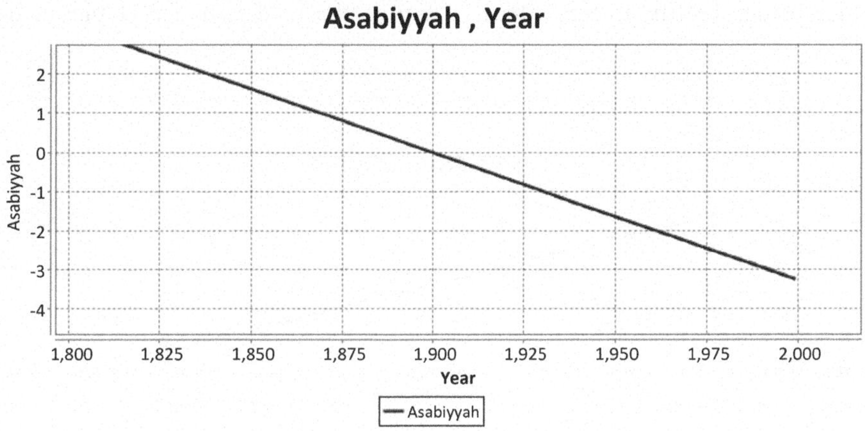

Note: Dashes are upper and lower confidence intervals.

Fig. 6.2 Bivariate linear regression of *asabiyyah* over time (AD 1800–1999)

As depicted in Fig. 6.2, GDP per capita data from AD 1800 to 1999 were obtained for the United Kingdom, the United States, Canada, New Zealand, and Australia, from the Maddison Project database (Bolt, Inklaar, de Jong, & van Zanden, 2018), a repository curated by the Groningen Growth and Development Centre (GGDC). MLM residuals were then exported for both GDP per capita and the unitary *asabiyyah* factor and used for subsequent general linear modeling. MLM residuals were thus statistically adjusted for the logarithmic effect of time as well as of any single-lagged heterogeneous autoregressive serial dependencies among successive data prior to regression modeling, thus circumventing this potential problem as a threat to the validity of correlational analysis. It was especially important to statistically control for the effects of time to ascertain that any association was not a simply coincidental one of GDP increasing ($r = 0.88$) and *asabiyyah* simultaneously but independently decreasing ($r = -0.97$) over the same period of time. The bivariate correlation of the time-adjusted MLM residuals of GDP per capita with those of *asabiyyah* was $r = -0.76$ (90% *CI*: −0.82, −0.70), $F(1198) = 276.21$, $p<0.0001$, empirically supporting Ibn Khaldun's (1377) hypothesis that declining *asabiyyah* is historically associated with increasing wealth, independently of the effects of time, as depicted in Fig. 6.3.

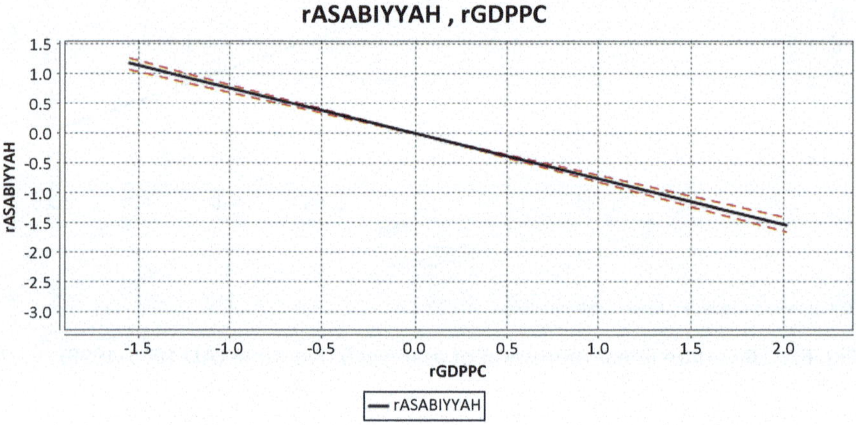

Note: Dashes are upper and lower confidence intervals.

Fig. 6.3 Time-adjusted MLM residuals of GDP per capita predicting MLM residuals of *asabiyyah* (AD 1800–1999)

6 Conclusions

Decline can come of falling behind advancing rivals, as seen in the waning Spanish Empire, or destroying the organic substructure of society through excessive change, as nearly happened in Petrine Russia. Similarly, decline can come just as well from subjects rebelling against their rulers as from rulers betraying their subjects. All such factors can precipitate decline or signal impending collapse to a lower level of societal aggregation. Distinct from these is the crowded genre of declinist literature, which Tainter (1988) refers to as *mystical*, wherein decline comes of *senility* and *decadence* or of waning *martial spirit*, *vigor*, and *virtue*. Plato and Polybius are listed as ancient forerunners, Gibbon and Montesquieu are Enlightenment exemplars, while Spengler and Toynbee are classed as famous modern examples of this *mystical* tradition of decline. Mystical explanations, Tainter argues, fail to account scientifically for decline or collapse. They are crippled by reliance on biological growth analogies, value judgments, and references to intangibles. Whether using virtue or spirit or any related term, one can ask from whence came its abundance and why was it lost. Though much methodological and measurement must ensue, conceptually replacing these various terms with measurable traits derived of the biological process of multilevel selection promises a way forward, marrying intuitive wisdom with rigorous science, as has been shown by example in the analyses presented above. These analyses show that concepts like *asabiyyah*, although perhaps "mystical" and intangible to historians, are well within the purview of modern psychological measurement when guided by appropriate scientific theory. Indeed, this allegedly *mystical* form of decline is qualitatively distinct from those signs and symptoms treated at the outset of this chapter, within Sects. 2 and 3. Instead of indicating a decline to hierarchically lower levels of group organization, authors within this genre seem to concern themselves with meta-population processes related to multilevel selection and its resultant traits, the decline of which is less easily reversible and often precipitates future collapse.

Notes

1. Just as societies may be characterized as having different mean life history speeds produced by differential *K*-selection among groups, societies may also be characterized as having varying mean strengths of social cohesion and integrity produced by the differential group selection to which they have been subjected.
2. Tainter (1988, p. 54) discusses a declinist genre referred to as *insufficient response to circumstances*, which is admired by Tainter for its recognition of external causes as mediated through internal characteristics: there is a pressure that does not mechanistically make for collapse but only leads to collapse for failure to adapt.
3. Having just read Chap. 4, one will see war serves equally as a stimulus to growth or an impetus to decline, depending on whether it comes in salutary or toxic doses; as was said, war has a curvilinear relationship with aggregation, and we are now, in this sixth chapter on decline, looking at the downside of the relationship.
4. Eventually, such was the prodigal outflow of treasure that Sir Robert Walpole attempted to convince the House of Commons to prop up the shell of the Spanish Empire in its twilight years, for it had become only a "canal" though which American silver passed before being distributed throughout Europe. The British among other nations had previously assisted in curtailing piracy on Spain's behalf when Spain was no longer able to defend her shipments of silver (Kamen, 2003).
5. Machiavelli can be mined for group selectionist reasoning, perhaps most evident in *The Discourses*, which, in comparison with *The Prince*, addresses itself more generally to national stability and power, of which the tenure of the leader is only one component. Ever didactic, *The Discourses* espouses maxim after maxim relevant to ingroup solidarity as it survives in the acid bath of outgroup competition.
6. The following is a description of the intellectual history of biological growth analogies such as those employed by Spengler (Farrenkopf, 2001):

   > In the late eighteenth century, biological analogies began to displace those derived from the mechanistic universe of Newtonian physics in historical and political thought. Herder, in exemplifying this trend, conceives of nations, within the flow of historical change, as organisms. They have a morphology; they are dynamic and alive. These

organisms are not rational in character; they are things in themselves and not means. Like a person, nations have characteristics: a life span and their own spirit. (p. 79)

7. As described in Chap. 4, these terms, especially when used in the declinist literature, are descriptively similar, some being nearly synonymous.
8. Please excuse this neologism invented for the sake of convenience. Of course, the term "group selection" refers to an evolutionary selective regime that selects for particular traits (such as altruism), and it is these traits that are actually the product of the group selection process.
9. We thank Maya Louise Bose for her excellent work in identifying these items from the original texts.
10. http://yourmorals.org/haidtlab/mft/index.php?t=questionnaires

References

Bolt, J., Inklaar, R., de Jong, H. and van Zanden, J. L. (2018). Rebasing "Maddison": New income comparisons and the shape of long-run economic development. *GGDC Research Memorandum, 174*, 1–67

Braddick, M. J. (2015). *The Oxford handbook of the English revolution*. New York: Oxford University Press.

Brody, M. (2000). *Mary Wollstonecraft: Mother of women's rights*. New York: Oxford University Press.

Brownson, O. A. (1865/2005). *The American Republic: Its constitution, tendencies, and destiny*. Delanco, NJ: The Classics of Liberty Library.

Burgess, J. W. (1915). *The reconciliation of government with liberty*. New York: Charles Scribner's Sons.

Burke, E. (1790/1992). *Reflections on the revolution in France and on the proceedings in certain societies in London relative to that event*. New York: The Classics of Liberty Library.

Casey, J. (2007). *Family and community in early Modern Spain: The citizens of Granada 1570–1739*. New York: Cambridge University Press.

Cressy, D. (2006). *England on edge: Crisis and revolution 1640–1642*. New York: Oxford University Press.

De Bruyn, F. (2004). Edmund Burke the political Quixote: Romance, chivalry, and the political imagination. *Eighteenth-Century Fiction, 16*(4), 695–734.

Elkins, S., & McKitrick, E. (1993). *The age of federalism*. New York: Oxford University Press.

Elliott, J. H. (1961). The decline of Spain. *Past & Present, 20*, 52–75.
Evans, R. J. W., & Von Strandmann, H. P. (Eds.). (2002). *The revolutions in Europe, 1848–1849: From reform to reaction*. New York: Oxford University Press.
Farrenkopf, J. (2001). *Prophet of decline: Spengler on world history and politics*. Baton Rouge, LA: Louisiana State University Press.
Figueredo, A. J., McKnight, P. E., McKnight, K. M., & Sidani, S. (2000). Multivariate modeling of missing data within and across assessment waves. *Addiction, 95*, 361–380.
Flynn, D. O. (1982). Fiscal crisis and the decline of Spain (Castile). *The Journal of Economic History, 42*(1), 139–147.
Gibbon, E. (1846). *The history of the decline and fall of the Roman Empire*. London: J. Murray.
Gorsuch, R. L. (1983). *Factor analysis* (2nd ed.). Hillsdale, NJ: L. Erlbaum Associates.
Haidt, J. (2012). *The righteous mind: Why good people are divided by politics and religion*. New York: Pantheon.
Hertler, S. C., Figueredo, A. J., Peñaherrera-Aguirre, M., Fernandes, H. B. F., & Woodley of Menie, M. A. (2018). *Life history evolution: A biological meta-theory for the social sciences*. New York: Palgrave Macmillan.
Hirst, F. W. (1935). *Liberty and tyranny*. London: Duckworth.
Ibn Khaldun, A. A.-R. (1377/1969). In N. J. Dawood (Ed.), *The Muqaddimah: An introduction to history*. Princeton, NJ: Princeton University Press.
Irwin, R. (2018). *Ibn Khaldun: An intellectual biography*. Princeton, NJ: Princeton University Press.
Kamen, H. (2003). *Empire: How Spain became a world power, 1492–1763*. New York: Harper Collins.
Lodge, H. C. (1898). *Alexander Hamilton*. New York/Boston: Houghton Mifflin & Co.
Machiavelli, N. (2010). *The prince and the discourses*. Rockville, MD: Gryphon Press.
Maltby, W. S. (2009). *The rise and fall of the Spanish Empire*. New York: Palgrave Macmillan.
Massie, R. K. (2012). *Peter the Great: His life and world*. New York: Random House.
Michel, Jean-Baptiste, Yuan Kui Shen, Aviva P. Aiden, Adrian Veres, Matthew K. Gray, The Google Books Team, Joseph P. Pickett, et al. (2011). Quantitative analysis of cutlure using millions of digitized books. *Science 331*(6014), 176–182.

Montesquieu, C. (1965). *Considerations on the causes of the greatness of Romans and their decline*. New York: Cambridge University Press.

Morgan, L. H. (1877). *Ancient society; or, researches in the lines of human progress from savagery, through barbarism to civilization*. New York: H. Holt and Company.

Morris, R. B. (1973). *Seven who shaped our destiny*. New York: Harper & Row Publishers.

Newman, P. D. (2004). *Fries's Rebellion: The enduring struggle for the American Revolution*. Philadelphia: University of Pennsylvania Press.

North, D. C., & Thomas, R. P. (1973). *The rise of the western world: A new economic history*. Cambridge, UK: Cambridge University Press.

Parsons, T. (2010). *The rule of empires: Those who built them, those who endured them, and why they always fall*. New York: Oxford University Press.

Philip, M. (2014). *Reforming ideas in Britain: Politics and language in the shadow of the French Revolution, 1789–1815*. New York: Cambridge University Press.

Pompa, L. (2010). *Vico: The first new science*. New York: Cambridge University Press.

Robertson, P. S. (1952). *Revolutions of 1848: A social history* (Vol. 1025). Princeton, NJ: Princeton University Press.

Sarraf, M. A., Woodley of Menie, M. A., & Feltham, C. (2019). *Modernity and cultural decline: A biosocial perspective*. London: Palgrave Macmillan.

Skocpol, T. (1979). *States and social revolutions: A comparative analysis of France, Russia, & China*. New York: Cambridge University Press.

Spengler, O. (1991). *The decline of the West*. Oxford: Oxford Paperbacks.

Tainter, J. A. (1988). *The collapse of complex societies*. New York: Cambridge University Press.

Toynbee, A. J. (1951). *A study of history*. New York: Oxford University Press.

Wimmer, A. (2018). *Nation building: Why some countries come together while others fall apart*. Princeton, NJ: Princeton University Press.

7

The Collapse and Regeneration of Complex Societies

Mateo Peñaherrera-Aguirre, Aurelio José Figueredo, and Steven C. Hertler

1 Introduction

The subject of societal collapse is a theme that, due to its political, social, economic, and ecological implications, still generates heated discussions. Researchers interested in developing a general theory of collapse face the challenge of identifying common patterns across human societies. This task is further complicated because multiple publications on the subject employ a case-by-case methodology, within which the causes of collapse are thought to be specific to each society. Such *historical particularism* persists to this day. Historical contingency is preferred to generalizable explanation. In response, some researchers have instead concentrated on examining how a society's internal dynamics predict the risk of collapse. For example, a society's institutional performance, macroeconomic yields, and level of collective action have been thought predictive of its structural integrity under adverse circumstances. Through this lens, external factors may lead to a sudden loss of sociopolitical complexity only when the system's capacity to address these conditions is compromised. Given variation in societies' level of cohesion and collective action, the case of

S. C. Hertler et al., *Multilevel Selection*, https://doi.org/10.1007/978-3-030-49520-6_7

societal collapse offers a unique glimpse into multilevel selection operating among social systems. This chapter describes critical elements developed in the collapse literature while providing an overview of the current multilevel selection perspectives on fluctuations in collective action. The present contribution also describes how institutional robustness and cultural innovations contribute to a society's regeneration capacity after experiencing a collapse.[1]

2 Defining Sociopolitical Collapse

Numerous publications have provided different descriptions of collapse. To circumvent the current overabundance of definitions in the literature, and because of its consonance with multilevel selection theory, this chapter will adopt J. A. Tainter's (1988) description featured in *The Collapse of Complex Societies*, wherein a society collapses when it features the following:

1. A decrease in social stratification and differentiation;
2. A decline in the society's economic specialism;
3. Lower regulation and integration among political and economic classes;
4. A fall in the allocation of resources to cultural phenomena such as art and architecture, among others;
5. Restricted exchange of information (e.g., at the level of individuals and groups and between the polity's core and its outskirts);
6. Limited trade and distribution of resources;
7. A loss of coordination within the system;
8. The emergence of smaller autonomous polities and a reduction in territory size.

In general, historical treatments of collapse frequently describe how external factors significantly and irreversibly disrupt a society's complexity. Natural catastrophes, invaders, resource depletion, and pandemics (Diamond, 2005; Fagan, 2009; Haug et al., 2003; Huntington, 1922; Kaniewski, Guiot, & Van Campo, 2015; Kennett et al., 2012; McNeill, 1998) are just a few of the commonly mentioned causes. Alternatively, other perspectives have suggested that collapse is instead the product of internal disruptions, including instances of interclass confrontations, accusations of maladministration, and other forms of legitimation crises, including using the state to attain economic, social, and political benefits at the expense of the rest of the population (Eisenstadt, 2017). The main limitation of these perspectives is that many societies repeatedly encountered these pressures without suffering from a sudden loss in sociopolitical complexity (Tainter, 1988; Turchin, 2003).

In response to some of the limitations of traditional collapse theories, Tainter (1988, 2004, 2006) developed a macroeconomic perspective wherein collapse resulted from the dynamic interplay between a society's complexity and the amount of energy required to sustain its organization. Tainter's (1988, 2006) theory rests on four principles: (1) human societies act as problem-solving systems; (2) these societies depend on a constant influx of energy to preserve their structural integrity; (3) as societies increase their level of sociopolitical complexity, they also experience rising per capita costs associated with preserving this organization; and (4) the benefits obtained from problem-solving institutions reach a point of diminishing returns over time, wherein the return on investment follows a negative quadratic trajectory.

Hence, for Tainter (1988), a sudden loss in sociopolitical complexity is caused by the system's internal economic dynamics. His model of marginal productivity claimed that the society's subsystems inescapably face a critical point of marginal returns where costs outweigh the benefits (Tainter, 1988, 2006). This dynamic permeates across social, political, and economic institutions. As social systems rise their level of complexity, they became increasingly dependent on the influx of information and material resources. To maintain an adequate flow of information and energy, the administration creates hierarchical institutions, coordinating and regulating how these resources spread within the system. The preservation and centralization of these connections rely on the labor of specialists. For example, in state-level societies, public institutions emerge to guarantee the production and distribution of goods and services to the population and to secure them against foreign and national threats (Tainter, 1988). Specialism ineluctably increases bureaucratization, with public servants proliferating as they specialize. In case any of its components are compromised, a society may also increase the number of redundant institutions. Tainter's (1988) theory also considered the challenge faced by academic and information processing institutions in producing new scientific and technological knowledge. For example, specialized information replaces general knowledge over time, a feature that subsequent publications identified in terms of evolved cognitive abilities (Woodley of Menie et al., 2017).

3 Multilevel Dynamics and Collapse

In addition to debates concerning the nature and the causes of collapse, another point of contention in the literature concerns the possibility of collapse operating sequentially over different social, political, and cultural units (Middleton, 2017) (Fig. 7.1).

Just as societies increase their level of sociopolitical complexity over time (e.g., band → tribe→ chiefdom→ state), so they can revert to a more primary level (Currie & Mace, 2011). As suggested by current perspectives on political evolution, collapse operates on different levels of

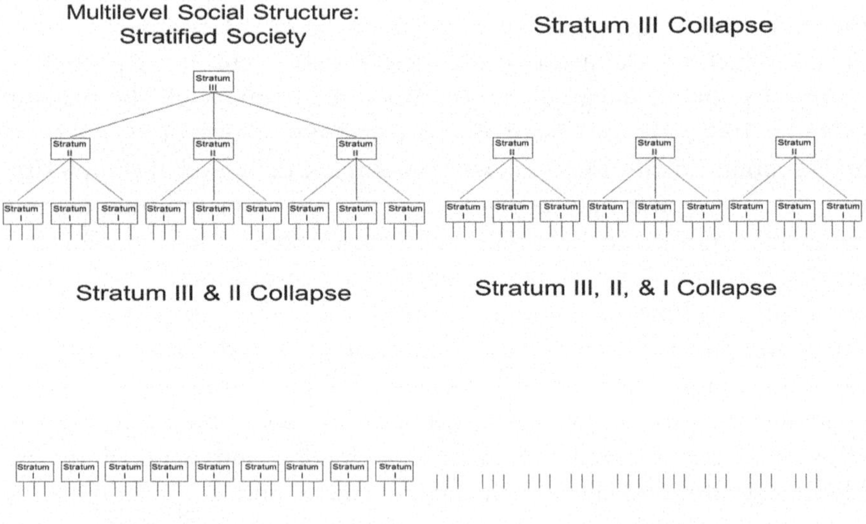

Fig. 7.1 Sequential multilevel social collapse operating on three different strata (I, II, and III)

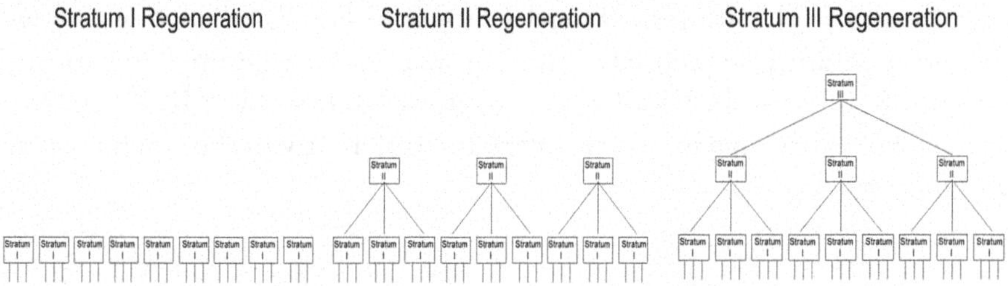

Fig. 7.2 Sequential multilevel social regeneration of strata I, II, and III

aggregation. Although complex societies often contain more than three strata (e.g., dyad/nuclear family → extended kin → neighborhood → district → city → prefecture→ province→ state), for the sake of illustration, consider a stratified society comprised of three hierarchical levels. As seen below in Fig. 7.2, the highest level, stratum III, corresponds to a fully autonomous superstructure featuring clear geopolitical boundaries. In this example, this highest stratum of the organization is governed by a group of individuals coordinating the efforts of lower-order components, including the distribution of information and resources. Although the subcomponents of stratum II can collect resources for local investment and spending, and are allowed other autonomous functions, they often respond to the instructions provided by stratum III.

In addition to implementing its own functions, stratum II acts as an intermediary between stratum III and stratum I. Indivisible elements[2] comprise stratum I, the lowest level of organization, as depicted in Fig. 7.1. In terms of collapse, the disappearance of stratum III does not imply the immediate elimination of the lower-level strata. If the disruption alters the integrity of stratum II, leading to its eventual collapse, the lowest level can persist despite the higher tiers experiencing considerable perturbations. Strata II and I could avoid collapse as long as their capacity to extract, process, and distribute resources remains relatively unaltered. Consistent with previous models, in this representation, collapse is not an outcome of internal or external perturbations, but rather the product of the strata's capacity to efficiently extract, process, and distribute resources in the face of changes and challenges. The survival of lower-level organizations also allows the eventual reemergence of higher-level organizations. This simplified model shown in Fig. 7.2 then views *regeneration*[3] operating sequentially on each level, with stratum II reliant on stratum I and stratum III reliant on stratum I and stratum II. Of course, each element within a level can contribute differently to the emergence of a higher-order level.

4 Cycles of Decline

From describing multilevel social organizations, we move on to consider multilevel temporal cycles. Traditionally, history is described as a series of events bound to contextual contingencies. Although such an approach provides insightful information into the particular features of a historical event, it also precludes the identification of general biohistorical patterns across contexts. In the last two decades, ecologists and demographers have begun working on such a general framework, displaying a renewed interest in those broad biohistorical dynamics initially broached by Toynbee, Spengler, and Huntington early in the twentieth century. Through this lens, historical events are no longer viewed as random occurrences but as parts of cycles embedded in a multilevel structure. Holling and Gunderson (2002), for example, theorized that complex ecological organizations experience a series of transformations through time. This theory viewed socioecological changes progressing from *exploitation* to *conservation*, to *release*, and on to *regeneration*, with each of these four stages, respectively, represented by the four mathematical functions: r, K, Ω, and α. Each stage differed depending on its level of intrinsic potential, the accumulated bioenergetic resources in the system, as well as on the subsystems' interconnectedness. The first stage, exploitation, is characterized by its low levels of interconnectedness and inherent potential. Under these circumstances, relatively unrestricted access to resources allows the population to increase in size (r). Eventually, the system expands its potential and connectedness, reaching the conservation phase with its accrual and storing of bioenergetic resources (K).[4] At some point, the accumulation of biomass and over-connectedness increases the system's vulnerability. Factors such as droughts, famines, pests, or fires release the structure's accumulated biomass (Ω). During the release phase, the system's connectedness declines while its intrinsic potential remains high. The end of this phase is followed by a period of reorganization. The remaining available resources are restructured to avoid unnecessary resource loss (α), generating the required conditions for the cycle to begin anew.

Other models emphasize destabilization as it occurs on one level and redounds to others. For instance, building on Holling and Gunderson (2002), Holling, Gunderson, and Peterson (2002) claimed that adaptive cycles operate within a nested organization of *economic*, *ecological*, and *institutional* systems, rather than occurring as isolated phenomena. Their view of collapse not only emphasizes interconnected systems but incorporates elements of chaos theory and is explicitly antihierarchical, both of which features are notable in their operationalization of *panarchy*, the title of their book, and the term describing their theory of *transformations in human and natural systems*. For Holling et al. (2002), each level in a system is linked to every other level through *panarchical* connections. Interconnectedness amplifies perturbations through all the system's levels. Consider three cycles: The first level is small and fast, the second presents moderate speed and size, and the third is large and slow. The first connection links the release phase of the small cycle to the conservation phase of the intermediate cycle and so on. These connections imply that disturbances at a lower level can impact a higher level. The likelihood of collapse then is proportional to the level's vulnerability and rigidity.[56]

In addition to the panarchy theory, a recent perspective inspired by multilevel selection theory describes biohistorical changes as *wheels within wheels* (Turchin, 2007), wherein a historical hierarchy is divided into four levels: (1) *business cycles*, (2) *father-son cycles*, (3) *secular cycles*, and (4) *asabiyyah cycles*. The lowest level in the hierarchy, business cycles, describes the macroeconomic fluctuations experienced by a society (Turchin, 2007) with each cycle lasting between five and twelve years (Korotayev & Tsirel, 2010; Turchin, 2007). Traditionally, business cycles are divided into *expansion* and *recession* phases, with the former referring to an increase in salaries, prices, and employment, whereas the latter refer to decreases in these same variables. The next level, father-son cycles, lasts approximately twenty-five years. It is manifested in the manner in which individuals from two contiguous generations differ in their involvement and responses to internal strife. The generation exposed to intense and often lethal intragroup competition, as through regional feuds and civil wars, eventually become exhausted by the atrocities perpetrated during the crisis. Survivors implement institutional safeguards to decrease the current conflict and avoid experiencing similar conditions in the future (Turchin, 2007). These sociopolitical changes generate a period of peace and intragroup cooperation. Under these conditions, a new generation arises. Sons are oblivious to the conflict experienced by their parents. The investment allocated to institutions involved in mediating intragroup confrontations is then redirected to other political or economic endeavors. Eventually, social tensions escalate into feuds and lethal confrontations, thus restarting the sequence. Secular and *asabiyyah* cycles, for their relevance and complexity, merit further discussion.

Located above father-son cycles, secular cycles last between 150 and 300 years. Turchin and Nefedov (2009) claimed that a population reaching an ecosystem's carrying capacity (*K*) experiences a series of social and economic *sequelae*. For instance, societies with higher population density approaching *K* are prone to experiencing food and land shortages, higher unemployment, and lower wages and display a decline in the consumption of goods and services (Turchin, 2007). Accelerated population growth also increases rents and property prices. These economic fluctuations reverberate within the system, with commoners and peasants partially or fully forfeiting their properties. The combined effects of downward social mobility, overpopulation, and the inability to store enough food increase the vulnerability of rural areas to famines. These dire conditions promote rural migration into urban centers (Turchin, 2003; Turchin & Nefedov, 2009). Although, initially, urban employers hire rural immigrants as craftsmen and traders, the continuous influx of peasants to cities eventually expands the number of unemployed immigrants (Turchin & Nefedov, 2009). Overcrowding and malnutrition generate the necessary conditions for the spread of pathogens. Demographic consequences ensue, with economic stagnation raising mortality rates, decreasing fertility rates, and decelerating population growth (Turchin, 2003). According to Turchin and Nefedov (2009), elites are not immediately affected by periods of resource scarcity, but often instead profit from the early stages of economic *stagflation*, which is an economic concept combining the phenomena of stagnation and inflation (Turchin, 2007; Turchin & Nefedov, 2009). In the absence of state regulation, landlords could raise rents or demand an increase in the rate of resource extraction without facing any significant opposition. Although peasants could rebel against their employers, the absence of alternative sources of income, as well as the threat of punishment, forces peasants to accept elite demands. During the early stages of the demographic-fiscal crisis, aristocrats experience a boost in wealth acquisition allowing an increased consumption rate (Turchin, 2007). Since the numbers of elites grow in conjunction with the rest of the population, resource scarcity eventually begins to affect the upper strata. Famines, epidemics, and violence interfere with the elite's revenues. Despite imminent financial crisis, the aristocracy's level of conspicuous consumption persists (Turchin & Nefedov, 2009). Decreasing profits and growing debts force elites to sell their properties to preserve their social standing. It is not uncommon for nobles to obtain loans from financial institutions or even request the state to provide financial assistance. These solutions temporarily buffer the nobility's financial crisis. However, the elite's inability or unwillingness to pay taxes forces the state to partially or fully halt any investment in public goods and services (Turchin & Nefedov, 2009). Among these services is the state's ability to provide internal and external security. Legal, penal, and military institutions operate at sub-optimal levels. *Intra-elite* confrontations escalate from minor disagreements to lethal violence; feuding, dueling, and civil wars increase in frequency (Turchin, 2007). This state of internal turmoil leads to a drop in the numbers of aristocrats, diminishing the intensity of intra-elite competition and restarting the cycle. It is worth noting that even though secular cycles are described as an autocatalytic process,[7] they can be moderated by external factors (Turchin, 2007). The impact of these exogenous forces will depend on the timing of their occurrence. An event like a drought or an epidemic coinciding with a society's integrative phase may alter some demographic parameters, such as the rate of population growth, but will not impact the society's organization to the point of collapse (Turchin, 2007).

Finally, *asabiyyah* cycles rest above secular cycles, ranging between 1000 and 2000 years (Turchin, 2007). This sequence covers fluctuations in a society's collective action and social solidarity. Quantitative examinations of this concept have recently provided additional support to historical perspectives considering the role of declining collective efforts in the collapse of complex societies. Turchin (2003), for example, described the change of society's *asabiyyah* in the following formula (Formula 7.1)[8]:

$$\dot{S} = r(A)S(1-S) \tag{7.1}$$

where r is a relative growth rate, A is the society's area or territory size, and S is the society's average level of collective solidarity (a value that ranges from 0 to 1, with 0 indicating a system unable to cooperate and 1 indicating the maximum-level collective action). This formula also assumes that some individuals will act selfishly or altruistically depending on the number of free riders and altruists in the group. Hence, the society must reach a critical point for altruism to spread. For Turchin (2003), this autocatalytic process[9] is best represented as a logistic function. Like any mathematical model, Formula 7.1 is necessarily simplified and thus faces a trade-off between sacrificing its external generalizability or risk being mathematically intractable. Hence, it is worth noting that this representation is limited to a single cycle, excluding instances of multiple collapses and regenerations. Similarly, it does not take into consideration the influence of exogenous elements, such as the presence of rival societies.

Though later rectified,[10] these models initially failed to explicitly accommodate the possibility that levels of *asabiyyah* might differ among hierarchical levels within a multistratum society, such as those depicted in Figs. 7.1 and 7.2. For example, Ibn Khaldun (1377) was clearly describing the *asabiyyah* of the Arabs (formerly Bedouin), who constituted the elite stratum III of the Islamic Empire that they created. He was not referring to the *asabiyyah* of conquered stratum II national polities, such as Egypt, Syria, or Persia. Although Ibn Khaldun did not explicitly address this issue, it is self-evident that higher levels of *asabiyyah* on the part of the natives of these conquered territories would be inimical to the survival of the empire, whereas higher levels of *asabiyyah* among the dominant Arab elites would doubtlessly help to preserve it as described. We might thus imagine antagonistic dynamism and difficulty maintaining prevailing hierarchies as consequences of a multistratum society expressing differential levels of *asabiyyah*. This was certainly true of the so-called "Byzantine" (Eastern Roman) Empire, which had previously sought to dominate those same national polities and encountered much nationalistic resistance from them. At that time, Egypt and Syria were both stratum II provinces that hosted alternative versions of Christianity, such as the Monophysite and Nestorian "heresies" (Khouri, 2007), in opposition to the "Orthodox" Christianity of Constantinople. Under the contemporaneous Sasanian Dynasty, Persia represented a competing stratum III empire in a state of near-perpetual war with the Byzantine, practicing an often militant Zoroastrianism in complete opposition to Christianity (Shapur Shahbazi, 2005).[11]

Socioecological conditions, such as being close to an ethnolinguistic boundary, favor the development of institutions enforcing intragroup cooperation (Turchin, 2003). Higher levels of social integration not only allow the group to defend the area from foreign incursions, but it also facilitates the annihilation or annexation of territories occupied by rival groups (Soltis, Boyd, & Richerson, 1995). As territorial expansion proceeds, threat of foreign rivals invading the core dissipates. Central areas, relatively invulnerable to invasions, experience an increase in intragroup competition. The decline of cooperation within the society brings the territorial expansion to a halt (Turchin, 2003). Lower levels of societal coordination interfere with the group's ability to defend its borders and the physical area held by the society contracts. The level of social coordination could potentially stabilize or even slightly increase due to the presence of invaders and the smaller territory size. However, this change is not sufficient to counter the effects of an even swifter dynamic. According to Turchin (2003), the amount of available resources decreases as the society's territory diminishes, which in turn accelerates institutional decline, causing further territorial contraction. A society's inability to halt this feedback loop generates the necessary conditions for collapse (Turchin, 2003). Although intersocietal competition can have a considerable impact on the structural integrity of societies, its occurrence does not imply the inevitable collapse of the factions involved. Instead, the degree to which institutions adequately coordinate collective efforts and enforce cooperation influences the likelihood that societies currently involved in a conflict will eventually collapse. Thus, Turchin's *wheels within wheels* metaphorically represent cycles within cycles, with collapse coming reliably from a coinciding downturn in secular and *asabiyyah cycles*, leading to the eventual loss of sociopolitical complexity of the state (Turchin, 2007).

5 Cultural Innovations, Institutional Robustness, and Resource Availability

Contemporary evolutionary theories no longer describe the evolutionary process as a unidirectional dynamic. Instead, evolution operates as a bidirectional phenomenon wherein selective pressures influencing the frequency of phenotypes in a population interact with the actions of individuals and collectives, both of which are able to alter local environmental conditions through the process of *niche construction*[12] (Laland & O'Brien, 2011; Laland, Odling-Smee, & Feldman, 2001, 2005; Odling-Smee, Laland, & Feldman, 2003). Even though in the long run such modifications can have lasting fitness-enhancing effects, altering local ecologies can also have considerable fitness-reducing consequences (Odling-Smee et al., 2003). In human societies, niche construction is facilitated by our species' ability to accumulate socially transmitted knowledge while employing this information to adequately respond to environmental conditions, as specifically denoted by the term *cultural niche construction* (Odling-Smee et al., 2003). To illustrate the interplay between cultural selection, genetic selection, and ecology, consider the example of individuals inhabiting a collective, who have at their disposal an array of cultural variants obtained from other individuals within the group (Hoppitt & Laland, 2013). A subcluster of these variants provides some information regarding how the adopter should optimally interact with the ecology. At first, individuals, or collectives, use cultural information from the material culture to modify their ecology, resulting in changed selective pressures to which that individual or collective is subject, despite being the agent of change. In turn, these ecological alterations encourage further cultural evolution, allowing the system to respond to novel environmental circumstances, denoted by the phrase *culturally modified cultural selection*

(Odling-Smee et al., 2003). If the system is unable to generate cultural innovations and respond accordingly to these environmental alterations, then the selective pressures are predicted to operate on the system's gene pool (Odling-Smee et al., 2003).

As proposed in a recent book chapter on the work of Alfred Crosby (Hertler, Figueredo, Peñaherrera-Aguirre, Fernandes, & Woodley of Menie, 2018), a successful human biocultural group that has made an innovation in subsistence technology is expected to expand its geographic range to the limits of the ecological niche within which that novel subsistence technology confers a competitive advantage. This process occurs because human biocultural groups engage in active *niche construction*, coevolving with other species of animals and plants to create *symbiotic portmanteau assemblages* (SPAs) that support higher carrying capacities within the same physical habitat. These SPAs are limited in ecological hyperspace to the specific range of ambient conditions under which such niches can be constructed out of the raw materials provided by the native ecologies being encroached upon. During such SPA expansions, the human biocultural group may incorporate subordinate human biocultural groups and exchange nonhuman SPA elements with them, resulting in richer species assemblages within the constructed niche, just as the human groups often hybridize and augment the human genetic diversity of the new aggregate. This process of constructed niche enrichment assists in the further expansion of the culturally constructed species assemblages.

By so doing, however, the previously dominant groups wind up exporting their superior subsistence technology to the incorporated subordinate groups and, through trade and imitation, exchanging some SPA elements with the surrounding ones that have not been fully merged with the new complex. As a result, the dominant group may gradually lose its competitive advantage with respect to these other groups and, with that, its ability to dominate them socially. When innovations in subsistence technology were spread to the "barbarian" (originally meaning *foreign*) nations surrounding the late Roman Empire, these enabled them to eventually achieve a level of agricultural parity with Roman Italy itself. For example, the Roman introduction of agricultural technologies, such as the heavy wheeled mouldboard plough in the late third and fourth centuries AD that helped till otherwise difficult Northern European soils

(Margaritis & Jones, 2008), facilitated a new balance of economic power between the provinces and the capital. This generalization of the technologically constructed ecological niche thus sets the stage for the collapse of the dominant group as the central authority in the expanded sociopolitical structure.

Even though the theory of cultural niche construction by no means implies this phenomenon is the sole source of cultural evolution (cf. Chap. 3 of this volume), this framework counters alternative theories claiming that cultural innovations and ecology occur as independent phenomena. Demographic evidence, mathematical models, and computational simulations support the fact that the rate of technological innovations is constrained in part by socioecological conditions. As indicated by global paleodemographic and macroeconomic reconstructions, the temporal trajectory of world GPD per capita remained relatively horizontal and unaltered throughout most of human history after the Neolithic Revolution (De Long, 1998). Reconstructions of global population growth reflect a similar trend, with occasional fluctuations without evidencing any sudden expansion (Artzrouni & Komlos, 1985). Before the 1800s, any increase in macroeconomic growth led to a corresponding enlargement in population size, restricting any increment in the global productivity per capita (Currie et al., 2016).

More recently, Currie et al. (2016) developed a mathematical model examining the coevolution between population size (*N*) and the total amount of useful knowledge in a society (meaning the society's technologies and institutions: *T*). The model tracked fluctuations in *N* and *T* through a period of 10,000 years. Even though *N* remained initially under the ecology's carrying capacity (*K*), ultimately *N* reached *K*. The authors modeled *T* as an autocatalytic process, wherein the creation of new technologies and institutions depended on the preexisting amount of *T*. Assuming this to be the case, the authors derived a mathematical equivalency between *K* and *T*. The simulation showed *N* generally outgrowing *T*, even as, over most of the model's timespan, *N* and *T* grew at a slow rate. This imbalance occurred due to the small number of innovators[13] in a population already at carrying capacity. In the model, population growth also consumed any additional benefits attained through the slow accumulation of novel technologies. This interaction allowed the authors to assess the amount of surplus based on the ratio *T*/*N*, an indicator of per capita consumption. When *N* outgrew *T*, the population experienced lower fertility rates, higher mortality rates, and an overall decline in size (i.e., Malthusian regime; Currie et al., 2016). Although throughout most of the model's time, the covariance between *N* and *T* complied with Malthus' predictions, eventually, the society's *T* managed to surpass *N*, allowing it to escape from this Malthusian trap, a process that took at least 9500 years.

Independent examinations have reached similar conclusions. Turchin and Nefedov (2009) modeled the association between population size, total production of resources, and the amount of surplus generated by pre-industrial agricultural societies. The authors' model identified a linear relation between population growth and the resources allocated to subsistence. In contrast, surplus production followed a curvilinear trend with larger populations producing smaller quantities of surplus. In this representation, carrying capacity equaled the intersection between the subsistence line and the surplus curve depending on population growth. In a subsequent model, the authors determined that the quantity of surplus is zero when population density is zero and when the system reaches the ecology's carrying capacity. In a previous publication, Turchin (2003) concluded that

> Unless the population size can somehow be prevented from crossing the N_{crit} threshold, the state's expenses will inexorably grow beyond its means, and the state will inevitably become insolvent. Once this point is reached, increasing the tax rate or cutting expenses on nonessentials like court luxuries, can at most be a short-term term solution. (p. 126)

Although other forces are involved in cultural evolution, ecological and population factors constrain a society's cultural evolution and the diversity of its institutional repertoires. These restrictions have clear implications for the system's capacity to address either internal or external pressures based on the socioecological conditions necessary for cultural and institutional evolution. For example, Bednar (2016) refers to an institution's *robustness* as the ability to retain its functionality despite the occurrence of perturbations. Through this lens, Bednar and Page (2016) consider that robust systems are: (1) *diverse*, exhibiting a constant influx of novel information; (2) *modular*, with information compartmentalized and processed by the system's components; and (3) *redundant*, in case one component is unable to function adequately, another component may act as a fail-safe, performing the same task. Institutional robustness, however, is not equivalent to stability. For instance, Bednar and Page (2016) clarified that the former concept addresses the system's transformation capacity, whereas the latter refers to the system's consistency across contexts. Cultures, and not only institutions, exhibit robustness. Hence, cultures are robust as long as the introduction or modification of cultural variants or the emergence of social agents does not alter the overall integrity of the existing culture (Bednar & Page, 2016). Furthermore, from an empirical point of view, the degree to which a cultural variant remains unaltered during transmission could be employed as a proxy for robustness. In addition to learning biases influencing the likelihood of adopting cultural variants (cf. Chap. 3 of this volume), institutions and cultures spread between groups depending on the adoptees' existing institutional and cultural systems (Spolaore & Wacziarg, 2016).

Bednar and Page's (2016) description is consistent with the definition of robustness employed in evolutionary and developmental biology. For example, Nijhout (2002) defines robustness as a weak correlation between a trait's variation and either genetic or environmental variation, implying a variation in its fitness consequences. Similarly, Bateson and Gluckman (2012) defined this concept as an organism's characteristics remaining unaltered by environmental and genetic changes. According to Bateson and Gluckman (2012), phenotypic robustness is often achieved through various mechanisms. For instance, the system's responsivity is dependent on its ability to detect environmental change. The higher the *insensitivity*, the lower the likelihood the alternation will impact the system's organization. The presence of barriers also buffers the system against external perturbations. Evolutionary, developmental, and ecological constraints also increase robustness. For example, after reaching a particular stage of differentiation, it is not feasible for the system to revert to previous steps without compromising its overall integrity (Bateson & Gluckman, 2012). Relative to older structures, more recent innovations are more susceptible to experience further modifications. Although Bateson and Gluckman

(2012) did not consider the degree of responsiveness of social systems, their descriptions can be productively generalized to this level of analysis. Hence, societies may experience greater difficulty modifying older and more foundational institutions, such as those associated with a subsistence economy. A robust system could also rely on its elasticity, wherein a structure will temporarily alter its form when confronted with external pressures, regaining its original form once the force is removed (Bateson & Gluckman, 2012). For instance, as discussed in the previous chapter, to this day, republics can decree a temporary state of exception, limiting civil liberties in response to an internal or external threat (Giorgio & Kevin, 2005).

6 Summary

The academic controversy associated with the study of collapse continues unabated. To some extent, theoretical integration provides a way forward. Instead of exclusively concentrating on the impact of external factors, such as climatic fluctuations or natural catastrophes, contemporary theories of collapse seek to understand the role of institutional, cultural, and economic dynamics. Similarly, some cyclical models inspired by multilevel selection theory claim that societies experience multiple autocatalytic cycles with collapse resulting from fluctuations in collective action and social solidarity. At the core of these theories lies the assumption that human societies feature considerable levels of behavioral flexibility allowing them to address socioecological challenges. Societies alter their ecology to sustain their level of sociopolitical complexity. In time, however, returns on investment decline, compromising institutional performance. The system's inability to continue to modify its environment reverberates across society. Without either incentives or threat of punishment, coordination and cooperation decline, and this increases the likelihood of further defection. With suboptimal levels of institutional performance and waning collective action, a sociopolitical system is prostrate, exposed to both higher- and lower-order threats. Higher-order threats descend from other complex sociopolitical systems retaining optimal levels of institutional performance and collective action, while lower-order threats emerge from less complex substrata, often showing higher asabiyyah, kinship, and cohesion. These dynamics explain the cyclical nature of history, while the superior competitive abilities of robust complex systems in Red Queen competition with simpler societies explain observable linear trends toward growing complexity through time. Both cyclical and linear trends are best synthesized and understood through the lens of multilevel selection.

Notes

1. Even though multilevel selection theory did not inspire some of the perspectives covered by this chapter, contributions have direct implications for understanding social collapse under a multilevel selection lens.
2. Even though it would be tempting to view this level as individuals, little consensus exists concerning whether the individual is the most elemental social unit. For the purposes of this representation, the dyad/nuclear family is considered as the simplest form of social organization.
3. Regeneration concerns the reconstruction of urban and economic systems, ideologies, and institutions following a decline in political and social centralization (Schwartz, 2010).
4. Holling and Gunderson (2002) viewed exploitative organisms as *r*-selected strategists, exhibiting fast population growth, while organisms corresponding to the conservation phase as *K*-selected strategists, displaying a deaccelerated growth rate.
5. In addition to this bottom-up approach, Holling et al. (2002) also proposed a "remember" connection bridging the conservation phase of the largest cycle to the reorganization phase of the intermediate cycle. Resources accumulated at a higher-order level can be allocated to the reorganization phase of a lower level of the panarchy.
6. Previous perspectives reached a similar conclusion: societies featuring overspecialism, institutional rigidity, or mismatched with its local ecology have a higher likelihood of collapsing (e.g., Flannery, 1972; Renfrew, 1979; Service, 1975).
7. This is a metaphor taken from chemistry where the products of a chemical reaction subsequently act to promote the same reaction that produced them.
8. Turchin (2003) also derived the following formula for the average polity *asabiyyah*:

$$r(A) = \frac{1}{A}\int_{A}^{0} r(x)\,dx = r_0\left(1 - \frac{A}{2b}\right)$$

where A is the size of the polity, $r(x)$ the relative growth rate of *asabiyyah*, and b= corresponds to the breadth of the polity's border,

9. As a complement to the current perspective, future studies could consider *asabiyyah* as a latent construct loading into clear indicators of institutional activity. This approach confers two significant advantages. First, it provides an avenue to empirically quantify the amount of a group's *asabiyyah* beyond the realm of mathematical modeling and computational simulations. Second, it avoids falling into traditional mystical perspectives often associated with fluctuations in a society's collective solidarity though time.
10. To be fair, Turchin (2007) did subsequently address the possible effects upon imperial dissolution of the *asabiyyah* of opposing biocultural groups along meta-ethnic frontiers in Chapter 14, *The End of Empire?*
11. Centuries later, a resurgent Arab nationalism also undermined the hegemony of the Ottoman Empire (Choueiri, 2000), despite religious similarities but exacerbated by ethnolinguistic differences.
12. Niche construction is not limited to an organism's actions reversing or neutralizing environmental pressures (*counteractive* niche construction; Odling-Smee et al., 2003), but it also occurs when an organism initiates a sequence of actions in the environment leading to its permanent or definite alteration (*inceptive* niche construction; Odling-Smee et al., 2003).
13. Previous research has concluded that demographic factors also predict cultural complexity and innovations (Bettencourt, Lobo, Helbing, Kühnert, & West, 2007; Kline & Boyd, 2010; Powell, Shennan, & Thomas, 2009). Similarly, population size predicts cultural extinction rates (Henrich, 2004). Consider, for example, two groups, group A which comprised ten individuals and group B containing one hundred. Suppose in both societies one-tenth of the individuals (i.e., models) are copied by other group members. While the death of one model in B could have no impact in the persistence of a cultural variant, as group B still contains nine other models, the elimination of the only remaining model in A represents the extinction of the cultural variant in the group.

References

Artzrouni, M., & Komlos, J. (1985). Population growth through history and the escape from the Malthusian trap: A homeostatic simulation model. *Genus, 41*(3-4), 21–39.

Bateson, P., & Gluckman, P. (2012). *Plasticity and robustness in development and evolution*. Oxford, UK: Oxford University Press.

Bednar, J. (2016). Robust institutional design: What makes some institutions more adaptable and resilient to changes in their environments than others? In D. S. Wilson & A. Kirman (Eds.), *Complexity and evolution* (pp. 201–238). Cambridge, MA: MIT Press.

Bednar, J., & Page, S. E. (2016). Complex adaptive systems and comparative politics: Modeling the interaction between institutions and culture. *Chinese Political Science Review, 1*(3), 448–471.

Bettencourt, L. M., Lobo, J., Helbing, D., Kühnert, C., & West, G. B. (2007). Growth, innovation, scaling, and the pace of life in cities. *Proceedings of the National Academy of Sciences, 104*(17), 7301–7306.

Choueiri, Y. (2000). *Arab nationalism–a history: Nation and state in the Arab world*. Oxford, UK: Wiley-Blackwell.

Currie, T., Turchin, P., Bednar, J., Richerson, P. J., Schwesinger, G., Steinmo, S., et al. (2016). Evolution of institutions and organizations. In D. S. Wilson & A. Kirman (Eds.), *Complexity and evolution* (pp. 201–238). Cambridge, MA: MIT Press.

Currie, T. E., & Mace, R. (2011). Mode and tempo in the evolution of socio-political organization: Reconciling 'Darwinian' and 'Spencerian' evolutionary approaches in anthropology. *Philosophical Transactions of the Royal Society B: Biological Sciences, 366*(1567), 1108–1117.

De Long, J. B. (1998). *Estimates of world GDP: One million B.C.-present*. Berkley, CA: University of California Press.

Diamond, J. (2005). *Collapse: How societies choose to fail or succeed*. New York: Penguin.

Eisenstadt, S. N. (2017). *The political systems of empires*. New York: Routledge.

Fagan, B. (2009). *Floods, famines, and emperors: El Niño and the fate of civilizations*. New York: Basic Books.

Flannery, K. V. (1972). The cultural evolution of civilizations. *Annual Review of Ecology and Systematics, 3*(1), 399–426.

Giorgio, A., & Kevin, A. (2005). *State of exception*. Chicago: The University of Chicago Press.

Haug, G. H., Günther, D., Peterson, L. C., Sigman, D. M., Hughen, K. A., & Aeschlimann, B. (2003). Climate and the collapse of Maya civilization. *Science, 299*(5613), 1731–1735.

Henrich, J. (2004). Demography and cultural evolution: How adaptive cultural processes can produce maladaptive losses—The Tasmanian case. *American Antiquity, 69*(2), 197–214.

Hertler, S. C., Figueredo, A. J., Peñaherrera-Aguirre, M., Fernandes, H. B. F., & Woodley of Menie, M. A. (2018). *Life history evolution: A biological meta-theory for the social sciences*. New York: Palgrave Macmillan.

Holling, C. S., & Gunderson, L. H. (2002). Resilience and adaptive cycles. In L. H. Gunderson & C. S. Holling (Eds.), *Panarchy: Understanding transformations in human and natural systems* (pp. 25–62). Washington, DC: Island Press.

Holling, C. S., Gunderson, L. H., & Peterson, G. D. (2002). Sustainability and panarchies. In L. H. Gunderson & C. S. Holling (Eds.), *Panarchy: Understanding transformations in human and natural systems* (pp. 63–102). Washington, DC: Island Press.

Hoppitt, W., & Laland, K. N. (2013). *Social learning: An introduction to mechanisms, methods, and models*. Princeton, NJ: Princeton University Press.

Huntington, E. (1922). *Civilization and climate*. New Haven, CT: Yale University Press.

Kaniewski, D., Guiot, J., & Van Campo, E. (2015). Drought and societal collapse 3200 years ago in the Eastern Mediterranean: A review. *Wiley Interdisciplinary Reviews: Climate Change, 6*(4), 369–382.

Kennett, D. J., Breitenbach, S. F., Aquino, V. V., Asmerom, Y., Awe, J., Baldini, J. U., et al. (2012). Development and disintegration of Maya political systems in response to climate change. *Science, 338*(6108), 788–791.

Khaldun, I. (1377/1969). In N. J. Dawood (Ed.), *The Muqaddimah: An introduction to history*. Princeton, NJ: Princeton University Press.

Khouri, R. O. (2007). Heresies in the early Byzantine Empire: Imperial policies and the Arab conquest of the Near East. *Collectanea Christiana Orientalia, 4*, 109–117.

Kline, M. A., & Boyd, R. (2010). Population size predicts technological complexity in Oceania. *Proceedings of the Royal Society B: Biological Sciences, 277*(1693), 2559–2564.

Korotayev, A. V., & Tsirel, S. V. (2010). A spectral analysis of world GDP dynamics: Kondratieff waves, Kuznets swings, Juglar and Kitchin cycles in global economic development, and the 2008–2009 economic crisis. *Structure and Dynamics, 4*(1), 3–57.

Laland, K. N., & O'Brien, M. J. (2011). Cultural niche construction: An introduction. *Biological Theory, 6*(3), 191–202.

Laland, K. N., Odling-Smee, J., & Feldman, M. W. (2001). Cultural niche construction and human evolution. *Journal of Evolutionary Biology, 14*(1), 22–33.

Laland, K. N., Odling-Smee, J., & Feldman, M. W. (2005). On the breadth and significance of niche construction: A reply to Griffiths, Okasha, and Sterelny. *Biology and Philosophy, 20*(1), 37–55.

Margaritis, E., & Jones, M. K. (2008). Greek and Roman agriculture. In J. P. Oleson (Ed.), *The Oxford handbook of engineering and technology in the classical world* (pp. 158–174). Oxford, UK: Oxford University Press.

McNeill, W. H. (1998). *Plagues and peoples*. New York: Anchor.

Middleton, G. D. (2017). *Understanding collapse: Ancient history and modern myths*. Cambridge, UK: Cambridge University Press.

Nijhout, H. F. (2002). The nature of robustness in development. *BioEssays, 24*(6), 553–563.

Odling-Smee, F. J., Laland, K. N., & Feldman, M. W. (2003). *Niche construction: The neglected process in evolution (MPB-37)* (Vol. 61). Princeton, NJ: Princeton University Press.

Powell, A., Shennan, S., & Thomas, M. G. (2009). Late Pleistocene demography and the appearance of modern human behavior. *Science, 324*(5932), 1298–1301.

Renfrew, C. (1979). Systems collapse as social transformation: Catastrophe and anastrophe in early state societies. In C. Renfrew & K. L. Cooke (Eds.), *Transformations: Mathematical approaches to culture change* (pp. 481–506). New York: Academic Press.

Schwartz, G. M. (2010). From collapse to regeneration. In G. M. Schwartz & J. J. Nichols (Eds.), *After collapse: The regeneration of complex societies* (pp. 3–18). Tucson, AZ: The University of Arizona Press.

Service, E. R. (1975). *Origins of the state and civilization*. Toronto, ON: W.W. Norton.

Shapur Shahbazi, A. (2005). Sasanian dynasty, *Encyclopædia Iranica*, online edition. Retrieved from: http://www.iranicaonline.org/articles/sasanian-dynasty. Accessed 23 Nov 2019.

Soltis, J., Boyd, R., & Richerson, P. J. (1995). Can group-functional behaviors evolve by cultural group selection?: An empirical test. *Current Anthropology, 36*(3), 473–494.

Spolaore, E., & Wacziarg, R. (2016). The diffusion of institutions. In D. S. Wilson & A. Kirman (Eds.), *Complexity and evolution* (pp. 147–166). Cambridge, MA: MIT Press.

Tainter, J. (1988). *The collapse of complex societies*. Cambridge, UK: Cambridge University Press.

Tainter, J. A. (2004). Plotting the downfall of society. *Nature, 427*(6974), 488–489.

Tainter, J. A. (2006). Social complexity and sustainability. *Ecological Complexity, 3*(2), 91–103.

Turchin, P. (2003). *Historical dynamics: Why states rise and fall*. Princeton, NJ: Princeton University Press.

Turchin, P. (2007). *War and peace and war: The rise and fall of empires*. New York: Penguin.

Turchin, P., & Nefedov, S. A. (2009). *Secular cycles*. Princeton, NJ: Princeton University Press.

Woodley of Menie, M. A., Figueredo, A. J., Sarraf, M. A., Hertler, S., Fernandes, H. B. F., & Aguirre, M. P. (2017). The rhythm of the West: A biohistory of the modern era, Ad 1600 to present. *Journal of Social, Political and Economic Studies Monograph Series, 37*.

8

Chimpanzee Intercommunity Conflict: Fitness Outcomes, Power Imbalances, and Multilevel Selection

Mateo Peñaherrera-Aguirre, Aurelio José Figueredo, and Steven C. Hertler

1 Introduction

This chapter is the first of five comprising Part III. Though, as in Chap. 6, we have previously allowed some lexical analyses to interpolate Part II's historical-empirical thrust, Part III is predominately statistical-empirical, even as it continues to review relevant literature and history. Although, consistent with the mandate of this monograph, we aim ultimately to establish the reality of human group selection, this initial chapter alone treats the subject of intergroup conflict in chimpanzees. To thoroughgoing evolutionists, the relevance will be self-evident; we only add that establishing evidence of multilevel selection in such a highly related species foundationally supports the empirical argument for human multilevel selection, as presented in the four subsequent chapters constituting Part III of this volume (for behavioral differences between these taxa see Chap. 9 in this volume, as well as Glowacki, Wilson, & Wrangham, 2017).

Contrary to some theoretical perspectives claiming that lethal violence between groups is a phenomenon restricted to contemporary nation-states, current literature indicates that aggressive intergroup competition

S. C. Hertler et al., *Multilevel Selection*, https://doi.org/10.1007/978-3-030-49520-6_8

is quite common across nonhuman clades (Kitchen & Beehner, 2007). In the last four decades, researchers have reported several instances of intergroup killings in gray wolves, cheetahs, hyenas, spider monkeys, and lions (Wrangham, 1999). Presumably due to the implications for understanding the evolutionary origins of warfare, these reports have generated heated debates concerning the nature of lethal intergroup aggression in nonhuman species. This is especially the case for publications addressing the various socioecological correlates of lethal intercommunity interactions in common chimpanzees (*Pan troglodytes*).[1] Although comparative data indicate that between-group killings emerged as an evolutionary adaptation, little agreement exists regarding the potential benefits obtained by attackers (for example, territorial expansion, recruitment of females, and elimination of sexual competitors). Similarly, some researchers have hypothesized that the social organization of chimpanzee communities enables attackers to raid and eliminate rivals without facing onerous costs. Despite disagreements, explanatory models reliably neglect multilevel selection, describing chimpanzee intercommunity aggression as an adaptation in terms of traditional, individual selectionism. This restrictive theoretical assumption, however, has limited the possibility of examining the persistence of chimpanzee intergroup competition due to multilevel selective pressures. Hence, in addition to providing the reader with an overview on this subject, the present chapter offers empirical evidence of multilevel selection operating on the number of male chimpanzee patrols.

2 Chimpanzee Intercommunity Conflict from an Adaptationist Perspective

During the early 1970s, the Kasekela community at Gombe, Tanzania, underwent a demographic fission (Feldblum, Manfredi, Gilby, & Pusey, 2018). While the original group remained in the north, the newly formed community of Kahama, comprising six mature males, one adolescent male, and three females, occupied the southern valley (Goodall, 1986; Williams et al., 2008). Intercommunity tolerance was short-lived. In 1974, Kasekela initiated a series of attacks against Kahama, reducing the southern group from a range of 10 km^2 to 1.8 km^2 (Goodall, 1986).

During this time, Kahama also suffered from the incursions of Kalande, another chimpanzee community in the south (Goodall, 1986). Three years after the first attack of Kasekela, Kahama finally collapsed. Kasekela immediately seized the abandoned regions, expanding its range to 17 km^2. Though this could have been the beginning of a period of further territorial expansion for Kasekela, this trend eventually reversed its course. Kahama served as a buffer between Kasekela and the southern communities (Goodall, 1986). Even though the identity of the aggressors is not known (perhaps chimpanzees from the Kalande community), Kasekela became the target of several attacks leading to multiple casualties (Goodall, 1986). The southern conflict led to a considerable reduction in Kasekela's territory, decreasing it to 9.6 km^2 in 1981 (Goodall, 1986). Kasekela and Kahama's territorial fluctuations evidenced the impact of intercommunity conflict on population stability. Detailed examinations of Kasekela's mortality patterns across forty-seven years concluded that out of eighty-six deaths with known causes, seventeen were the product of intraspecific aggression, with eight of these resulting from lethal intergroup interactions (Williams et al., 2008; Wilson, 2013). Moreover, half of the twelve Kahama and Kasekela males (aged between twenty and thirty) who died during the observation period were known or suspected to have been killed during intercommunity attacks (Williams et al., 2008). Albeit it is presumed that some of the females who disappeared during this time could have also been the target of foreign chimpanzees raiding the territory, only two such attacks were directly observed (Williams et al., 2008).[2]

Although intergroup killings were thought at first to be exclusive of the chimpanzees at Gombe, independent observations conducted at Kibale National Park, Uganda (Watts et al., 2006), provided yet another detailed account of intercommunity aggression and territorial expansion. Between the years of 1999 and 2008, chimpanzees at the Ngogo community eliminated eighteen foreign rivals (Mitani, Watts, & Amsler, 2010). Most of these attacks ($n = 13$) occurred during patrols close to the northeastern border of the Ngogo community. Even though Mitani et al. (2010) did not have an exact count of the number of individuals in the Northeastern community, assuming the targeted group was equivalent in size to other chimpanzee unit-groups (e.g., 47 individuals), the Northeastern faction experienced a death rate of 2790 per 100,000 per year, according to their calculations. Presuming Ngogo and Northeastern communities had similar group sizes (~150 chimpanzees), this number still implies a rate of 867 per 100,000 per year, an estimate that exceeds the killing rate experienced by some small-scale human societies (Mitani et al., 2010). Echoing Gombe's intergroup killings, the confrontations at Ngogo also generated significant territorial changes. By 2009, Ngogo chimpanzees acquired 6.4 km^2 from their Northeastern rivals, representing a 22% territorial expansion.[3]

In contrast to Ngogo, another community at Kibale, Kanyawara, underwent a 46.8% range contraction, falling from 29.5 km^2 to 13.8 km^2 in 8 years (Wilson, Kahlenberg, Wells, & Wrangham, 2012). Except for the suspected death of three adult males[4] (Wilson et al., 2014; Wrangham, Wilson, & Muller, 2006), over 80% of the 120 intercommunity contacts between 1992 and 2006 were limited to acoustic displays (Wilson et al., 2014). These events often occurred at the borders of Kanyawara's territory, within a range of 288 to 4406 meters away from the community's center (Wilson et al., 2012). Despite the low death rate, the threat of between-group conflict eventually forced Kanyawara chimpanzees to avoid regions where they tended to encounter the opposing parties (Wilson & Glowacki, 2017).

Within the fields of primatology and physical anthropology, reports of lethal aggression in chimpanzees, such as the cases of Kahama, Kasekela, Ngogo, and Kanyawara, generated an array of responses ranging from scientific curiosity to skepticism (Power, 1991; Sussman, 2013). Although their arguments varied, the views of skeptics can be classed under the general umbrella of *human impact hypotheses* (HIH), an array of perspectives denying the adaptive function of lethal aggression, and instead explaining these killings as a product of human activities, such as food provisioning or habitat degradation. According to Wilson et al. (2014), the premises of HIH can be summarized by the following predictions:

1. The killing rates should not significantly differ between chimpanzees and bonobos, with the severity of ecological impact superseding any cladistic effect.
2. Due to the increasing rate of environmental disturbance over time, killings should also exhibit a positive temporal trend.

3. There should be no significant difference in the frequency of lethal attacks perpetrated by males or females.
4. Both males and females should be equally represented as the victims of the attacks.
5. There should be no noticeable difference regarding the age of the victims.
6. Genetic relatedness between the attackers and the victims should not have any influence on the rate of killings.
7. There should be no numerical asymmetries between the attackers and the victims during the attack.

Other researchers, remaining unconvinced that chimpanzee intercommunity aggression was the product of human disturbance, suggested that this behavior instead evolved as an adaptation to natural conditions. Following Wilson et al. (2014), the predictions of the *adaptive strategies hypotheses* (ASH) can be summarized as follows:

1. Chimpanzees should display higher killing rates relative to bonobos.
2. Even if human ecological encroachment increases over time, killing rates should not be affected by these temporal changes.
3. Males should be the perpetrators of the attacks more often than females.
4. Males should be the victims of attacks more often than females.
5. Relative to adults, younglings are predicted to be at higher risk of being killed.
6. Attackers are expected to preferentially kill individuals more distantly related to them, such as members of other communities.
7. Perpetrators are expected to outnumber the victims during attacks.

Despite the preponderance of evidence tending to disconfirm the HIH, researchers supporting this alternative hypothesis remained unconvinced. Due to persistence of the HIH in the literature, Wilson et al. (2014) collected data on intergroup killings from eighteen chimpanzee and four bonobo communities from various online databases and publications.[5] Their study also gathered information on the dimensions of the protected area, the presence of ecological disturbance, the number of adult males in the community, the number of animals per km^2 (not associated with human environmental disturbance), the location of the community in Eastern or Western Africa, and whether or not the researchers artificially provisioned the apes. The authors reported 152 killings, including inferred and suspected fatal attacks, in 15 of the 18 chimpanzee communities. Model comparisons determined that those including community density and number of males as predictors best fitted the data. More detailed examinations concluded that even though females occasionally killed other individuals, males were more often the perpetrators of lethal attacks. Similarly, a generalized linear mixed model estimated that infants and adult males had a higher probability of being victims of these attacks. With respect to intercommunity conflict, over 60% of the 99 victims were killed by members of a different community, suggesting an inclination toward targeting either unrelated or distantly related individuals. Hence, the data favors ASH over HIH across a variety of sites.

3 Intercommunity Conflict and Individual Fitness Outcomes

Even though cross-regional examinations indicate that intercommunity killings are not a product of anthropogenic factors, there is little agreement among researchers endorsing ASH on the fitness benefits attained from intercommunity incursions. According to Wilson (2013), some of the hypothesized benefits include restricting foreign males from copulating with resident females, accessing feeding grounds, defending themselves or others from an attack, and encouraging the migration of foreign females into the community. Evaluating the *female acquisition hypothesis*, lethal and nonlethal intercommunity aggression seems to encourage females to abandon their group and migrate into the attacker's community. The case of the K-group community at Mahale Mountains in Tanzania offers compelling evidence. In seventeen years, the K-group community went from being demographically stable to experiencing the systematic disappearance of its males (Nishida, Hiraiwa-Hasegawa, Hasegawa, & Takahata, 1985). This demographic change encouraged all cycling females to associate with males from a rival community (M-group; Nishida et al., 1985). Eventually, at least by 1983, the K-group was reduced to three females and one adolescent male. Although M-group males were not observed killing K-group males (Mitani et al, 2010; Nishida et al., 1985), demographic estimations for the 1966–1999 period indicate that 3.8% of the total number of deaths could have been the product of intergroup conflict (Nishida et al., 2003; Wilson, 2013). In addition to offering a unique glimpse into female dispersal patterns after male disappearance, K-group's fate provides evidence of community extinction occurring without the complete elimination of the chimpanzee population (Nishida et al., 1985). Besides the latter case, support for the female recruitment hypotheses emerges from the observed differential treatment of foreign females. Researchers have reported that females exhibiting signs of sexual receptivity, such as anogenital swellings, are less vulnerable to intergroup aggression (Nishida et al., 1985; Williams, Oehlert, Carlis, & Pusey, 2004). In contrast, non-swollen females were more likely to suffer from intercommunity attacks.[6] Furthermore, individual differences, such as the female's age as well as her offspring number, increased the risk of experiencing an aggressive encounter (Williams et al., 2004). It is worth noting that socioecological factors could mediate these dynamics. For instance, mathematical modeling has predicted that males inhabiting groups with low reproductive skew should be more inclined to attack foreign females (Pradhan, Pandit, & Van Schaik, 2014).

Communities could also benefit from territorial expansion by accessing coveted feeding grounds (Wilson, 2013). Resource acquisition could impact the life history of females and the group's social dynamics. Researchers at Gombe analyzed data collected over eighteen years of observation to determine the association between community range size and several demographic indicators (Williams et al., 2004). Although community range size did not have any influence on the total number of adult males nor on the number of adult females (for a more recent take on the association between territory size, group size, and number of males, see Lemoine et al., 2020a; likewise, see Lemoine et al., 2020b, for a study on the variation in female reproductive success due to between-group differences in number of males), it did predict the time that males interacted in mixed-sex parties, as well as the size of mixed-sex groups (Williams et al., 2004). Furthermore, these analyses concluded that a larger home range decreased the females' interbirth intervals (Williams et al., 2004). A more recent perspective, the *group augmentation hypothesis*, also argued in favor of indirect as well as direct benefits obtained by individuals. According to Langergraber, Watts, Vigilant, and Mitani (2017), males could be more inclined to patrol, depending on the influence of several sociodemographic factors. Although no immediate benefits are obtained if group size increases, such augmentation could provide a positive effect on the males' future reproduction (Langergraber et al., 2017). Relying on a generalized linear mixed model, the authors examined the effects of paternity success, dominance rank, age, maternal relatedness, and male group size, on the total number of male patrols. The model detected that only paternity success and the males' rank positively predicted patrol participation (Langergraber et al., 2017). In terms of long-term reproductive success, Langergraber et al. (2017) identified that most males who did not have any offspring when they joined patrol parties would eventually sire offspring. These results further support the hypothesis that males could obtain delayed fitness benefits by providing immediate service to the community.

4 Intergroup Killings and Power Imbalances in Chimpanzees

Concentrating exclusively on the benefits of intercommunity killings provides a partial perspective of the adaptive nature associated with this behavior. Some authors have argued that the accrued toll associated with attacking another group will also influence the prevalence of this behavioral phenotype. According to the imbalance of power hypothesis, raiding a rival community imposes several costs to attackers, including those entailed from spending time and energy patrolling the territory to the risk of suffering injury and death during the incursion (Manson & Wrangham, 1991; Wilson, 2013; Wrangham, 1999). Raiding parties, however, decrease exposure to lesions and death by targeting individuals who are either foraging or traveling alone (Manson & Wrangham, 1991; Pandit et al., 2016; Wilson, 2013). Furthermore, animal species living in societies governed by fission-fusion dynamics are expected to be vulnerable to lethal intergroup conflict (Wrangham, 1999). The outcome of the raid depends on the numerical asymmetry between the number of attackers and the number of defenders (Wilson, 2013; Wilson et al., 2014). The hypothesis, however, does not imply that a numerical asymmetry will decrease the cost to zero, as raiders will still spend a portion of their bioenergetic budget reaching the target (Amsler, 2010; Wilson, 2013); instead, it offers a framework for studying the likelihood that the raiding party will physically engage a target once they encounter it.

Empirical evidence agrees with the imbalance of power hypothesis. A field experiment conducted with three chimpanzee communities in the Taï National Park in Cote d'Ivoire detected significant variation in vocal and locomotor behavior in response to a recording simulating the presence of a nearby chimpanzee (Herbinger, Papworth, Boesch, & Zuberbühler, 2009). Researchers played three simulated *pant-hoots*, the first recorded from one of the males in the focal community; a second recorded in a neighboring community; and a third recorded in a community 70 km away. Focal chimpanzees in the *member* condition had a higher frequency of pant-hoot vocalization relative to both the *neighbor* and the *stranger* conditions. Herbinger et al. (2009) determined that the number of males present influenced the likelihood of responding to the recordings. The chimpanzees' reactions extended to other behavioral dimensions; for instance, patrolling increased during the playback of strangers and neighbors (Herbinger et al., 2009).

In a similar experiment at Kanyawara, the probability of vocalizing in response to a recording increased with the number of males in the party (Wilson, Hauser, & Wrangham, 2001). The number of males present also increased the probability of approaching the speaker. No significant relation existed between the male's agonistic rank and the mean approach rank to the speaker. Although further examinations should consider the role of individual differences in counter-calling and patrolling, current evidence indicates that both high-ranking and low-ranking males are more likely to respond to the presence of a foreign rival depending on the number of males in the party (Wilson et al., 2001). Inspired by Lanchester's theory of conflict (1916),[7] researchers examined the relevance of chimpanzee numerical assessment during intercommunity conflict (Wilson et al., 2002). By adding the data collected during the playback experiments into derived equations, these researchers estimated that a party of adult male chimpanzees (A) would engage a rival group (B) if A is 1.5 times larger than B (Wilson et al., 2002).

Observational data at Kanyawara provides further evidence. Several logistic regression models analyzed the effect of the number of adult males, the number of females in estrus, number of infants, the distance from the range center, and the food value of the disputed resource (as indicated by the proportion of forage time spent in the location where the encounter occurred) on the probability of counter-calling, as well as on the probability of approaching rivals (Wilson et al., 2012). Male chimpanzees at Kanyawara were more likely to vocalize toward foreign rivals, depending on the number of adult males in the group (Wilson et al., 2012). Model-averaged parameter estimates identified that the number of males present had a significant positive effect on the probability of approaching foreign rivals. The number of infants in the group did not have a significant effect. Neither the distance from the center nor the food value had any significant effect on the probability of approaching the intruders. The local conditions of the encounter area also seemed to be unrelated to engaging rivals. Instead, the numerical asymmetry has a significant influence on the direction and escalation of the conflict. Moreover, the number of estrous females had a negative effect on engaging rivals. According to Wilson et al. (2012), adult males face a trade-off: either defend the range or mate-guard females with sexual swellings. Mate guarding not only reduces the likelihood that the female will copulate with males from the neighboring groups, but also limits the risk of copulations between the female and interloping males from her own group (Wilson et al., 2012). This pattern, however, does not generalize across chimpanzee communities. Mitani and Watts (2005) scrutinized the influence of various socioecological indicators on patrolling through a series of logistic regressions. The analyses included the size of the male party, the presence of estrous females, the availability of fruit, reports of chimpanzee hunting behavior during the patrol, and if they made vocal or visual contact with competing factions. While male party size and fruit availability positively predicted male patrols, the presence of estrous females did not have a significant effect.

5 Chimpanzee Intercommunity Conflict and Multilevel Selection

Before we proceed with our examination of multilevel selection in chimpanzee intercommunity conflict, it is essential to provide the reader with a review of key concepts and methodologies associated with multilevel selection theory (MLS), only alluded to previously. Damuth and Heisler (1988) distinguished between two types of multilevel selection: MLS1 and MLS2. According to these authors, MLS1 can be said to occur when:

1. Group selection is operationalized as the effects of group membership on individual-level fitness;
2. Fitness is defined as limited to individuals;
3. Characters or traits are defined as restricted to individuals;
4. Populations are comprised of individuals, and are classified into groups;
5. Explicit evolutionary inferences are limited to the observed variations in frequency among different types of individuals in the population.

Alternatively, MLS2 is characterized by:

1. Group selection is operationalized as the variations in frequency among different types of groups;
2. Groups exhibit differential fitness;
3. Groups feature variations in characters or traits;
4. Populations contain groups, which in turn are comprised of individuals;
5. Explicit evolutionary inferences referring to changes in the frequencies of different types of groups within a population.

More recently, authors such as Okasha (2006) further explored the distinction between MLS1 and MLS2 by reconsidering how fitness is defined. For Okasha, the group's fitness is calculated in MLS1 as the average individual fitness across all individuals within the group. Alternatively, MLS2 operationalizes the collective's fitness as the number of offspring *groups* sired by each group. Even though, at first glance, the MLS1/MLS2 distinction seems to mirror the difference between aggregate and emergent properties, this is not always the case. In MLS1, for example, it is feasible that in addition to a group's *aggregate trait*, which is computed as an average based on the individual-level data, a group's *emergent property*, which denotes a trait that is irreducible to individual characteristics, could significantly predict the individuals' fitness (Okasha, 2006). Similarly, according to Okasha (2006), MLS2 is not limited to the covariation between emergent group-level traits and the number of offspring groups. Aggregate traits may also have a significant effect on the probability of group reproduction.

Distinguishing between types of MLS is just the first step before conducting an examination based on MLS theory. To explore MLS hypotheses, it is necessary to employ an analytic procedure design to detect both individual- and group-level effects. Developed in 1987 by Heisler and Damuth, *contextual analysis* emerged as an extension of regression models wherein individual- and group-level traits are viewed as predictors of individual-level outcomes. Group traits, also known as contextual characters, may differ depending on the estimation procedure. While an aggregate character is often calculated from the individual data (e.g., as an average), a global character refers to a unique property of the collective irreducible to the characteristics of individuals within the group. Hence, according to these authors, MLS1 assumes that selection operates not only upon the individual but also on contextual characters. Contextual analysis is best represented by the following regression equation (Eq. 8.1):

$$w_{ij} - w.. = \beta_I \left(z_{ij} - z.. \right) + \beta_C \left(z_{i.} - z.. \right) + \varepsilon_{ij} \tag{8.1}$$

where w_{ij} is the individual level of fitness, $w..$ is the average fitness across all individuals, z_{ij} corresponds to the individual-level trait, $z..$ is the average trait value across all individuals, z_i is the average trait value for each group, and ε_{ij} is the equation's error term. Hence, β_I is the partial regression coefficient between the individual-level trait and the individual-level fitness, after controlling for the average phenotype at the level of the group. Alternatively, β_C is the partial regression coefficient between the mean phenotype at the level of the group and the individual-level fitness, after controlling for individual-level effects. Heisler and Damuth argued that any evidence of group selection requires β_C to be significantly different from zero. The flexibility of contextual analyses allows the inclusion of multiple predictors into the equation. For example, a model with two traits z_1 and z_2 generates the following equation (Eq. 8.2):

$$\begin{aligned} w_{ij} - w_{..} = \beta_{I1}\left(z_{1ij} - z_{1..}\right) + \beta_{I2}\left(z_{2ij} - z_{2..}\right) \\ + \beta_{C1}\left(z_{1i.} - z_{1..}\right) + \beta_{C2}\left(z_{2i.} - z_{2..}\right) + \varepsilon_{ij} \end{aligned} \tag{8.2}$$

It is worth noting, however, that analogous to other statistical procedures employed to examine multilevel selection, contextual analyses operate under a set of specifiable conditions. Okasha (2004) summarizes these elements as follows:

1. Group selection does not require fitness variation between groups, meaning that *soft selection* can occur as long as β_C is nonzero;
2. Group selection depends on global trait variation between groups;
3. Group selection does not rely on the nonrandom formation of groups;
4. Group selection does not require individual fitness to be group-dependent;
5. Individual-level selection depends on within-group fitness variation.

Critics have argued that the lack of group reproductive isolation and the occurrence of migration between groups violate the conditions that are presumably necessary for group selection to occur (Wrangham & Glowacki, 2012). Though this assumption was indeed one of the tenets of naïve group selection theory, researchers favoring a more contemporary multilevel selection view of evolution argue that the *trait-group* rather than the *deme* is the operative level of group selection, such that demographic isolation is no longer considered a necessity for group selection to operate (Sober & Wilson, 1998). Similarly, MLS1 describes fitness as a feature of individuals rather than groups (Okasha, 2006). Moreover, even when fitness is estimated at the level of the group, it is equal to the average fitness of all individuals within that collective. The distinction between MSL1 and MLS2 has not only theoretical but methodological implications. For example, the multilevel selection hypotheses generated for this chapter were developed under the premise that chimpanzee groups provide a context for individuals to replicate their genes (MLS1). Having provided this theoretical and methodological overview, we can proceed to describe the contextual analyses conducted on male chimpanzee patrolling behavior. For an illustration in this chapter, we used Langergraber et al.'s (2017, 2018) Ngogo database of chimpanzee territorial behavior to assess whether the aggregate frequency of patrolling in male chimpanzees predicted the observed variation in the reproductive success of male chimpanzees (providing evidence of MLS1). This online resource contains data on the participation of males above 13 years of age across 284 patrols observed between the years of 1996 and 2015. The dataset also includes information on (1) the male's age (estimate computed as a quadratic term); (2) his dominance rank, calculated from the outcome of agonistic interactions; (3) his maternal relatedness, estimated from the number of genetic relatives alive at the time of the patrol; and (4) his paternity success, computed as the sum of the male's genetic relatedness to his offspring that were alive at the time of the patrol.[8]

Prior to conducting the analyses, the participation dataset was reshaped into a transposed matrix[9] with males as columns and patrols as rows.[10] In turn, this array was transformed into a polychoric correlation matrix to be subsequently examined with a principal axis factor analysis. The factor analysis identified thirteen main factors. Horn's parallel analysis supported this number of dimensions. After classifying each male chimpanzee into one of these groups, we proceeded to compute the various variance component terms defined in (Eq. 8.1) (see Table 8.1 for a list of the equations estimated along with the corresponding description).

A general linear model (using Type II sums of squares) revealed that the aggregate number of male participations in patrols ($Patrols_C$) significantly predicted the individual relative fitness ($\beta = 0.456$, $p = 0.012$). In contrast, the number of patrols conducted by each individual ($Patrols_I$) had no significant effect on the individual relative fitness (*Relative* w_I; $\beta = 0.176$, $p = 0.318$). The overall model explained 36% of the variance ($p < 0.000$). A Linear Mixed Model (LMM) with REML, variance components, and Group as a random factor reached similar conclusions ($Patrols_I$: $\beta = 0.176$, $p = .306$; $Patrols_C$: $\beta = 0.447$, $p = 0.018$). A model comparison recommended the inclusion of both individual and

Table 8.1 Glossary of variables names and equations employed in these analyses

Term	Equation	Description
Relative w_I	*Z(wij-w..)*	Standardized *individual's* relative fitness
$Patrols_I$	*Z(zijpatrols-z..patrols)*	Standardized number of patrols at the *individual* level[a]
$RPatrols_I$	*Z(zijRpatrols-z..Rpatrols)*	Standardized number of patrols at the *individual* level[a] relative to the total number of opportunities
Age^2_I	$Z(zijAge^2\text{-}z..Age^2)$	Standardized quadratic transformation of age at the *individual* level[a]
$Dominance_I$	*Z(zijav_rank -z..av_rank)*	Standardized dominance rank at the *individual* level[a]
Maternal r_I	*Z(zijMtrnl_Rlt -z..Mtrnl_Rlt)*	Standardized value for the males' coefficient of maternal relatedness at the *individual* level[a]
$Patrols_C$	*Z(zi patrols -z..patrols)*	Standardized number of patrols at the *aggregate* level[b]
$RPatrols_C$	*Z(zi Rpatrols -z..Rpatrols)*	Standardized number of patrols at the *aggregate* level[b] relative to the total number of opportunities
Age^2_C	$Z(ziAge^2\text{-} z..Age^2)$	Standardized quadratic age at the *aggregate* level[b]
$Dominance_C$	*Z(ziav_rank -z..av_rank)*	Standardized dominance rank at the *aggregate* level[b]
Maternal r_C	*Z(ziMtrnl_Rlt -z..Mtrnl_Rlt)*	Standardized coefficient of maternal relatedness at the *aggregate* level[b]

[a]The *individual* level is defined as the difference between the individual score and the grand mean of each trait, as in a general linear model

[b]The *aggregate* level is defined as the difference between group mean and the grand mean of each trait, as in a general linear model

contextual information, (Only $Patrols_I$: Δ *BIC* = 4.140, *weight* = 0.112; $Patrols_I$ and $Patrols_C$: Δ *BIC* = 0.000, *weight* = 0.888). It was pertinent to further explore the latter results based on males' relative patrol participation ($RPatrols_I$), which are the number of times a male joined a patrol party divided by the total number of patrol opportunities available. The results remained unaltered. The aggregate estimate of relative patrolling ($RPatrols_C$) significantly predicted the individual-level relative fitness (*Relative* w_I; β = 0.582, p = 0.004). In contrast, the relative individual-level values of patrolling ($RPatrols_I$) had no significant effect on the relative fitness (*Relative* w_I; β = −0.056, p = 0.769). The overall model accounted for 29% of the variance (p = 0.000). The LMM detected the same pattern ($RPatrols_I$: β = −0.056, p = 0.755; $RPatrols_C$: β = 0.570, p = 0.007). Fit comparison favored the full model (Only $RPatrols_I$: Δ *BIC* = 5.987; *weight* = 0.048, $RPatrols_I$ and $RPatrols_C$: Δ *BIC* = 0.000, *weight* = 0.952). To determine whether these results were a statistical artifact of the method, the same procedure was employed to examine the association between the individual ($Dominance_I$) and the aggregate ($Dominance_C$) values for the male's dominance rank. In contrast to the results of the previous analyses, the individual level of dominance rank significantly predicted the individual-level relative fitness (*Relative* w_I; β = 0.716, p = 0.000), whereas the aggregate value of male dominance ($Dominance_C$) did not have any significant effect (*Relative* w_I; β = 0.033, p = 0.853). The model explained 55% of the variance (p < 0.000). Multicollinearity diagnostics were computed for each GLM ($Patrols_I$ and $Patrols_C$: *VIF* = 2.327; Condition Index for min Eigenvalue = 2.677; Variance proportion = 0.88; $RPatrols_I$ and $RPatrols_C$: *VIF* = 2.511; Condition Index for min Eigenvalue = 2.814; Variance proportion = 0.89; $Dominance_I$ and $Dominance_C$: *VIF* = 3.411; Condition Index for min Eigenvalue = 3.399; Variance proportion = 0.92).

A sequential canonical analysis (SEQCA), the results of which are displayed in Table 8.2, examined a cascade model of the association between age,[11] dominance rank, maternal relatedness, and the number of patrols at both the *individual* and *aggregate* levels. The model accounted for 72% of the variance (p = 0.0001). $Patrols_C$ was positively and significantly predicted by Age^2_I and $Dominance_I$; similarly, Age^2_C and $Dominance_C$ positively predicted the *aggregate*-level number of patrols. In the next step of the cascade,

Table 8.2 Sequential Canonical Analysis of individual level number of patrols (*Patrols*$_I$), aggregate level number of patrols (*Patrols*$_C$), and individual level relative fitness (*Relative* w_I)

Variables	Effect size	C.I. (LB)	C.I. (UB)	*F*-ratio	df1/df2	*p-value*
Overall ($V = 1.553$)	$E = 0.72$	0.00	1.00	8.05	18/135	<0.0001
Y variable: *Patrols*$_C$						
Age^2_I	$sR = 0.60$	0.38	0.75	87.51	1/45	<0.0001
Dominance$_I$	$sR = 0.47$	0.22	0.67	55.07	1/45	<0.0001
Maternal r_I	$sR = -0.14$	−0.41	0.14	5.08	1/45	0.03
Age^2_C	$sR = 0.37$	0.10	0.59	33.88	1/45	<0.0001
Dominance$_C$	$sR = 0.28$	0.00	0.52	18.85	1/45	<0.0001
Maternal r_C	$sR = 0.01$	−0.27	0.29	0.04	1/45	0.84
Multiple	$R = 0.90$	0.81	1.00	33.40	6/45	<0.0001
Residual: Mean = 0.00; SD = 0.43; Skew/Kurtosis = 0.40/−1.31; Range = −0.65–0.65						
Y variable: *Patrols*$_I$						
Prior Y variables						
Patrols$_C$	$sR = 0.76$	0.60	0.85	167.22	1/44	<0.0001
X variables						
Age^2_I	$sR = 0.03$	−0.25	0.31	0.30	1/44	0.58
Dominance$_I$	$sR = 0.33$	0.06	0.56	32.42	1/44	<0.0001
Maternal r_I	$sR = 0.25$	−0.03	0.50	18.40	1/44	0.00
Age^2_C	$sR = -0.18$	−0.44	0.10	9.58	1/44	0.00
Dominance$_C$	$sR = -0.26$	−0.51	0.02	20.23	1/44	<0.0001
Maternal r_C	$sR = -0.06$	−0.34	0.22	1.10	1/44	0.30
Multiple (Xs only)	$R = 0.53$	0.46	0.60	13.67	6/44	<0.0001
Residual: Mean = 0.00; SD = 0.64; Skew/Kurtosis = 0.73/0.34; Range = −1.38–1.50						
Y variable: *Relative* w_I						
Prior Y variables						
Patrols$_I$	$sR = 0.52$	0.28	0.70	63.42	1/43	<0.0001
Patrols$_C$	$sR = 0.30$	0.02	0.53	20.94	1/43	<0.0001
X variables						
Age^2_I	$sR = 0.56$	0.33	0.73	73.03	1/43	<0.0001
Dominance$_I$	$sR = -0.10$	−0.37	0.18	2.55	1/43	0.12
Maternal r_I	$sR = 0.03$	−0.25	0.31	0.25	1/43	0.62
Age^2_C	$sR = -0.29$	−0.53	−0.01	20.25	1/43	<0.0001
Dominance$_C$	$sR = -0.18$	−0.44	0.10	7.66	1/43	0.01
Maternal r_C	$sR = 0.12$	−0.17	0.38	3.19	1/43	0.08
Multiple (Xs only)	$R = 0.68$	0.60	0.78	17.82	6/43	<0.0001
Residual Mean = 0.00; SD = 0.54; Skew/Kurtosis = 0.47/0.83; Range = −1.20–1.52						

Notes: $N = 52$. As measures of effect size, the symbol *sR* represents the semi-partial correlation coefficient (statistically controlled for all prior "X" predictor and "Y" criterion variables) and upper-case *R* represents the combined multiple correlation for all predictor ("X") variables within each equation, while *E* is the multivariate eta (η) or "trace" correlation. For the 90% confidence intervals (C.I.), *LB* lower bound, *UB* upper bound

$Patrols_C$ positively and significantly predicted $Patrols_I$. $Dominance_I$ and $Maternal\ r_I$ also positively predicted the *individual*-level number of patrols. Moreover, Age^2_C and $Dominance_C$ were negatively associated with $Patrols_I$. Finally, in the last step of the cascade, $Patrols_I$ and $Patrols_C$ positively predicted the individual-level relative fitness; Age^2_I and Age^2_C also significantly predicted the relative w_I, featuring a positive and negative effect respectively.

6 Conclusions

Evidence collected by Jane Goodall convinced primatologists and evolutionists that neither Westernization nor Modernization could capably explain lethal intergroup conflict. Thereafter, data gathered from multiple African sites now suggests that lethal intergroup conflict in chimpanzees is a natural adaptation. Some unyielding critics continued investing in, and insisting on, anthropogenic explanations. Most primatologists and evolutionists, now satisfied with the adaptive function of lethal intergroup conflict, instead debated whether its direct and indirect fitness benefits were derived from the recruitment of females from targeted communities, territorial expansion, the elimination of sexual rivals, or shortened interbirth intervals.[12] Multilevel selection theory provides an overarching framework within which to consider lethal intergroup competition's potential adaptive advantages. Some researchers remained skeptical, preferring simpler explanations, such as selfish gene theory over multilevel selection theory. Overemphasizing the principle of parsimony in relation to that of explanatory power, however, ignores the basic scientific principle that the rejection of a hypothesis should not be conducted prior to its empirical examination.[13] From this present exploration, we gain evidence for multilevel selection theory. The contextual analysis and subsequent statistical tests (GLMs and SEQCA) supported the claim that in addition to individual-level attributes, aggregate-level traits have significant effects on the relative fitness of individuals. The results of this chapter complement, rather than contradict, the information collected in the last four decades on chimpanzee intercommunity competition. As per the results of the SEQCA, both at the individual and aggregate levels, the number of patrols independently predicted the individuals' relative fitness. These data offer foundational evidence that multilevel selection is operative in a closely related animal model, allowing us to better contextualize human lethal intergroup competition within multilevel selection theory in the next chapter.

Notes

1. Chimpanzees inhabit multimale-multifemale communities characterized for their fission-fusion dynamics, within which individuals decrease the intensity of intragroup competition by foraging or exploring in subgroups (Aureli et al., 2008; Lehmann & Boesch, 2004; Lehmann, Korstjens, & Dunbar, 2007). Chimpanzees are polygynandrous, with males and females copulating with multiple individuals (Van Schaik, 2016). Although male reproductive skewness has been reported, males also employ an array of alternative mating tactics such as sperm competition (Dixson, 2012; Muller & Pilbeam, 2017) and collective mate guarding (Watts, 1998). In contrast to other primates living in polygynous societies, where a single male has a reproductive monopoly, chimpanzees display an attenuation in their sexual dimorphism (e.g., in canine size; Plavcan, 2001, 2012; Plavcan, Van Schaik, & Kappeler, 1995). Chimpanzees exhibit sex-biased dispersal, with females abandoning their natal group after reaching sexual maturity (Langergraber, Mitani, & Vigilant, 2009; Mitani, Watts, & Muller, 2002; Pusey, 1980). Male philopatry has considerable social *sequelae* such as the development of intracommunity coalitions and alliances (Chapais, 2009; Gilby et al., 2013; Wilson & Glowacki, 2017).
2. While Madame Bee fell during a Kasekela attack to Kahama in 1975 (Goodall, 1986; Williams et al., 2008; Wilson et al., 2014), Patti died during a conflict between the Mitumba and Kasekela in 2005 (Williams et al., 2008; Wilson et al., 2014).
3. The annexation of this area generated several behavioral changes. For example, Ngogo chimpanzees spent over 30% of the observation time foraging and socializing in the captured region, a pattern that lasted for at least five months (Mitani et al., 2010).
4. Julian, Badfoot, and Light Brown, in 1991, 1998, and 2001

5. A similar approach was adopted in previous publications. For example, Wrangham et al. (2006) generated a comprehensive cross-site database describing instances of intracommunity and intercommunity lethal aggression from five sites and nine communities. Following traditional epidemiological procedures, the authors estimated the sites had a median mortality rate of 69 per 100,000 per year, based on observed and inferred cases, and 287 per 100,000 per year, including suspected cases (Wrangham et al., 2006). Communities also exhibited noticeable differences in killing rates. Even though the total rate across communities ranged from 125 to 306 per 100,000 per year, some communities experienced higher rates than others (Wrangham et al., 2006). Kahama, for example, reached a value of 12,000 per 100,000 per year, while Sonso did not experience any attack (Wrangham et al., 2006). Across communities, adult and adolescent males were often the victims of these attacks, with a rate of 355 per 100,000 per year, followed by infants and juveniles with 92, and adult and adolescent females with 28. This value stands in contrast to the frequency of intracommunity lethal aggression, with infants and juveniles displaying a median rate of 429 per 100,000 per year, as compared to adult and adolescent males with 254 per 100,000 per year (Wrangham et al., 2006).
6. Though this pattern generalizes to other chimpanzee communities, regional differences exist between Eastern and Western communities. In contrast to chimpanzees from Gombe, female chimpanzees at the Taï National Park experience less severe, life-threatening attacks (Boesch et al., 2008). Moreover, the frequency of sexual interactions between neighboring communities at Taï is five times greater than that of Eastern communities (Boesch et al., 2008). These behavioral differences could be attributed to socioecological variations. Taï chimpanzees forage in larger parties and exhibit more social cohesion (Boesch, 1991; Boesch et al., 2008). Higher levels of gregariousness allow vulnerable individuals to be rescued by nearby supporters during intercommunity encounters, a phenomenon uncommon in Eastern communities (Boesch et al., 2008). Taï's larger group size could also be attributed to higher predation rates (Boesch, 1991). It is worth noting, however, that even under circumstances of greater social cohesion, intercommunity killings do occur (Boesch et al., 2007, 2008).
7. Lanchester's "linear law" predicts that the largest group will not deploy all its units in a battle. Victory will depend on the relative difference in fighting force between the factions (Wilson, Britton, & Franks, 2002).

According to the "square law," if one of the groups outnumbers the other, the largest group should allocate all its units in a concentrated attack. The numerical advantage will influence the outcome of the conflict (Wilson et al., 2002).

8. Given that previous publications have used reproductive success as a proxy for fitness in chimpanzees (Gilby et al., 2013), the present study employed a similar approach by using the males' paternity success as a surrogate for the individual's fitness.
9. We would like to thank Robyn Stea for help in coding these data and JohnMichael Jurgensen for his feedback on this chapter.
10. Even though traditional factor analyses rely on the extraction of latent variables from observable indicators across individuals (an R-type matrix), it is also statistically feasible to determine underlying groups by examining the correlations between individuals across occasions (an S-type matrix; Gorsuch, 2015).
11. Langergraber et al. (2017) squared this variable. The present chapter retained this transformation.
12. Researchers have also argued that the persistence of lethal intercommunity competition arises from the low costs accrued by raiding males when targeting vulnerable or solitary individuals in the rival group. Even though the experimental and observational evidence endorses perspectives concentrating on examining the low fitness costs of ambushes and incursions, such as the imbalance of power hypothesis, these results should not discourage researchers from further examining the fitness benefits obtained by raiding males. Similarly, future studies should consider the role of multilevel selection in the evolution of chimpanzee intercommunity competition.
13. See Sober and Wilson (1998), for a detailed overview of the logic inconsistencies associated with the parsimony argument.

References

Amsler, S. J. (2010). Energetic costs of territorial boundary patrols by wild chimpanzees. *American Journal of Primatology: Official Journal of the American Society of Primatologists, 72*(2), 93–103.

Aureli, F., Schaffner, C. M., Boesch, C., Bearder, S. K., Call, J., Chapman, C. A., et al. (2008). Fission-fusion dynamics: New research frameworks. *Current Anthropology, 49*(4), 627–654.

Boesch, C. (1991). The effects of leopard predation on grouping patterns in forest chimpanzees. *Behaviour, 117*(3–4), 220–242.

Boesch, C., Crockford, C., Herbinger, I., Wittig, R., Moebius, Y., & Normand, E. (2008). Intergroup conflicts among chimpanzees in Tai National Park: Lethal violence and the female perspective. *American Journal of Primatology: Official Journal of the American Society of Primatologists, 70*(6), 519–532.

Boesch, C., Head, J., Tagg, N., Arandjelovic, M., Vigilant, L., & Robbins, M. M. (2007). Fatal chimpanzee attack in Loango National Park, Gabon. *International Journal of Primatology, 28*(5), 1025–1034.

Chapais, B. (2009). *Primeval kinship: How pair-bonding gave birth to human society*. Cambridge: Harvard University Press.

Damuth, J., & Heisler, I. L. (1988). Alternative formulations of multilevel selection. *Biology and Philosophy, 3*(4), 407–430.

Dixson, A. F. (2012). *Primate sexuality*. New York: Oxford University Press.

Feldblum, J. T., Manfredi, S., Gilby, I. C., & Pusey, A. E. (2018). The timing and causes of a unique chimpanzee community fission preceding Gombe's "four-year war". *American Journal of Physical Anthropology, 166*(3), 730–744.

Gilby, I. C., Brent, L. J., Wroblewski, E. E., Rudicell, R. S., Hahn, B. H., Goodall, J., et al. (2013). Fitness benefits of coalitionary aggression in male chimpanzees. *Behavioral Ecology and Sociobiology, 67*(3), 373–381.

Glowacki, L., Wilson, M., & Wrangham, R. (2017). The evolutionary anthropology of war. *Journal of Economic Behavior and Organization*. https://doi.org/10.1016/j.jebo.2017.09.014

Goodall, J. (1986). *The chimpanzees of Gombe: Patterns of behavior*. Cambridge, MA: Belknap Press of Harvard University Press.

Gorsuch, R. L. (2015). *Factor analysis*. New York: Psychology Press.

Heisler, I. L., & Damuth, J. (1987). A method for analyzing selection in hierarchically structured populations. *The American Naturalist, 130*(4), 582–602.

Herbinger, I., Papworth, S., Boesch, C., & Zuberbühler, K. (2009). Vocal, gestural and locomotor responses of wild chimpanzees to familiar and unfamiliar intruders: A playback study. *Animal Behaviour, 78*(6), 1389–1396.

Kitchen, D. M., & Beehner, J. C. (2007). Factors affecting individual participation in group-level aggression among non-human primates. *Behaviour, 144*(12), 1551–1581.

Lanchester, F. W. (1916). *Aircraft in warfare: The dawn of the fourth arm*. London: Constable limited.

Langergraber, K., Mitani, J., & Vigilant, L. (2009). Kinship and social bonds in female chimpanzees (Pan troglodytes). *American Journal of Primatology: Official Journal of the American Society of Primatologists, 71*(10), 840–851.

Langergraber, K. E., Watts, D. P., Vigilant, L., & Mitani, J. C. (2017). Group augmentation, collective action, and territorial boundary patrols by male chimpanzees. *Proceedings of the National Academy of Sciences, 114*(28), 7337–7342.

Langergraber, K. E., Watts, D. P., Vigilant, L., & Mitani, J. C. (2018). Data from: Group augmentation, collective action, and territorial boundary patrols by male chimpanzees. *Dryad, Dataset.* https://doi.org/10.5061/dryad.kk33f

Lehmann, J., & Boesch, C. (2004). To fission or to fusion: Effects of community size on wild chimpanzee (Pan troglodytes verus), social organization. *Behavioral Ecology and Sociobiology, 56*(3), 207–216.

Lehmann, J., Korstjens, A. H., & Dunbar, R. I. M. (2007). Fission–fusion social systems as a strategy for coping with ecological constraints: A primate case. *Evolutionary Ecology, 21*(5), 613–634.

Lemoine, S., Boesch, C., Preis, A., Samuni, L., Crockford, C., & Wittig, R. M. (2020a). Group dominance increases territory size and reduces neighbour pressure in wild chimpanzees. *Royal Society Open Science, 7*(5), 200577. https://doi.org/10.1098/rsos.200577

Lemoine, S., Preis, A., Samuni, L., Boesch, C., Crockford, C., & Wittig, R. M. (2020b). Between-group competition impacts reproductive success in wild chimpanzees. *Current Biology, 30*(2), 312–318.

Manson, J. H., & Wrangham, R. W. (1991). Intergroup aggression in chimpanzees and humans. *Current Anthropology, 32*(4), 369–390.

Mitani, J. C., & Watts, D. P. (2005). Correlates of territorial boundary patrol behaviour in wild chimpanzees. *Animal Behaviour, 70*(5), 1079–1086.

Mitani, J. C., Watts, D. P., & Amsler, S. J. (2010). Lethal intergroup aggression leads to territorial expansion in wild chimpanzees. *Current Biology, 20*(12), R507–R508.

Mitani, J. C., Watts, D. P., & Muller, M. N. (2002). Recent developments in the study of wild chimpanzee behavior. *Evolutionary Anthropology: Issues, News, and Reviews, 11*(1), 9–25.

Muller, M. N., & Pilbeam, D. R. (2017). The evolution of the human mating system. In M. N. Muller, R. W. Wrangham, & D. R. Pilbeam (Eds.), *Chimpanzees and human evolution* (pp. 383–426). Cambridge, MA: Harvard University Press.

Nishida, T., Hamai, M., Hasegawa, T., Hiraiwa-Hasegawa, M., Hosaka, K., Hunt, K. D., et al. (2003). Demography, female life history, and reproductive profiles among the chimpanzees of Mahale. *American Journal of Primatology, 59*(3), 99–121.

Nishida, T., Hiraiwa-Hasegawa, M., Hasegawa, T., & Takahata, Y. (1985). Group extinction and female transfer in wild chimpanzees in the Mahale National Park, Tanzania. *Zeitschrift für Tierpsychologie, 67*(1–4), 284–301.

Okasha, S. (2004). Multilevel selection and the partitioning of covariance: A comparison of three approaches. *Evolution, 58*(3), 486–494.

Okasha, S. (2006). *Evolution and the levels of selection*. Oxford: Oxford University Press.

Pandit, S. A., Pradhan, G. R., Balashov, H., & Van Schaik, C. P. (2016). The conditions favoring between-community raiding in chimpanzees, bonobos, and human foragers. *Human Nature, 27*(2), 141–159.

Plavcan, J. M. (2001). Mating systems, intrasexual competition and sexual dimorphism in primates. In P. C. Lee (Ed.), *Comparative primate socioecology* (pp. 241–269). New York: Cambridge University Press.

Plavcan, J. M. (2012). Sexual size dimorphism, canine dimorphism, and male-male competition in primates. *Human Nature, 23*(1), 45–67.

Plavcan, J. M., Van Schaik, C. P., & Kappeler, P. M. (1995). Competition, coalitions and canine size in primates. *Journal of Human Evolution, 28*(3), 245–276.

Power, M. (1991). *The egalitarians-human and chimpanzee: An anthropological view of social organization*. New York: Cambridge University Press.

Pradhan, G. R., Pandit, S. A., & Van Shaik, C. P. (2014). Why do chimpanzee males attack the females of neighboring communities? *American Journal of Physical Anthropology, 155*(3), 430–435.

Pusey, A. E. (1980). Inbreeding avoidance in chimpanzees. *Animal Behaviour, 28*(2), 543–552.

Sober, E., & Wilson, D. S. (1998). *Unto others: The evolution and psychology of unselfish behavior*. Cambridge: Harvard University Press.

Sussman, R. W. (2013). Why the legend of the killer ape never dies: The enduring powers of cultural beliefs to distort our view of human nature. In D. P. Fry (Ed.), *War, peace, and human nature: The convergence of evolutionary and cultural views* (pp. 97–111). New York: Oxford University Press.

Van Schaik, C. P. (2016). *The primate origins of human nature*. Hoboken: Wiley.

Watts, D. P. (1998). Coalitionary mate guarding by male chimpanzees at Ngogo, Kibale National Park, Uganda. *Behavioral Ecology and Sociobiology, 44*(1), 43–55.

Watts, D. P., Muller, M. N., Amsler, S. J., Mbabazi, G., & Mitani, J. C. (2006). Lethal intergroup aggression by chimpanzees in Kibale National Park, Uganda. *American Journal of Primatology, 68*(2), 161–180.

Williams, J. M., Lonsdorf, E. V., Wilson, M. L., Schumacher-Stankey, J., Goodall, J., & Pusey, A. E. (2008). Causes of death in the Kasekela chimpanzees of Gombe National Park, Tanzania. *American Journal of Primatology: Official Journal of the American Society of Primatologists, 70*(8), 766–777.

Williams, J. M., Oehlert, G. W., Carlis, J. V., & Pusey, A. E. (2004). Why do male chimpanzees defend a group range? *Animal Behaviour, 68*(3), 523–532.

Wilson, M. L. (2013). Chimpanzees, warfare, and the invention of peace. In D. P. Fry (Ed.), *War, peace, and human nature: The convergence of evolutionary and cultural views* (pp. 361–388). New York: Oxford University Press.

Wilson, M. L., Boesch, C., Fruth, B., Furuichi, T., Gilby, I. C., Hashimoto, C., et al. (2014). Lethal aggression in Pan is better explained by adaptive strategies than human impacts. *Nature, 513*(7518), 414–417.

Wilson, M. L., Britton, N. F., & Franks, N. R. (2002). Chimpanzees and the mathematics of battle. *Proceedings of the Royal Society of London. Series B: Biological Sciences, 269*(1496), 1107–1112.

Wilson, M. L., & Glowacki, L. (2017). Violent cousins: Chimpanzees, humans, and the roots of war. In M. N. Muller, R. W. Wrangham, & D. R. Pilbeam (Eds.), *Chimpanzees and human evolution* (pp. 464–508). Cambridge, MA: Harvard University Press.

Wilson, M. L., Hauser, M. D., & Wrangham, R. W. (2001). Does participation in intergroup conflict depend on numerical assessment, range location, or rank for wild chimpanzees? *Animal Behaviour, 61*(6), 1203–1216.

Wilson, M. L., Kahlenberg, S. M., Wells, M., & Wrangham, R. W. (2012). Ecological and social factors affect the occurrence and outcomes of intergroup encounters in chimpanzees. *Animal Behaviour, 83*(1), 277–291.

Wrangham, R. W. (1999). Evolution of coalitionary killing. *Yearbook of Physical Anthropology, 42*, 1–30.

Wrangham, R. W., & Glowacki, L. (2012). Intergroup aggression in chimpanzees and war in nomadic hunter-gatherers. *Human Nature, 23*(1), 5–29.

Wrangham, R. W., Wilson, M. L., & Muller, M. N. (2006). Comparative rates of violence in chimpanzees and humans. *Primates, 47*(1), 14–26.

9

Lethal Intergroup Competition in Non-State Societies: From Small-Scale Raids to Large-Scale Battles

Mateo Peñaherrera-Aguirre, Aurelio José Figueredo, and Steven C. Hertler

1 Introduction

This chapter will provide a synthesis of the current evolutionary literature concerning lethal coalitional aggression in small-scale societies. Attacks, raids, skirmishes, ambushes, and other forms of intergroup aggression present significant risk of injury or death, irrespective of group size, though the means of differentiation among groups, like the mechanisms of ensuring coordination within groups, change as a function of group size. Human coalitional violence is often explained via *kin selection* and *reciprocal altruism*, such that an individual's assumption of risk is compensated by fitness-enhancing benefits to relatives and allies. These explanations become increasingly inapplicable in progressing from bands and tribes to chiefdoms and states. The growth of larger social aggregations compelled the emergence of institutions enforcing intragroup cooperation above and beyond the effects of underlying social networks based on kinship and direct reciprocity. Perspectives reviewed herein, such as *cultural group selection*, consider the cultural evolution of such institutions in generating between-group variance and facilitating lethal intergroup

S. C. Hertler et al., *Multilevel Selection*, https://doi.org/10.1007/978-3-030-49520-6_9

competition. According to these theories, cultural transmission, group differentiation, symbolic ornamentation, punishment of defectors, and ethnocentrism are integral components of intergroup competition, with lethal coalitional aggression being an extreme manifestation of between-group rivalry. Furthermore, due to the significant fitness costs imposed upon defeated factions, the study of lethal coalitional aggression in small-scale societies provides fertile ground for examining the interaction between group-level and individual-level selective pressures.

2 Intergroup Violence in Warlike Societies

To this day, cross-cultural studies on human violence have provided a powerful framework for examining how socioecological and cultural factors may influence the persistence of human intragroup and intergroup aggression. Fabbro (1978), for example, reviewed seminal ethnographic papers describing peaceful societies. The author identified seven social organizations allegedly lacking physical interpersonal violence: (1) the Siriono of Bolivia, (2) the !Kung of the Kalahari Desert, (3) the Semai of Malaya, (4) the Mbuti of Equatorial Africa, (5) the Hutterites of North America, (6) the inhabitants of Tristan da Cunha in the South Pacific, and (7) Canada's Copper Eskimo. However, contrary to Fabbro's predictions, Knauft (1987) and Kelly (2000) established the reality of lethal intragroup violence among the Semai, the !Kung, the Mbuti, and the Copper Eskimo, groups respectively experiencing homicide rates of 30.3, 41.9, 39.7, and 419[1] per 100,000 annually. Margaret Mead's *Coming of Age in Samoa* exemplifies the studied mischaracterizations of the prevailing Boasian anthropological paradigm. These inaccuracies extend equally to intragroup and intergroup violence, such that small-scale societies are imagined to be at peace among themselves and with their neighbors. This, too, is fantastical. As extensively reviewed by Ellingson (2001), Pinker (2000, 2012), and our chapter on Lawrence Keeley's anthropological legacy (Hertler, Figueredo, Peñaherrera-Aguirre, Fernandes, & Woodley of Menie, 2018), lethal intergroup violence has been associated with high civilization, at least since Rousseau imagined "noble savages" subsequently corrupted by cultural institutions.[2] The absence of war among these purportedly peaceful primitives, perforce, suggested that lethal violence was a byproduct of complex civilization. As we will see, the ethnographic data says otherwise.

The archaeological and ethnographic record suggests that non-state societies frequently experience raids, ambushes, and massacres (Beckerman et al., 2009; Gat, 2008, 2015; Guilaine & Zammit, 2008; Keeley, 1997; Lahr et al., 2016, LeBlanc & Register, 2003; Pinker, 2012; Soltis, Boyd, & Richerson, 1995; Wrangham & Glowacki, 2012). Examining the frequency of warfare across 50 hunter-gatherer societies, Ember (1978) calculated that 64% of social systems waged war at least once every two years, in contrast to the previously presumed general absence of small-scale intergroup conflict (Lee & Devore, 1968). While 26% experienced war *occasionally*, only 10% *rarely* or *never* exercised any form of intergroup conflict.[3] Adopting a similar methodology, Boehm (2013) collected ethnographic information on 49 bands of foragers from Africa, the Arctic, Asia, Australia, and the Americas. These societies were characterized by their nomadic and egalitarian lifestyle, as well as by their economic autonomy, meaning that they are not dependent on economic exchanges with neighboring horticulturalists, fur traders, sedentary foragers, or equestrian cultures. Boehm referred to them as *Late Pleistocene Appropriate* (LPA), due to their hypothesized resemblance with behaviorally modern prehistoric cultures (45,000 years ago). Boehm's database on LPA foragers allowed him to reconstruct the approximate frequency, in terms of central tendencies, of inter-band conflict in the past. Boehm's analyses revealed that 59% of LPA bands experienced at least one form of intergroup conflict, such as revenge killings, raids, or intense warfare (Boehm, 2013). Although far from suggesting the universality of lethal intergroup aggression, this estimate provides additional evidence against the prevalence of peace among foragers. In terms of conflict resolution, Boehm determined that bands employ various mechanisms to temporarily or permanently finalize confrontations. Even though highly mobile, only 35% of LPA foragers used avoidance as an avenue to resolve conflicts (Boehm, 2013). Moreover, bands attempted to negotiate conflicts in some way (59%) or reach a temporary truce (27%). In only 16% of the cases, combatants attended formal peace meetings.

In a similar vein, Otterbein (1989) classified 100 societies according to their frequency of warfare (*continuous*, *frequent*, *rare/never*) and their level of political organization (states, chiefdoms, tribes, and bands). Otterbein detected a curvilinear trend between the percentages of societies experiencing continuous warfare and the level of sociopolitical complexity: bands 33.3%, tribes 80%, chiefdoms 50%, and states 40%. This pattern also replicated the associations between different types of subsistence economy and continuous warfare: foraging 20%, animal husbandry 88.9%, shifting cultivation 85.7%, and intensive agriculture 47.1%. Hence, although intergroup conflict exists in bands of hunter-gatherers, it intensifies in tribal societies dependent on the practices of animal husbandry or horticulture. Subsequent analyses by Wrangham, Wilson, and Muller (2006) also concluded that killing rates in farmers (595 per 100,000 per year) far exceed those of hunter-gatherers (164 per 100,000 per year).

Additional cross-cultural comparisons further supported the hypothesis that small-scale societies were subject to intense lethal intergroup interactions. Keeley (1997) reviewed the historical, archaeological, and ethnographic literature, generating a detailed database on deaths due to warfare. In contrast to the percentage of US and European males killed in combat during the twentieth century (less than 1%), tribal societies, such as the Jivaro in lowland Amazonia, were subject to considerably more intense competition (59% of male deaths being due to warfare). Keeley concluded that this pattern extended to prehistoric societies.[4] Expanding on Keeley's dataset, Pinker (2012) calculated that, across 21 prehistoric archaeological sites (14,000 BC to AD 1770), 15% of deaths were due to warfare, with estimates ranging from 0% (Gobero, Niger, 14,000–6200 BC) to 60% (Crow Creek archaeological site, South Dakota, AD 1325). The sample included a variety of hunter-gatherers and horticulturalists endemic to Asia, Africa, the Americas, and Europe, suggesting that lethal intergroup violence was not limited to any geographical region or type of subsistence economy. Furthermore, these metrics were within the ranges displayed by contemporary small-scale societies, indicating a degree of consistency across past and present non-state societies. According to Pinker's (2012) database, extant hunter-gatherers (n = 8) reached an average of 14% of war deaths, with values ranging from 4% (the Anbara in Australia) to 30% (the Ache in Paraguay).

The average for contemporary Amazonian, New Guinean, and European tribal societies ($n = 10$) was close to 25%. Pinker (2012)[5] estimated that 0.7% of individuals died in battles during the twentieth century. Although critics could argue that war deaths are not limited to direct combat casualties, Pinker's (2012) evidence suggests that, even after adding deaths due to wartime famines, epidemics, or genocides, the percentages of war-related deaths remain considerably lower (3%) relative to that of small-scale societies.

Similarly, Walker and Bailey (2013) analyzed the degree of lethal conflict across a sample of 44 lowland Amazonian societies. The authors estimated that violence accounted for 30% of all adult deaths, with the majority corresponding to male victims (69%). Cross-cultural comparisons also revealed considerable variation in mortality estimates, from 6% in the Tsimane to 56% in the Waorani. Walker and Bailey thereafter delineated cumulative violent deaths into three categories: (1) *within-village homicides*; (2) *internal warfare*, in which rival factions are part of the same ethnolinguistic group; and (3) *external warfare*, in which rival factions differ in their ethnolinguistic group of origin. Internal warfare occurred more frequently (55% of events in the database), though subsequent analyses demonstrated that external warfare killed more people (Walker & Bailey, 2013). This difference has been attributed to the lack of significant social connections between groups, such as affinal or consanguineal kinship (Ellsworth & Walker, 2014). Despite the lethality of these confrontations, attackers died in only 2% of the incursions.[6]

Unique among researchers reviewed in this section, Walker and Bailey relate observed mortality estimates in small-scale Amazonian societies explicitly to multilevel selection. As per Walker and Bailey, in contrast with other populations around the globe, Amazonian organizations exhibit considerable levels of genetic variation between groups, as indicated by their high autosomal *Fst* values[7] and low heterozygosities (Wang et al., 2007). The level of genetic differentiation observed in eastern South America is attributable to genetic drift and *assortative fissioning*, in which the daughter groups produced by a split become more homogeneous by recruiting subsets of similar individuals. If individuals marry between allied communities, and if these groups then cooperate against a common rival, selection can operate along the boundary between the marriage-trade cluster and the rival outgroup faction without unduly eroding between group genetic heterogeneity. Walker and Bailey (2013) also suspected that reproductive leveling, such as the absence of land and livestock inheritance, the restriction of polygyny, and an increase in shared paternity, could also promote multilevel selection in Amazonia.

3 Revenge and Residence Patterns

As mentioned in the previous chapter, lethal killings are not a uniquely human phenomenon. Evidence indicates that chimpanzee intercommunity raids and human deadly intergroup aggression feature numerous similarities, including (1) groups practicing collective territoriality, (2) males establishing coalitions and alliances, (3) attackers experiencing low risk of injury or death, and (4) raiders systematically weakening the rival group's cohesion (Manson et al., 1991; Wilson, 2013). Nevertheless, noticeable differences also exist. For instance, although revenge and treachery feature predominantly in lethal confrontations across small-scale human societies (Beckerman et al., 2009; Kelly, 2005; Valentine & Beckerman, 2008; Walker & Bailey, 2013), there is no evidence these behaviors facilitate chimpanzee intercommunity competition. This section then focuses on revenge, residence patterns, and their relation to lethal intergroup conflict.

In small-scale human societies, revenge killings often operate differently depending on the identity of the killer. Rather than interpreting a within-village homicide as a collective affair, the group typically views the attack as a personal loss (Kelly, 2005). This distinction avoids dragging other members of the community into the conflict, circumscribing the dispute between the murderer on the one hand and the victim's immediate kin and allies on the other (Kelly, 2005). Moreover, grievers have at their disposal an array of alternatives destined for dealing with the killer and settling disputes (Boehm, 1999). Mourners, for example, could ask for *weregild*, *éraic*, *galanas*, or any related form of blood money compensation (Dunbar, Clark, & Hurst, 1995) or demand the murderer's expulsion from the group (Boehm, 1999); these terms are commonly accepted by the murderer's kin and allies. However, if these solutions are deemed unacceptable, the only recourse is to kill the murderer (Boehm, 1999). Although grievers could eliminate the killer without first consulting other members of the community, such an action could bring escalating revenge cycles. To circumvent these risks, it is not uncommon for executioners to meet with influential people in the community (so-called big men) as well as with the killers' relatives and friends (Boehm, 1999). Hence, revenge takes the form of capital punishment, in which only the murderer is considered liable (Kelly, 2005).

In contrast to these within-village revenge homicides, where a victim's allies, or kin, precisely dispense lethal aggression toward the murderer, the attackers in *intergroup retaliatory raids* instead direct lethal aggression against *any* member of the rival group, a style of vengeance termed *social substitutability* (Kelly, 2005). Hence, social substitutability can generate additional grievances, rather than settling the dispute. In turn, retaliatory incursions can thereafter create new grievances, leading to future killings (Gat, 2010). The underlying political structure can also create the necessary conditions for a continuing state of intergroup violence. Without a dispassionate Hobbesian *Leviathan* to justly mediate conflict, families, lineages, and clans can be dragged into chronic clashes by retaliating disproportionately, injudiciously, or unjustly (Gat, 2017). In consequence, small-scale societies organized as multilevel systems are more prone to revenge cycles.

Further still, there is covariation between warfare and sociodemographic factors such as residency patterns (patrilocality vs. matrilocality). At first glance, it would be expected that relative to patrilocal societies, matrilocal organizations would rarely wage war due to their absence of consanguineous bonds between males. This, however, is not the case. Matrilocal societies, such as the Waorani, also engage in lethal intergroup competition (Erickson, 2008). With both patrilocal and matrilocal societies displaying between-group killing, researchers have concentrated instead on the influence of residence patterns and migration on internal and external warfare.[8] Divale, for example, suggested (1974) that patrilocal communities can transition to matrilocality after migrating into a territory already occupied by another society. Matrilocality, Divale hypothesized, could disrupt fraternal interest groups, limiting the frequency of internal warfare among the immigrant communities. Greater cooperation between migrant groups reorients the war effort toward defeating native rivals. As per analyses of Divale's sample of 43 societies, groups that migrated within 500 years were more likely to be matrilocal. In terms of warfare, matrilocal organizations tended to wage only external warfare, rather than either internal warfare alone or any combination of internal and external warfare (Divale, 1974). These associations remained significant after statistically controlling for the degree of sociopolitical complexity. Divale also suspected that the coevolution of matrilocality with external warfare was the product of communities losing their young males at an accelerating rate.[9] These circumstances forced the heads of the households to establish coalitions and alliances with men outside of their communities (Divale, 1974; Otterbein, 2004). Otterbein (2004) further expanded on the association between war, kinship, and fraternal interest groups. Tribal warfare, unlike intergroup conflicts between bands of hunter-gatherers, relied on the "recruitment" of males from nearby villages based on their respective kinship lineages. Since patrilineages contain fraternal interest groups, the risk of internal conflicts pervades tribal organizations (Otterbein, 1968, 2004). The threat of external warfare generally forces internal disputes to be suspended or resolved until rival groups are defeated. Regarding mobilization in matrilineal/matrilocal societies, Otterbein (2004) concurs with Divale (1974). During wartime, matrilineal villages featuring recurrent peaceful interactions with each other often dispatched their warriors as part of their contribution to the war effort (Otterbein, 2004).

4 Parochial Altruism and Group Differentiation

Having considered some relevant ethnographic and archaeological evidence, it is now crucial to review explicitly evolutionary explanations of non-state warfare, which emphasize the interplay between ingroup altruism and outgroup antagonism.

Bowles and Gintis (2013) generated a mathematical model for the evolution of altruism under multilevel selection dependent on five assumptions: (1) individuals inhabiting partially isolated groups can either be altruists or non-altruists; (2) altruists incur a cost when they generate a benefit collected by other group members; (3) a member of a group comprised exclusively of altruists attains greater gains relative to an individual living in a group of non-altruists; (4) within mixed groups, altruists fare worse compared with non-altruists; and (5) the expected fitness of the individual is equal to (a) the effects of variation in the frequency of the altruistic allele in the group, plus (b) the existence of the altruistic allele in the individual, and (c) the baseline replication rate (Bowles & Gintis, 2013). Using an adapted version of Price's equation[10,11] for this model, these considerations allow for the separate estimation of within-group and between-group selection. Bowles and Gintis assumed that the change in the frequency of altruists will be zero, if the absolute magnitude of between-group selection and within-group selection remains equal. Alternatively, the enlargement of groups containing altruists and the reduction in the size of groups with fewer altruists counterbalance the natural decline in the number of altruists (Bowles & Gintis, 2013). As an additional step, Bowles and Gintis also adapted Wright's inbreeding coefficient ($F_{ST;}$ 1935), referring to

> The ratio of the between-group variance in the fraction of altruists to the total population variance, which is the within-group plus the between-group variance of the fraction of altruists. (p. 55)

Altering this equation in terms of payoffs, Bowles and Gintis suggested that, if F_{ST} is larger than the ratio between the cost and the benefit, the proportion of altruists will rise, while if F_{ST} is lower, their proportions will decline. Given the costs associated with ingroup altruism and parochialism, it follows that selective pressures should act against the persistence of lethal intergroup conflict. Yet, as evidenced in the previous section, warfare remains a pervasive state among small-scale societies. Choi and Bowles (2007) simulated the potential interactions between four behavioral types: (1) tolerant altruists, (2) tolerant non-altruists, (3) parochial non-altruists, and (4) parochial altruists. Since combat requires hostility toward outsiders and the agent's willingness to accrue a cost, the authors limited this behavior to parochial altruists. In terms of individual gains, Choi and Bowles' model (2007) allowed parochial altruists to attain a direct benefit from warfare, assuming this gain would remain lower than the costs. In contrast, altruists accrued a cost to themselves by providing a public good whose value was to be distributed equally among adult members of the group.

The model also assumed the action of two types of selection: (1) *selective extinction*, the product of intergroup antagonism favoring parochial altruism; and (2) *within-group selection*, promoting tolerant non-altruists and selecting against parochial altruists (Choi & Bowles, 2007). The payoffs received by each adaptive strategy are dependent on the presence or absence of war. Under peaceful circumstances, tolerant individuals in each group attain a net benefit. Since parochial altruists cannot obtain any gains from these conditions, tolerance prevails (Bowles & Gintis, 2013; Choi & Bowles, 2007). The number of parochial altruists in each group predicts the likelihood of intergroup hostility. Intergroup antagonism can turn into warfare if one of the groups contains a higher number of warriors. Hence, group survival depends on the number of warriors per group. The conquering group randomly draws a set of individuals from its ranks and proceeds to replace the proportion of members "killed" in the other group. Reproduction occurs when individuals are randomly paired, with the number of resultant offspring being proportional to the breeding pair's percentage of the group's benefits. The simulation predicted an increase in the frequency of wars when parochial altruists comprised most of the population (Bowles & Gintis, 2013; Choi & Bowles, 2007). The analyses predicted two cutoff points in the frequency of parochial altruists leading to the outbreak of intergroup conflict. Simulations with fewer than 30% parochial altruists generated lower war frequencies due to the limited opportunities for hostile intergroup interactions. In contrast, settings with more than 80% fighters often predicted martial impasses where warriors refused to attack due to the balanced fighting ability between groups. Similarly, the simulated frequency of war and parochial altruism was dependent on population parameters such as migration rates and group sizes. An increase in these parameters decreased the simulated magnitude of between-group variation (Choi & Bowles, 2007).

As mentioned before, a critical assumption of this model rests on partial isolation between groups. According to Bowles (2006), most empirical estimates of genetic F_{ST} are higher than 0.02, a threshold indicating F_{ST} is at equilibrium[12] and interdemic selection unfeasible. In subsequent years, for example, Bowles and Gintis (2013) assembled a list of F_{ST} values in a sample of extant hunter-gatherer populations. The authors classified the data based on three indices. Hence, while F_{DG} provided information on the genetic differentiation between demes within the same ethnolinguistic cluster, F_{GT} and F_{DT} referred to the magnitude of between-group and between-deme variance within the same metapopulation. Across all indices, the mean differentiation value was 0.080. Removing F_{DG} from the analyses slightly increased this estimate (0.087; Bowles & Gintis, 2013). These metrics are inconsistent with traditional descriptions of isolated bands of hunter-gatherers displaying group-level cooperation due to reciprocal altruism, or kin selection.

The work of Bowles and Gintis cited above represents one interpretation, though the degree of genetic differentiation in small-scale societies remains inconclusive. For example, Langergraber et al. (2011) computed F_{ST} scores based on autosomal microsatellite genotypes collected from hunter-gatherers and food-producing organizations. The authors generated pair-wise comparisons between food producers (FP-FP), between hunter-gatherers and food producers (HG-FP), and between hunter-gatherers (HG-HG). The average F_{ST} for FP-FP (0.015), HG-FP (0.011), and HG-HG (0.005) were below 0.02. These inconsistencies led some researchers to consider cultural evolutionary dynamics as the driver of human intergroup variation (Richerson et al., 2016).[13] For instance, Zefferman and Mathew (2015) reviewed the literature collecting information on the *genetic* and *cultural* F_{ST} values of small- and large-scale human societies. The authors then proceeded to compare these estimates with the genetic F_{ST} of chimpanzee communities and Argentine ant supercolonies. Though differing from the average genetic F_{ST} estimate *between* different Argentine ant supercolonies, F_{ST} estimates between human societies were relatively similar to the values *within* Argentine ant supercolonies, which are generally composed of multiple related nests within about one hectare, and also similar to F_{ST} estimates between chimpanzee communities. Alternatively, the human cultural F_{ST} between-groups estimates for both small-scale and state societies were considerably higher than the genetic F_{ST} values in human societies, chimpanzee communities, and Argentine ant supercolonies. As per Zefferman and Mathew, these results suggest that cultural F_{ST} values are more likely to be of sufficient magnitude than genetic F_{ST} estimates to allow for the evolution of large-scale lethal intergroup conflict in humans.

Recent examinations have also compared the difference in magnitude between genetic and cultural F_{ST}. Bell, Richerson, and McElreath (2009), for example, collected data from the *World Values Survey*, an online database describing various beliefs and values. The authors limited their analyses to geographically adjacent polities, assuming neighboring societies would compete more frequently. Cultural F_{ST} scores were contrasted to genetic F_{ST} estimates previously published in *The History and Geography of Human Genes* (Cavalli-Sforza, Menozzi, & Piazza, 1994). Average cultural F_{ST} scores (0.080) differed from average genetic F_{ST} (0.005). Employing the left component of Equation 9.1,[14] Bell et al. calculated the genetic and cultural group benefits of altruistic traits, with results suggesting that genetic benefits (mean = 437) considerably exceeded the cultural benefits (mean = 16) of altruism. According to the authors, this difference provides evidence that selection can promote the persistence of cultural traits associated with group-oriented behaviors.

While considering the relative magnitudes of genetic and cultural F_{ST} values in multilevel selection, it is important to keep in mind that trait-group selection theory, as covered in previous chapters of this volume, does not require the more stringent assumptions of naïve group selection theory, such as reproductive isolation or an absence of migration between groups. Dispensing with these unnecessarily restrictive requirements renders these models more tractable. Additionally, individually selected selfishness does not undermine mean group altruism due to altruistic punishment, as discussed in previous chapters. Evidence of altruistic punishment is provided by Mathew and Boyd (2011), who collected information on 88 raids conducted among Turkana communities, a pastoralist culture located in East Africa. Warriors accrued severe costs, suffering from injuries or death during these confrontations. Due to the acephalic political structure of Turkana's communities, men are not directly instructed to participate in a raid. However, refusal to join a raiding party must be adequately justified. Since an incursion often involves multiple participants, it is not uncommon for some individuals to defect at any point in time. Desertions occurred in 43% of the attacks. Men also defected by staying behind, refusing to engage the enemy, and fleeing the area. Acts of cowardice occurred in 45% of raids. Even though defections are a pervasive phenomenon, free riders rarely go unpunished. Indeed, Mathew and Boyd estimated that other group members punished at least one deserter in 47% of the cases of desertion. Individuals also punished at least one defector in 67% of cases of cowardice. In both instances, sanctions ranged from public recriminations and financial penalties to corporal punishment.

5 Ultrasociality and the Evolution of Large-Scale Warfare

In contrast to small-scale societies such as bands or tribes, large-scale societies, such as chiefdoms and states, feature considerable levels of *ultrasociality*, in which genetically unrelated (or only distantly related) individuals cooperate regularly (Richerson & Boyd, 1998; Turchin, 2010, 2013; Turchin, Currie, Turner, & Gavrilets, 2013). Ultrasocial norms and institutions[15] allowed some small-scale societies to defeat rival neighbors (Turchin, 2016). In modeling the impact of intergroup warfare and the diffusion of military technology, Turchin et al. (2013) predicted that the outcome of an attack during warfare rested on the attackers' average level of ultrasociality.[16] Social systems displaying a higher capacity for collective endeavors were more likely to defeat their competitors. Employing an agent-based simulation, Turchin's mathematical model matched historical data quite closely, predicting over 65% of the variance, paralleling the spread of large-scale societies in Africa and Eurasia (1500 BC to AD 1500).

Geopolitical factors also have a significant effect on the severity of warfare. Turchin (2010), for example, examined the lethality of warfare in steppe societies, comparing culturally similar groups to that of rival factions differing in their meta-ethnic affiliation.[17] As per Turchin, atrocities frequently occurred as part of external confrontations occurring close to steppe borders. For example, the likelihood of genocide was 1.4% during internal conflicts, whereas it reached 63% on steppe frontiers. In addition to the presence of meta-ethnic frontiers, as per Turchin, some biogeographical regions exhibited the necessary conditions for the evolution of *mega empires*, which are defined as social organizations comprising a population of at least 10 million inhabitants and occupying an area of at least 1 million Km^2 (Turchin, 2013). Besides the influence of specific biomes, the presence of draft animals (such as *perissodactyl* or *artiodactyl* ungulates) enabled the rise and spread of these complex conglomerates. Turchin gathered information on large-scale polities from various published databases.[18] As predicted, mega-empires emerge more frequently in arid and transitional zones featuring domesticates, such as horses or camelids. Analyses revealed that over 90% of mega-empires appeared on steppe frontiers.

The coevolution of large-scale warfare within sociopolitically complex systems extends beyond monarchical mega-empires in Afroeurasia. Truly, warfare is exclusive to neither antiquity nor monarchy. In fact, contrary to the *democratic peace theory*, self-governing peoples are sometimes aggressors and can act belligerently in their own interest, as per the predictions of multilevel selection theory. Though not pacific, democracies are exceptional in terms of martial success. Reiter and Stam (2002), for example, gathered historical information on interstate wars (from AD 1819 to 1990) involving more than a thousand casualties. After classifying each faction as a *dictatorship*, an *oligarchy*, or a *democracy*, Reiter and Stam determined that, when democracies initiated the attack, they won in 93% of the cases, as compared to success rates for oligarchies and dictatorships, which were, respectively, 58% and 60%. Similarly, when a polity was the target of aggression, democracies defeated invaders 63% of the time, relative to dictatorships and oligarchies, which, respectively, defeated invaders in 34% and 40% of instances. In attempting to explain these results, Reiter and Stam (2002) surmised that democratic leaders more judiciously decided when to initiate war, perhaps suggesting that democracies are less often pressed unwisely into war. More than this, public opinion and political accountability also had a significant effect, with democratic leaders facing continuous scrutiny during war. Hence, rather than pinnacles of pacifism, democratic regimes successfully play the part of both passive and active belligerents (Reiter & Stam, 2002), perhaps because group interests are more carefully considered when deciding whether to wage war and because group solidarity is increased when individuals perceive themselves as citizens rather than subjects.

6 Conclusions

Although researchers have examined instances of human intergroup competition in an array of economic and religious contexts, warfare remains the best-documented facet of collective human behavior emerging in competitive settings. Rather than being limited to nation-states, lethal intergroup aggression occurs across non-state societies, as we have seen in this chapter, and has precedents in comparative primatology, as we have seen in the previous chapter. We have furthermore observed how local raids between small-scale societies became large-scale battles requiring the collective action of myriads of relatively unrelated individuals against rivals having different cultures, institutions, and languages. Showing multiple ways in which human warfare was elaborated alongside the growth of groups, *cultural group selection theory* emphasizes the evolution of symbolic markers, allowing groups to cooperate with neighbors and compete against factions lacking these identifiers. In addition to the transmission and persistence of symbolic markers within groups, residence patterns also influenced intergroup aggression. The theories reviewed herein, inclusive of their methods and findings, are relevant to multilevel selection for their ability to reconstruct group formation and fractionalization. As we have also noted, the *theory of parochial altruism* (Bowles & Gintis, 2013) and the *meta-ethnic frontier theory* (Turchin, 2003, 2007) are pointed elaborations of multilevel selection theory, which view the evolution of warfare as the product of ingroup prosociality and outgroup hostility. Part III's first chapter used review and analysis to establish intergroup conflict's biological precursors, while this second chapter used ethnography and modeling to explain the cultural elaboration of those precursors with respect to lethal intergroup conflict in small-scale societies. Hence, we are positioned to apply multilevel selection theory to Ancient Rome and Modern Europe, presenting analyses of representative large-scale societies of antiquity and modernity.

Notes

1. Kelly (2000) considers the homicide rate of Copper Inuit to be of the same order of magnitude to the one reported for the Gebusi (419 per 100,000).
2. Possibly from a mixture of industrialized warfare, high absolute death tolls, and ideological motivations, Modern Western nations are assumed especially bellicose. These attributions are doubly incorrect, as violence decreases with civilization and is restricted in the Modern West.
3. Removing cultures dependent on equestrian or fishing economies from the analysis did not alter the overall results, with 12% of societies living peacefully with other groups (Ember, 1978).
4. For example, attackers killed close to 50% of males at the Nubian site of Djebel Sahaba (12,000–10,000 BC; Keeley, 1997; Wendorf, 1968).
5. In addition to the percentage of deaths, other metrics such as standardized rates provide additional information by taking into consideration the number of living individuals in the population. Pinker calculated that the average rate for 27 non-state societies, including hunter-gatherers and horticulturalists, was of 524 war deaths per 100,000 individuals per year. Alternatively, twentieth-century states such as Germany and the United States suffered lower annual death rates (144 and 3.7 per 100,000, respectively; Pinker, 2012). Adding all deaths due to genocides, purges, battles, and war-related famines during the twentieth century generated a rate of 60 per 100,000 per year, close to 9 times lower than the average of non-state societies (Pinker, 2012).
6. Several factors have been attributed to the raiders' relative low rate of injury or death. First, although attackers can use shock weapons to inflict blunt trauma during hand-to-hand combat, projectiles, such as spears, arrows, and darts, enable raiders to injure or kill their rivals from a safe distance (Keeley, 1997). Second, raiders tend to choose solitary victims or smaller groups that are unlikely to successfully fend off an attack (Glowacki, Wilson, & Wrangham, 2017). Third, if competing groups

share an ethnolinguistic background and maintain a minimum level of open communication (as in lowland Amazonian societies, in which the frequency of treachery tactics was 9.5 times higher during internal conflicts relative to external clashes), the killing party could host a meeting or feast with the sole intention of eliminating their rivals (Walker & Bailey, 2013).

7. Wang et al. (2007) examined autosomal microsatellite data collected from 24 Amerindian populations. The analyses concluded that the F_{ST} values for Amerindian samples located in eastern South America ($F_{ST} \times 100 = 14.7$) were considerably higher relative to the worldwide estimate ($F_{ST} \times$ x100 = 7.1).
8. See Turchin and Korotayev (2006) for a mathematical examination of population pressure and internal warfare in non-state societies.
9. For a complementary take, see Ember and Ember (1971).
10. In 1970, George Price developed a mathematical expression describing the change in the frequency of an allele A nested within a group (Δp). Price (1970, 1972) proposed that Δp was equal to the covariance between the individual gene frequency and the number of offspring sired by each partner (assuming each offspring is the product of half of each parent's contributions).
11. This adaptation, however, made no distinction between MLS1 and MLS2. Okasha (2009), for example, demonstrated that Price's equation differs depending on the type of multilevel selection. The author derived the following formula for MLS1:

$$cov(w_i, z_i) = cov(W_k, Z_k) + E_k(cov_k(w_{jk}, z_{jk}))$$

wherein the overall covariance between trait and fitness is equal to the group-level covariance (mean fitness and mean trait) and the average of intragroup covariance (individual trait and individual fitness; Okasha, 2009). This formulation, however, assumes that the group's fitness is equal to the average individual fitness, and the group trait equals the average individual character (Okasha, 2009). In this case:

$$\overline{w}\Delta\overline{z} = cov(w_i, z_i)$$

Concerning MLS2, the focal units are the individuals and the groups (Okasha, 2009). Extending Price's equations to MLS2, Okasha proposed the following:

$$\bar{Y}\Delta\bar{Z} = cov(Y_K, Z_k) + E(Y_k \Delta Z_k)$$

The change in average group trait is equal to the covariance between the trait of the group and group's absolute fitness (number of offspring groups) and the average of the total fitness multiplied by the groups' transmission fidelity.

12. According to Bowles and Gintis (2013), the equilibrium level of genetic differentiation is represented by the following formula:

$$F_{ST}^{*} = \frac{1}{1 + 4m_e + N_e}$$

where m_e is the rate of migration between groups and N_e is the group size.

13. Some critics of the parochial altruism hypothesis claim that self-sacrificial behavior rose due to the cultural effects of warfare rather than as the outcome of its evolutionary selective forces (Wrangham, 2019).
14. In Equation 9.1, the numerator represents an increase in the average group's fitness based on an increase in the number of altruists in the group.

$$\frac{\beta(w_g, p_s)}{\beta(w_{ig}, p_{ig})} > \frac{1 - F_{ST}}{F_{ST}}$$

Alternatively, the denominator indicates a decline in fitness in an individual featuring an altruist allele (Bell et al., 2009).

15. Institutions involved in maintaining cooperation in large-scale societies; these social subsystems attain collective or higher-order benefits by imposing a cost on individuals or other lower-level components (Turchin, 2013).
16. The authors employed the following formula to model the association between ultrasociality and the outcome of a confrontation between polities:

$$\bar{U}_{att} = \frac{\sum_j \sum_i U_{ij}}{S_{att}}$$

In this equation the term $\bar{U}_{att}$ represents the success of the attack, where S_{att} is equal to the polity's size, and U_{ij} corresponds to an ultrasocial trait, ranging from 0 to 1, and present in the *i*th particle within the *j*th group.

17. The author assigned the outcome of each confrontation a value ranging from 0 to 10, where 0 represented a non-violent takeover and 10 indicated that over 50% of the defeated population was killed or enslaved. The study examined events from AD 1 to 1700.
18. For example, Chase-Dunn, Hall, & Turchin, 2007; Taagepera, 1997; Turchin, Adams, & Hall, 2006

References

Beckerman, S., Erickson, P. I., Yost, J., Regalado, J., Jaramillo, L., Sparks, C., et al. (2009). Life histories, blood revenge, and reproductive success among the Waorani of Ecuador. *Proceedings of the National Academy of Sciences, 106*(20), 8134–8139.

Bell, A. V., Richerson, P. J., & McElreath, R. (2009). Culture rather than genes provides greater scope for the evolution of large-scale human prosociality. *Proceedings of the National Academy of Sciences, 106*(42), 17671–17674.

Boehm, C. (1999). *Hierarchy in the forest: Egalitarianism and the evolution of human altruism.* Cambridge, MA: Harvard University Press.

Boehm, C. (2013). The biocultural evolution of conflict resolution between groups. In D. P. Fry (Ed.), *War, peace, and human nature: The convergence of evolutionary and cultural views* (pp. 315–340). New York: Oxford University Press.

Bowles, S. (2006). Group competition, reproductive leveling, and the evolution of human altruism. *Science, 314*(5805), 1569–1572.

Bowles, S., & Gintis, H. (2013). *A cooperative species: Human reciprocity and its evolution.* New York: Princeton University Press.

Cavalli-Sforza, L., Menozzi, P., & Piazza, A. (1994). *The history and geography of human genes.* Princeton, NJ: Princeton University Press.

Chase-Dunn, C., Hall, T., & Turchin, P. (2007). World-systems in the biogeosphere: Urbanization, state formation and climate change since the Iron Age.

In A. Hornborg & C. Crumley (Eds.), *The world system and the earth system: Global socioenvironmental change and sustainability since the Neolithic* (pp. 132–148). New York: Columbia University Press.

Choi, J. K., & Bowles, S. (2007). The coevolution of parochial altruism and war. *Science, 318*(5850), 636–640.

Divale, W. T. (1974). Migration, external warfare, and matrilocal residence. *Behavior Science Research, 9*(2), 75–133.

Dunbar, R. I., Clark, A., & Hurst, N. L. (1995). Conflict and cooperation among the Vikings: Contingent behavioral decisions. *Ethology and Sociobiology, 16*(3), 233–246.

Ellingson, T. (2001). *The myth of the noble savage* (Vol. 1). Berkeley, CA: University of California Press.

Ellsworth, R. M., & Walker, R. S. (2014). Sociobiology of lethal violence in small-scale societies. In M. DeLisi & M. G. Vaughn (Eds.), *The Routledge international handbook of biosocial criminology* (pp. 85–102). New York: Routledge.

Ember, C. R. (1978). Myths about hunter-gatherers. *Ethnology, 17*(4), 439–448.

Ember, M., & Ember, C. R. (1971). The conditions favoring matrilocal versus patrilocal residence. *American Anthropologist, 73*(3), 571–594.

Erickson, P. I. (2008). Revenge, bride capture, and gender strategies for survival among the Waorani. In S. Beckerman & P. Valentine (Eds.), *Revenge in cultures in lowland South America* (pp. 93–116). Gainesville, FL: University of Florida Press.

Fabbro, D. (1978). Peaceful societies: An introduction. *Journal of Peace Research, 15*(1), 67–83.

Gat, A. (2008). *War in human civilization*. New York: Oxford University Press.

Gat, A. (2010). Why war? Motivations for fighting in the human state of nature. In P. M. Kappeler & J. B. Silk (Eds.), *Mind the gap* (pp. 197–220). New York: Springer.

Gat, A. (2015). Proving communal warfare among hunter-gatherers: The quasi-Rousseauian error. *Evolutionary Anthropology: Issues, News, and Reviews, 24*(3), 111–126.

Gat, A. (2017). *The causes of war and the spread of peace: But will war rebound?* New York: Oxford University Press.

Glowacki, L., Wilson, M. L., & Wrangham, R. W. (2017). The evolutionary anthropology of war. *Journal of Economic Behavior and Organization*. https://doi.org/10.1016/j.jebo.2017.09.014. Retrieved from: https://www.sciencedirect.com/science/article/pii/S016726811730255X

Guilaine, J., & Zammit, J. (2008). *The origins of war: Violence in prehistory*. New York: John Wiley & Sons.

Hertler, S. C., Figueredo, A. J., Peñaherrera-Aguirre, M., Fernandes, H. B. F., & Woodley of Menie, M. A. (2018). *Life history evolution: A biological meta-theory for the social sciences*. New York: Palgrave Macmillan.

Keeley, L. H. (1997). *War before civilization*. New York: Oxford University Press.

Kelly, R. C. (2000). *Warless societies and the origin of war*. Ann Arbor: University of Michigan Press.

Kelly, R. C. (2005). The evolution of lethal intergroup violence. *Proceedings of the National Academy of Sciences, 102*(43), 15294–15298.

Knauft, B. M. (1987). Reconsidering violence in simple human societies: Homicide among the Gebusi of New Guinea. *Current Anthropology, 28*(4), 457–500.

Lahr, M. M., Rivera, F., Power, R. K., Mounier, A., Copsey, B., Crivellaro, F., et al. (2016). Inter-group violence among early Holocene hunter-gatherers of West Turkana, Kenya. *Nature, 529*(7586), 394–398.

Langergraber, K., Schubert, G., Rowney, C., Wrangham, R., Zommers, Z., & Vigilant, L. (2011). Genetic differentiation and the evolution of cooperation in chimpanzees and humans. *Proceedings of the Royal Society B: Biological Sciences, 278*(1717), 2546–2552.

LeBlanc, S. A., & Register, K. E. (2003). *Constant battles: The myth of the peaceful, noble savage*. New York: Macmillan.

Lee, R., & Devore, I. (1968). *Man the hunter*. Chicago: Adelaide.

Manson, J. H., Wrangham, R. W., Boone, J. L., Chapais, B., Dunbar, R. I. M., Ember, C. R., et al. (1991). Intergroup aggression in chimpanzees and humans [and comments and replies]. *Current Anthropology, 32*(4), 369–390.

Mathew, S., & Boyd, R. (2011). Punishment sustains large-scale cooperation in prestate warfare. *Proceedings of the National Academy of Sciences, 108*(28), 11375–11380.

Okasha, S. (2009). *Evolution and the levels of selection*. New York: Oxford University Press.

Otterbein, K. F. (1968). Internal war: A cross-cultural study 1. *American Anthropologist, 70*(2), 277–289.

Otterbein, K. F. (1989). *The evolution of war: A cross-cultural study* (3rd ed.). New Haven, CT: HRAF.

Otterbein, K. F. (2004). *How war began* (Vol. 10). College Station, TX: Texas A&M University Press.

Pinker, S. (2000). The blank slate, the noble savage, and the ghost in the machine. *Tanner Lectures on Human Values, 21*, 179–210.

Pinker, S. (2012). *The better angels of our nature: Why violence has declined.* New York: Viking.

Price, G. R. (1970). Selection and covariance. *Nature, 227*, 520–521.

Price, G. R. (1972). Extension of covariance selection mathematics. *Annals of Human Genetics, 35*(4), 485–490.

Reiter, D., & Stam, A. C. (2002). *Democracies at war.* Princeton, NJ: Princeton University Press.

Richerson, P., Baldini, R., Bell, A. V., Demps, K., Frost, K., Hillis, V., et al. (2016). Cultural group selection plays an essential role in explaining human cooperation: A sketch of the evidence. *Behavioral and Brain Sciences, 39*, 1–68.

Richerson, P. J., & Boyd, R. (1998). The evolution of human ultrasociality. In I. Eibl-Eibesfeldt & F. K. Salter (Eds.), *Indoctrinability, ideology, and warfare: Evolutionary perspectives* (pp. 71–95). New York: Berghahn Books.

Soltis, J., Boyd, R., & Richerson, P. J. (1995). Can group-functional behaviors evolve by cultural group selection?: An empirical test. *Current Anthropology, 36*(3), 473–494.

Taagepera, R. (1997). Expansion and contraction patterns of large polities: Context for Russia. *International Studies Quarterly, 41*(3), 475–504.

Turchin, P. (2003). *Historical dynamics: Why states rise and fall.* Princeton, NJ: Princeton University Press.

Turchin, P. (2007). *War and peace and war: The rise and fall of empires.* New York: Penguin.

Turchin, P. (2010). Warfare and the evolution of social complexity: A multilevel-selection approach. *Structure and Dynamics, 4*(3), 1–37.

Turchin, P. (2013). How did large-scale complex societies evolve? In P. J. Richerson & M. H. Christiansen (Eds.), *Cultural evolution: Society, technology, language, and religion* (pp. 61–74). Cambridge, MA: MIT Press.

Turchin, P. (2016). *Ultrasociety: How 10,000 years of war made humans the greatest cooperators on earth.* Chaplin, CT: Beresta Books.

Turchin, P., Adams, J. M., & Hall, T. D. (2006). East-west orientation of historical empires and modern states. *Journal of World-Systems Research, 12*(2), 219–229.

Turchin, P., Currie, T. E., Turner, E. A., & Gavrilets, S. (2013). War, space, and the evolution of Old World complex societies. *Proceedings of the National Academy of Sciences, 110*(41), 16384–16389.

Turchin, P., & Korotayev, A. (2006). Population dynamics and internal warfare: A reconsideration. *Social Evolution and History, 5*(2), 112–147.

Valentine, P., & Beckerman, S. (Eds.). (2008). *Revenge in the cultures of lowland South America*. Gainesville, FL: University Press of Florida.

Walker, R. S., & Bailey, D. H. (2013). Body counts in lowland South American violence. *Evolution and Human Behavior, 34*(1), 29–34.

Wang, S., Lewis Jr., C. M., Jakobsson, M., Ramachandran, S., Ray, N., Bedoya, G., et al. (2007). Genetic variation and population structure in Native Americans. *PLoS Genetics, 3*(11), e185.

Wendorf, F. (1968). *The prehistory of Nubia*. Dallas, TX: Southern Methodist University Press.

Wilson, M. L. (2013). Chimpanzees, warfare, and the invention of peace. In D. P. Fry (Ed.), *War, peace, and human nature: The convergence of evolutionary and cultural views* (pp. 361–388). New York: Oxford University Press.

Wrangham, R. (2019). *The goodness paradox: The strange relationship between virtue and violence in human evolution*. New York: Pantheon.

Wrangham, R. W., & Glowacki, L. (2012). Intergroup aggression in chimpanzees and war in nomadic hunter-gatherers. *Human Nature, 23*(1), 5–29.

Wrangham, R. W., Wilson, M. L., & Muller, M. N. (2006). Comparative rates of violence in chimpanzees and humans. *Primates, 47*(1), 14–26.

Wright, S. (1935). Evolution in populations in approximate equilibrium. *Journal of Genetics, 30*(02), 257–266.

Zefferman, M. R., & Mathew, S. (2015). An evolutionary theory of large-scale human warfare: Group-structured cultural selection. *Evolutionary Anthropology: Issues, News, and Reviews, 24*(2), 50–61.

10

The Sociopolitical Integrity of the Roman State: Intragroup Competition, Intergroup Competition, and Economic Dynamics

Mateo Peñaherrera-Aguirre, Aurelio José Figueredo, and Steven C. Hertler

1 Introduction

As exemplified by the history of the Roman State, economic distress can decrease the level of intrasocietal cohesion even in the absence of external pressures. Since its inception, the Roman Republic faced numerous foreign threats, from rival cites in the Italian peninsula to tribal confederacies in Gaul. A chronic state of intergroup conflict favored the evolution of cultural variants necessary for sustaining large-scale cooperation. These innovations allowed the Republic to conquer its neighbors and expand outside the Italian peninsula. Even though the spoils of war created a climate of macroeconomic prosperity in Rome, rising income and wealth inequality, along with the elimination of ultrasocial institutions, increased popular discontent. It was during these times of turmoil that ambitious men dismantled the foundations designed to guard the State from the exploitation of autocrats. Debates concerning land and wealth distribution, at first restricted to senatorial rhetoric, escalated into street

S. C. Hertler et al., *Multilevel Selection*, https://doi.org/10.1007/978-3-030-49520-6_10

revolts and political assassinations. The mobilization of "private armies," as part of a series of civil wars, would eventually lead to the end of the Republic.

Past the quasi-legendary Roman Monarchy (with seven kings purportedly reigning from 753 to 509 BC), most of the history of the Roman State can be roughly divided into at least three major political periods, sometimes prefixed with the adjectives *early* or *late* as subdivisions: (1) the Republic (509–27 BC), (2) the Principate (27 BC–AD 284), and (3) the Dominate (AD 284–476). Although both latter periods are often collectively referred to as the Roman Empire, the Principate maintained the outward forms of Republican government while being ruled by a de facto autocrat, whereas the Dominate was the culmination of a gradual transition to outright despotism. The rise of the Principate temporarily reestablished peace after the civil wars that wracked the Late Republic; however, the continuous territorial expansion, and its corresponding economic growth, once again decreased intragroup cooperation, bringing the Empire close to the brink of collapse during the third century AD. Even though the Roman State would survive for centuries to come, territorial fragmentation between East and West would mark the end of classical Rome. The case of the Roman State evidences the fragility of within-group cooperation once outside rivals are neutralized or eliminated. Though multiple volumes exist on the history of Rome, fewer studies have quantitatively examined the association between Rome's economic distress and its decline in State cohesion.[1] Hence, in addition to providing a historical overview of the dynamic interaction between external warfare, civil warfare, and economic growth, we present empirical data on the covariation of these factors within classical Rome.

2 Roman Physical Ecology

The geographical greatness of Rome, its vast holdings and far-flung outposts spanning the 24th to the 56th parallel (Harper, 2017) and ranging longitudinally between 9° East and 38° West (Harper & McCormick, 2018), requires one to speak of multiple climates (Blasi et al., 2014; Blasi,

Filibeck, Frondoni, Rosati, & Smiraglia, 2004). Bounded by the *Atlantic* climate of the Northern and Western reaches of the Empire to the north, and the once-fertile deserts of Northern Africa to the south, the heartland of the Roman Empire was a Mediterranean climate assembling the promising combination of arid summers and humid winters under a temperate constant. In addition, an extended warm period provided the Roman State with unusually mild and agriculturally favorable climatic conditions from about 250 BC to AD 400.

For as long as this *Roman Climate Optimum* prevailed, so did its correlative parameters, among which can be counted solar output, volcanic activity, surface temperatures, and precipitation.[2] Except for one significant event in the first century BC, ice core samples and dendrochronology indicate that Rome also enjoyed an extended length of low volcanic activity, which was only interrupted by a cluster of significant activity sixty years after the fall of the Western Roman Empire, beginning around AD 536 (Harper & McCormick, 2018). Without gasses, dirt, and dust occluding the atmosphere, there were then correspondingly higher temperatures (Koutsoyiannis, Montanari, Lins, & Cohn, 2009), which were exceptionally stable between 60 BC and AD 90, as established by dinoflagellate cysts sampled at the Po River Delta (Chen, Zonneveld, & Versteegh, 2011). Keeping in mind its relevance, we then consistently see warm, wet summers give way to stably low rainfall throughout the latter part of the third century AD. Despite briefly rebounding under the reigns of Emperors Constantine and Valentinian, adequate precipitation did not return until after 476 (Büntgen et al., 2011).

During the Roman Climate Optimum, there was thus a spate of more than three centuries wherein the whole of the circum-Mediterranean region enjoyed an advantageous combination of relative stability and generous warmth (Harper & McCormick, 2018). Warmth and moisture rendered the southern reaches of Italy and the north of Africa sufficiently fertile to serve as net exporters of grain. It was upon such surpluses that the Roman Army was said to march. Rome was then a preindustrial, organic economy, deeply dependent on cereal crops for which reliable rains were necessary and providentially provided for a long spate (Harper, 2017) previous to decline and dissolution. In sum, the Romans leveraged the Roman Climate Optimum to erect an ever-expanding supranational State out of the societal entropy of the ancient world. Even without reference to the direction or absolute value of any climatic parameter, change imparts stress and strain on a society in and of itself. In progressing from the Roman Climate Optimum (200 BC–AD 150) to the Transitional Period (AD 150–450), as Harper (2017) explains, stable conditions gave way to severe variability and thereafter to frigid temperatures, which were brought about by an atmosphere occluded by volcanism, even as there was a concomitant reduction in solar output.[3]

3 Intergroup and Intragroup Competition in the Roman Republic

Though historians continue to debate the veracity of some events regarding Rome's history,[4] little disagreement exists in terms of the influence that the threat of invasion had on the political evolution of Rome (Duncan, 2017; Turchin, 2007). In its early years, the Roman Republic frequently clashed with tribal societies, such as the Umbrians and the Aequi (Oakley, 2004; Rawlings, 2007), as well as with Italian city-states, including those of the Etruscans and the Latins (Cornell, 2012; Keppie, 2002). A combination of diplomacy and military innovations allowed the Republic to expand its sphere of geopolitical influence beyond the Italian peninsula.[5,6] The political, mercantile, and military influence of Rome in the Mediterranean inevitably led to conflict with other polities in Eurasia and North Africa (Duncan, 2017). The elimination of the rival city-state of Carthage by the end of the Third Punic War generated a cascade of geopolitical and macroeconomic changes that would alter Western history for centuries to come. This event would also modify the internal organization of the Roman State.

In the first half of the 2nd Century BC, the treasury of the Roman Republic reached a steady growth. Current historical reconstructions suggest wealth accrual fluctuated during this period. For instance, the economic landscape of the Middle and the Late Republic indicates the fortunes of the aristocracy grew from 4–5 million *sestertii* in the second century BC to 25 million in the first century BC (Scheidel, 2017). The influx of goods and services facilitated this increase after the defeat of Carthage and Corinth (Duncan, 2017). For example, close to a third of the senatorial class augmented their fortunes, thanks to military campaigns (Scheidel, 2017). Rather than decreasing in subsequent decades, the wealth of the aristocracy grew to 200 million *sestertii* between 60 and 50 BC (Scheidel, 2017).

This level of prosperity, however, encountered a point of inflection. Current estimates suggest that external threats decreased the amount of wealth, forcing the Senate to either sell or melt valuable goods (Turchin & Nefedov, 2009). Even though military campaigns in the Late Republic initially represented opportunities for looting and raiding, the expenses required to sustain such confrontations in some circumstances exceeded the gains collected (Duncan, 2017). Hence, in contrast to the plunder acquired during conflicts, such as the war against Macedon (providing 120,000,000 *sestertii*; De Nardis, 2015a), confrontations against invading tribal groups, such as the Cimbri, did not fill the Republic's coffers (Duncan, 2017). Territorial reorganization, the implementation of new tax laws, as well as collecting tribute assuaged these financial shortfalls (Duncan, 2017). It was in this economic context that modifying the quality of the coins became a common alternative for supplementing the budget (Turchin & Nefedov, 2009). Metallurgical analysis suggests that during the Late Republic, senators and consuls implemented coin debasement as a tactic to counterbalance financial distress.[7] Debasing the currency was thought to both decrease the public debt and permit the continuation of expensive military campaigns (Turchin & Nefedov, 2009). Even though currency debasement temporarily balanced the state's finances, the reduction of metal content eventually increased inflation.

Resource competition (wealth accumulation and skewed distribution) was among the core causes of the various civil wars. Two factions dominated the political theater between 80 BC and 30 BC: the *Optimates*, interested in the preservation of the political and economic platform of the *patrician* aristocracy, and the *Populares*, agitating for improvement in the condition of the *plebian* lower classes by proposing wealth redistribution schemes, such as agricultural reforms based on land expropriation (Duplá, 2011). Although clashes between *Optimates* and *Populares* were initiated as heated senatorial debates, conflict escalation frequently led to expulsions, executions, and even assassinations (Scheidel, 2017). Between 90 and 80 BC, 291 senators died violently (Scheidel, 2017). The violent elimination of political competitors allowed senators to seize the wealth of other magistrates, further increasing the economic inequality and exacerbating the underlying competition among social classes (Scheidel, 2017). The imperial system emerged from chronic economic and political discontent, as well as the exacerbation of past grievances, which escalated to a series of internal wars during the Late Republic, including the (1) Social Wars, (2) Sullan Wars, (3) Servile Wars, (4) Caesar's Civil War, (5) Liberators' Civil Wars, and (6) the final civil war that ended the Roman Republic (Duncan, 2017).

To this day, the causes behind the end of the Republic remain open to discussion. Barton (2001), for example, argued that the last century of the Roman Republic witnessed the abandonment of codes of honor along with symbols, rituals, and gestures associated with these ultrasocial cultural institutions. Furthermore, traditions proscribing certain forms of intragroup competition became ignored or actively modified (Barton, 2001). The elimination of regulations promoting within-group cooperation enabled the rise of political upstarts who would continue to dismantle the remaining sociopolitical safeguards against the emergence of autocrats. Golden (2013) reached a similar conclusion. The author argued that the modification of institutions and roles associated with crisis management paved the way for the eventual end of the Republican organization. During the Early and Middle Republic, the State had at its disposal a series of institutional measures destined to preserve Rome's safety. When armed forces threatened the security of the State, the senate could declare a *tumultus*, or a *senatus consultum ultimum*, and thereby suspend some civil liberties. According to Golden, Roman politics in the Early Republic specified that executive leaders should be granted temporary dictatorial powers to address the urgency of the threat. As Rome expanded outside of Italy, however, the flexibility of the Roman political system phased out the office of the dictator to manage crises occurring abroad (Golden, 2013). Instead, the Roman Senate became the main governmental body responsible for addressing an emergency. This institutional reorganization allowed the Roman Republic to outlast the

Second Punic War. As per Golden, Rome reached a political impasse due to the growing social discontent between commoners and the aristocracy after the Punic Wars. This rivalry led some senators to employ crisis response institutions against political rivals, setting a precedent for future autocrats to follow.

Even though the Roman State exhibited sociopolitical stratification before the ignition of the crisis of the Late Republic, the presence of ancestral ultrasocial traditions decreased the likelihood of minor disagreements escalating into lethal outcomes (Barton, 2001). Concerning the association between territorial expansion and increased intragroup competition, Barton pondered that

> When, as a result of the imperial expansion of Rome, the spiritual walls around the city were irreparably breached, *urbanitas,* originally the ways and manners peculiar to those who lived within the walls of the city of Rome, took on the connotation of our modern "urbanity" or "cosmopolitanism." The citizen of Rome became a citizen of the world. And because, for the cosmopolite, limits, like definitions, had to be chosen, morality and adhesion to particular traditions and limits required a prodigious act of will. Preserving a sense of being, of identity, thus became a continuous—and ultimately exhausting—assault on the will. As a result, the enervating power of unrelieved good fortune became as common a theme in Roman literature as the annealing power of adversity. (p. 95)

Barton's argument gravitates around the impact of cultural heterogeneity eroding the persistence of Roman cultural variants, including those associated with intragroup cooperation. Hence, Roman settlements located close to the State's borders should have exhibited higher levels of within-group competition due to their exposure to cultural variants developed by outside groups. Although theoretically plausible, it is worth remembering that mathematical models suggest cultural group differences are more pronounced closer to ethnolinguistic boundaries (McElreath, Boyd, & Richerson, 2003). Similarly, subsequent models have also demonstrated that intragroup cooperation tends to decline closer to the center of the polity (Turchin, 2003). Hence, future studies are required to determine how Roman morality, and consequently intragroup cooperation, varied depending on its territorial expansion and its proximity to other cultures.

4 Intergroup and Intragroup Competition During the Principate and the Dominate

Military campaigns after the collapse of the Roman Republic allowed the Principate to reach its maximum geographical extension, such as at the time of the Marcomanni Campaigns or the Britannic Wars of Consolidation (Potter, 2014). This territorial growth generated a period of prosperity and peace,[8] as evidenced in the declining frequency and lethality of external wars (Rankov, 2015; Whatley, 2015). Imperial campaigns became asymmetrical confrontations wherein Roman legions frequently fought against tribal societies (Levick, 2002). Rather than encountering well-organized enemies during the Principate, the imperial army faced ambushes, raids, and other forms of guerilla tactics with higher frequency (Thorne, 2015a, 2015b). During the Dominate and the Late Empire, external wars continued to transpire. Despite noticeable political differences between the late republican and imperial structures, the Empire remained vulnerable to the occurrence of military clashes among generals by inheriting the military practices of the Late Republic. While earlier Republican morality enjoined the armies to protect the abstract notion of a unified state, the legions' allegiance during the Principate and the Dominate lay personally with their commanders.[9] (Drinkwater & Lee, 2015). Under these conditions, military capacity had considerable political repercussions[10] (Campbell, 2015). Hence, victories bolstered the political stance of the ruler, whereas military defeats could lead to the questioning of their commanding competency (Drinkwater & Lee, 2015).

The Roman State required a constant influx of wealth to continue with its military campaigns. The fiscal and monetary crises ensuing by the end of the Late Republic-Early Principate forced Emperor Augustus to implement radical institutional reforms to restore the Roman State to its former levels of political, social, and economic stability. Military expenditures, however, remained a lingering issue during the rest of Principate, with military campaigns consuming a sizeable portion of the state's budget

(Turchin & Nefedov, 2009). Some estimates suggest that military expenditures required over 500 million *sestertii* every year during the first century AD (De Nardis, 2015b). Private fortune accumulation, however, continued to grow during the Empire. Current reconstructions suggest that the wealth of the aristocracy in the Principate and the Late Empire ranged between 300 and 400 million *sestertii*, 80 times higher than the elites' wealth of the late second century BC (Scheidel, 2017). Augustus, for example, received over 1.4 billion *sestertii* from other aristocrats during his reign as *Princeps* (Scheidel, 2017). In response to the State's debt, Nero debased the denarius from 3.72 grams of silver to 3.14 (Turchin & Nefedov, 2009). Nero's[11] decision would set a precedent for future emperors who would devalue the currency repeatedly, considering this practice as an alternative to selling the palace's treasury (Turchin & Nefedov, 2009). During Commodus's reign,[12] the silver content per coin declined rapidly. The Severan dynasty temporarily halted this financial disaster (Turchin & Nefedov, 2009). However, climatic, political, and social instability during the third century once again generated a financial crisis that would plunge the Empire into a series of civil wars[13] (Harper, 2017).

Akin to the fate of the *denarius*, other currencies experienced similar devaluations. For example, the *sestertius* was frequently melted down and combined with other metals (Harper, 2017). Although the *antoninianus* was introduced to dampen the monetary crisis, this currency was also debased by becoming an alloy coin (Harper, 2017). Emperor Aurelian tried to contain the economic freefall; however, despite his efforts, inflation continued to escalate during his reign (Harper, 2017). Diocletian[14] also attempted to decrease financial stress by replacing the *denarius* with the *argenteus* (Burgess, 2015). By the turn of the fifth century, the devaluation of silver and bronze finally led to a complete halt in production (Burgess, 2015). During this time, inflation not only increased the cost of everyday goods, such as wheat (Harper, 2017), but also increased the price of gold (Burgess, 2015). In five centuries, the value of a pound of gold rose from 1500 *denarii* in 46 BC to 5184 billion in AD 400 (Burgess, 2015). Although there is no consensus concerning the association between wealth accumulation, currency debasement, and inequality, trend examinations across 9000 years of European history indicate that one of the highest peaks in inequality occurred between AD 125 and 425 in Rome (Scheidel, 2017).[15] These dire socioeconomic conditions became the catalysts for lethal internal strife, ranging from minor revolts and rebellions limited to a specific province, to full-fledged civil wars involving multiple regions.

In contrast to the Principate, civil wars during the Dominate and the Late Empire became considerably bloodier relative to external wars (except for the battle of Adrianople in AD 378; Whatley, 2015). These internal conflicts, in turn, decreased the ability of the Empire to defend its borders, and this led to chronic confrontations with tribes such as the Alamanni and the Goths (Stickler, 2007). In this state of political despair, the Empire opted for the recruitment of *foederati*, which temporarily decreased the frequency of conflict (Stickler, 2007). Unfortunately, the recurrent invasions of Vandals, Alans, and Huns overwhelmed these defenses, and this led to the abandonment of the Western Roman Empire's borders and facilitated its fall in AD 476 (Duncan, 2017). The military history of the Roman State depended on its treasury, with successful campaigns returning on its investment and defeats draining its coffers. Consequently, to sustain its expansion and defense of its borders, the Roman State opted for various economic practices, some of which would eventually lead to significant internal crises.

5 Empirical Examination: Methods

The present study examined the association between economic distress and cohesion in the Roman State. Biographical data of 190 *Roman Chief Executives* (RCEs) were collected from various historical sources, including (1) Broughton's *The Magistrates of the Roman Republic* volumes I (Broughton, 1951) and II (Broughton, 1952), (2) Livy's collection on *The History of Rome* (1982, 2006), (3) *The Roman Antiquities* (Cary, 1937), (4) Potter's *Emperors of Rome* (2014), (5) Matyszak's *Chronicle of the Roman Republic* (2008), (6) Scarre's *Chronicle of Roman Emperors* (1995), (7) *The Encyclopedia Britannica*, and (8) the *De Imperatoribus Romanis* online website. Some of the measures extracted from these sources included the RCE's age at the start of rule, the time of government, the age of death, and the cause of death. Mortality rates per 100,000 ruler-years were also estimated (ruler-years were used instead of person-years due to our interest in the risk period associated with their time of rule, rather than their overall lifetime). Following Eisner's publication on European regicide (Eisner, 2011), causes of death were classified as *suicide*, *accident*, *warfare*, *assassination*, *execution*, or *natural.* Killings were further classified based on the degree of doubt or certainty, which was determined based on the level of consistency across literary reports (Eisner, 2011). As the Republican political system differed from the dynastic rule observed in imperial times, the time for which each senator occupied the office of consul was used instead for Republican magistrates.

Data on two indicators of physical ecology, *solar irradiance*, and *precipitation* were collected from Steinhilber, Beer, and Fröhlich (2009) and Büntgen et al. (2011). According to Steinhilber et al. (2009), estimation of solar irradiance is based on cosmogenic radionuclide ^{10}Be extracted from ice cores. Büntgen et al. (2011) estimated the level of precipitation according to the tree-ring samples collected from Germanic locations. This measure has been found to predict social unrest and the assassinations of Roman emperors (Christian & Elbourne, 2018). Data from 504 external and civil war battles were counted.[16] The average weight (in grams) of silver and bronze coins per decade was collected from the *Coinage of the Roman Republic Online* (American Numismatic Society & The British Museum, 2018) and the *Online Coins of the Roman Empire* databases (American Numismatic Society & The Institute for the Study of the Ancient World, 2018). The frequency of battles in external and civil wars was computed based on the information provided by Don Taylor's compendiums on battles in the Roman Republic (Taylor, 2017) and Empire (Taylor, 2016).

Information on coin hoard frequency was gathered from the Coin Hoards of the Roman Republic (Lockyear, 2018) and the Coin Hoards of the Roman Empire (Thyssen-Bornemisza, 2018) databases. It is relevant to provide an additional explanation regarding the scientific relevance of coin hoards for examining economic fluctuations. According to Crawford, different from other forms of monetary collection, such as depositing coins in an *arcas*, or an *armariums* (Crawford, 1969), coin hoards reflect social circumstances in which the hoarder was forced, due to dangerous sociopolitical conditions, to hide his wealth in hopes of retrieving it sometime in the future (Crawford, 1974). For Crawford (1969), the presence of external and internal wars encouraged people to use caches, hoping others would not loot their wealth. In his examination of coin hoards from Italy, Corsica, Sardinia, and Sicily dating from 218 to 03 BC, he found the frequency of hoards increased during the Second Punic War, the Social War, Caesar's Civil War, and the Final War of the Republic[17] (Crawford, 1969). In contrast to these results, previous publications had found coin hoarding increased in the absence of any particular social disturbance (Aitchison, 1988). If this is indeed the case, coin hoards should not be used as the sole indicator of sociopolitical instability but in conjunction with other variables reflecting these crises. Coin hoards had also been interpreted as a proxy for monetary savings (Aitchison, 1988). Due to the fact hoarders are assumed to cache goods based on the value of the items at the time of their burial (Reece, 1988), the frequency of caches may also be considered as a metric of wealth accumulation (see Martin, 1995, for another perspective concerning the difficulties of reconstructing the monetary value associated with each hoard).

6 Empirical Examination: Analyses

Parametric examinations require the data to be statistically independent. This assumption, however, is not met when examining temporal patterns. The presence of serial autocorrelations in the data increases the likelihood of incurring Type I errors when exploring the covariation between two variables. Statistical methods, such as linear and nonlinear time series analyses, provide an alternative by circumventing these issues. In the current chapter, however, multiple linear mixed models with random intercepts, based on maximum likelihood estimation and specifying a residual covariance matrix of single-lagged heterogeneous serially autoregressive effects (*ARH1*), were computed with the decade as a predictor. This procedure extracted residualized values, after controlling for temporal and autoregressive effects, for the various indicators used in subsequent analyses for this chapter. The current study employed unit-weighted factor scoring to examine the underlying factorial structure of three factors evidencing: (1) intragroup competition; (2) intergroup competition; and (3) economic distress. The *intragroup competition factor* contained the standardized rate of RCEs killed in civil war battles per 100,000 ruler-years per decade, the standardized rate of RCEs assassinated per 100,000 ruler-years per decade, and the standardized number of civil war battles per decade. The *intergroup competition factor* included the standardized rate of RCEs killed in external battles per 100,000 ruler-years per decade and the standardized number of external battles fought per decade. Similarly, a higher-order factor containing the standardized intragroup competition factor scores (reverse scored) and the standardized intergroup competition factor scores was also estimated to explore the overall level of the State cohesion. In terms of the economic variables, a lower-order factor was calculated based on the standardized values of the weight loss of silver and bronze coinage per decade, relative to their baseline weight in 220 BC (the oldest coins in the dataset). In turn, the standardized values of this metallurgic factor, along with the standardized frequency of coin hoards per decade, served to compute a general *economic distress factor*.

7 Empirical Examination: Results

As explained above, all the following results are based on standardized MLM residuals adjusted for both the effects of time and serial autocorrelations among the successive decadal observations. The main measurement model is shown in Fig. 10.1. The *intra*state conflict factor loaded onto the frequency of civil wars per decade, the rate of RCEs civil war battle deaths per 100,000 years in office per decade, and the rate of RCEs assassinations per 100,000 years in office per decade. The *inter*state conflict factor loaded onto the frequency of external battles per decade and onto the rate of RCEs' external battle deaths in external warfare. This latent variable, in turn, loaded negatively onto the intrastate conflict factor and positively onto the intergroup conflict factor.

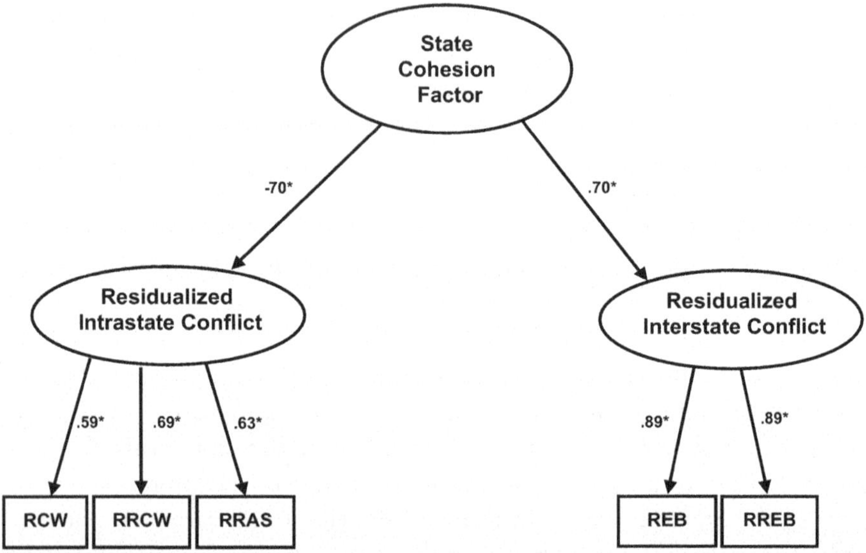

Fig. 10.1 The latent hierarchical structure of the general State cohesion factor (REB, residualized frequency of external battles per decade; RREB, residualized rate of RCEs' deaths in external battles per 100,000 ruler-years per decade; RCW, residualized frequency of civil war battles per decade; RRCW, residualized rate of RCEs' deaths in civil war battles per 100,000 ruler-years per decade; RRAS, residualized rate of RCEs' assassinations per 100,000 ruler-years per decade). (* $p < 0.05$)

An additional examination determined that the relative loss of coinage metal quantity had a positive correlation with the frequency of coin hoards per decade ($r = .315$, $p = .001$). A single economic distress factor thus loaded onto the coinage weight loss and the coin hoard indicators ($r = .811$, $p = .001$). As expected, this economic distress factor then negatively predicted the State cohesion factor ($r = -.256$, $p = .029$). In an alternative analysis, we also considered the ratio between the levels of intergroup and intragroup conflict in place of the additive estimate for the State cohesion factor. The economic distress factor, however, did not significantly predict this conflict ratio ($r = .068$, $p = 568$). Given that the Mediterranean location and Roman Climate Optimum were considerably important to the growth and productivity of the Roman State, we also decided to explore the possible confounding influence of climatic factors on State cohesion. A hierarchical general linear model examined

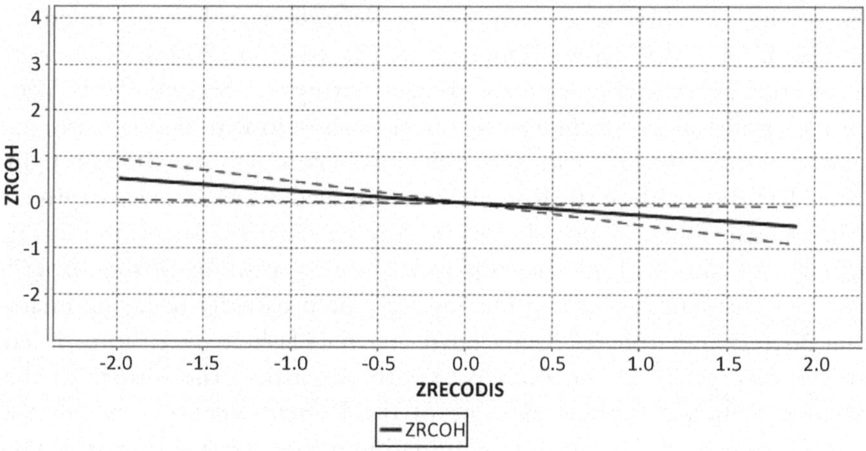

Fig. 10.2 Time-adjusted MLM-standardized residuals of State cohesion (Z-RCOH) predicted by time-adjusted MLM-standardized residuals of economic distress (Z-RECODIS), after accounting for the physical ecology factor

the effect a common factor of physical ecology,[18] the economic distress factor, and the interaction the two predictors had on the general State cohesion factor. The overall model accounted for 33% of the variance. The economic distress factor negatively predicted the level of State cohesion ($r = -.260$, $p = .02$; see Fig. 10.2) above and beyond the effects of physical ecology ($r = .080$, $p = .50$). The interaction between economic distress and physical ecology did not predict the State cohesion factor ($r = .140$, $p = .22$). These analyses demonstrated that the influence of economic distress on the integrity of the Roman State was not affected by climate.

8 Conclusions

The looming threat of external wars promoted the evolution of political and social institutions in Rome favoring large-scale endeavors, such as the defense of the State's territorial integrity (Turchin, 2007). The establishment of coalitions and alliances with other Italian city-states allowed

Rome and its allies to concentrate on defending against the frequent invasions of Celtic tribes (Duncan, 2017; Turchin, 2007). Moreover, even after experiencing repeated defeats during the Second Punic War, Rome's political and military organization allowed it to defeat Carthage and its allies in the long run. While the loss of competing polities brought considerable wealth to the Roman State (in the form of enslaved captives, tribute, and loot), it also eliminated any external pressures promoting intragroup cooperation. The Roman Republic transitioned from experiencing considerable levels of macroeconomic prosperity to facing financial distress and suffering from lethal internal clashes, which in turn led to the emergence of the Principate with Augustus. The history of the Roman Principate echoes the fate of the Roman Republic, wherein a period of territorial expansion, eliminating major external threats, is followed by an increase in magnicides, revolts, and civil wars. The numismatic, biographic, and military information collected for this chapter supports the hypothesized association between Rome's economic distress and its decline in State sociopolitical integrity.

Notes

1. The expression *State cohesion* herein refers to the integration of Stratum III, defined as per Chap. 7, as opposed to Stratum II regional polities that might have retained internal unity during civil wars and rebellions.
2. Concerning the solar output, Harper and McCormick (2018) presented descriptive data showing lows of below 1365 watts/per square meter, bracketing the Empire's rise and fall, with the Roman Climate Optimum enjoying relatively higher and less variable insolation.
3. According to Harper and McCormick (2018), circa AD 250 brought "a phase of instability and general cooling that lasts until the Medieval Climate Anomaly."
4. Including Rome's regal period, the expulsion of the last Tarquin king and the foundation of the Republic.
5. Albeit wars of conquest under the Middle Republic were detained by foreign invasions such as that of the Pyrrhic and the Punic Wars (Keppie, 2002), after the resolution of these conflicts, Roman borders extend beyond the Italian peninsula (Beard, 2015).

6. In the Late Republic, Carthage's defeat and the annexation of Corinth also saw the transformation of military campaigns from affairs circumscribed to the season cycle to prolonged endeavors lasting for years (Duncan, 2017). In later years, institutional changes, such as Marius' reforms, allowed the Roman State not only to defend its territorial integrity but also to wage wars at any point in time (Duncan, 2017).
7. For example, between 48 BC and 28 BC, Caesar and Mark Anthony, after him, debased the denarius to 92% of its original silver content (Turchin & Nefedov, 2009).
8. Except for the Parthian and Sasanian Wars (Farrokh, 2007).
9. Although the office of the princeps relied on the senate's legitimation (Hekster, 2015), most of the emperor's legitimacy came from the support of the army (Ando, 2007; Drinkwater & Lee, 2015).
10. For example, in the late imperial period, the imperial authority was mainly contested by imperial officers who grew in military might due to their defense of the borders but also by establishing patronage relationships with local *foederati* (Liebeschuetz, 2007).
11. Initially, the political reorganization of the Principate saw a decline in major internal conflicts, such as civil wars, though provincial revolts remained a pervasive nuisance (Cosme, 2015). This tranquility was eventually interrupted by intense domestic crises. The first one occurred after Nero's suicide, with the empire falling into its first major civil war in AD 68, since the Final War of the Roman Republic (Turchin & Nefedov, 2009). Following this period of instability, the ascension of the Nerva-Antonine Dynasty saw a period of political and social growth (Turchin & Nefedov, 2009).
12. After Commodus' murder in AD 192, once again, a period of political unrest between AD 193 and 197 ensued, leading to a civil war between Clodius Albinus, Septimius Severus, and Pescennius Niger (Cosme, 2015).
13. Interestingly, the frequency of assassinations occurring during this period decreased the regularity and intensity of civil wars, with rulers being killed before armies could face in battle (Drinkwater & Lee, 2015).
14. Even though the reign of figures such as Diocletian increased the level of political security, civil wars during the Dominate, such as the wars of the Tetrarchy, as well as those in the Late Empire, remained recurrent affairs (Drinkwater & Lee, 2015).

15. Although further analyses are required to determine the effect of inequality upon social disturbances in Antiquity, robust evidence across contemporary nation-states had found inequality indicators. The Gini coefficient, for example, is correlated with metrics of internal social upheaval, such as homicide rates (Daly, Wilson, & Vasdev, 2001; Figueredo et al., 2017; Peñaherrera-Aguirre et al., 2019). Hence, across societies, income, and wealth inequality covary with lethal competition within social organizations (Daly, 2017).
16. We would like to thank Maya Louise Bose, Garrett Dien, and Jonathan Revel for their help in collecting these data.
17. More recently, Turchin and Nefedov (2009) further examined Crawford's data and its correspondence with an instability index (based on the occurrence of rebellions, magnicides, civil wars, and wars with other nation-states and tribes), indicating an overlap between the frequency of hoards found on each decade and the degree of social disruption. Even though the association between coin hoarding and this index is noticeable during the Republic, it is less clear for the Principate. Turchin and Nefedov (2009) found an overlap between the frequency of coin hoards in Britain and Alexandria and the instability observed in the second half of the third century AD. This general trend, however, indicated considerable variation depending on where the hoards were found, suggesting local crises could also impact the frequency of hoarding. Although suggestive, the authors recognized this association might be confounded by the level of monetization of the economy, as well as how long the hoard was buried, with older caches having a higher risk of being found and looted (Turchin & Nefedov, 2009).
18. This factor loaded onto the average precipitation per decade and the average solar irradiance per decade ($r = 0.645$, $p = 0.0001$).

References

Aitchison, N. (1988). Roman wealth, native ritual: Coin hoards within and beyond Roman Britain. *World Archaeology, 20*(2), 270–284.

American Numismatic Society & The British Museum. (2018). *Coinage of the Roman Republic online*. Data retrieved from http://numismatics.org/crro/visualize/metrical

American Numismatic Society & The Institute for the Study of the Ancient World. (2018). *Online coins of the Roman Empire*. Data retrieved from http://numismatics.org/ocre/visualize/metrical

Ando, C. (2007). The army and the urban elite: A competition for power. In P. Erdkamp (Ed.), *A companion to the Roman Army* (pp. 359–378). Oxford, UK: John Wiley & Sons.

Barton, C. A. (2001). *Roman honor: The fire in the bones*. Berkeley, CA: University of California Press.

Beard, M. (2015). *SPQR: A history of ancient Rome*. New York: Liveright Publishing Corporation.

Blasi, C., Capotorti, G., Copiz, R., Guida, D., Mollo, B., Smiraglia, D., et al. (2014). Classification and mapping of the ecoregions of Italy. *Plant Biosystems-An International Journal Dealing with All Aspects of Plant Biology, 148*(6), 1255–1345.

Blasi, C., Filibeck, G., Frondoni, R., Rosati, L., & Smiraglia, D. (2004). The map of the vegetation series of Italy. *Fitosociologia, 41*(1), 21–25.

Broughton, T. S. (1951). *The magistrates of the Roman Republic 1: Volume 1: 509 B.C.- 100 B.C.* New York: American Philological Association.

Broughton, T. S. (1952). *The magistrates of the Roman Republic 2: Volume 2: 99 B.C.- 31 B.C.* New York: American Philological Association.

Büntgen, U., Tegel, W., Nicolussi, K., McCormick, M., Frank, D., Trouet, V., et al. (2011). 2500 years of European climate variability and human susceptibility. *Science, 331*(6017), 578–582.

Burgess, R. (2015). Numismatics: Late Empire. In Y. Le Bohec, G. Brizzi, E. Deschler-Erb, G. Greatrex, B. N. Rankov, & M. Reddé (Eds.), *The encyclopedia of the Roman Army* (pp. 673–704). Oxford, UK: Wiley Blackwell.

Campbell, B. (2015). Imperator: Principate. In Y. Le Bohec, G. Brizzi, E. Deschler-Erb, G. Greatrex, B. N. Rankov, & M. Reddé (Eds.), *The encyclopedia of the Roman Army* (pp. 509–510). Oxford, UK: Wiley Blackwell.

Cary, E. (1937). *Dionysius of Halicarnassus: Roman antiquities*. Cambridge, MA: Harvard University Press.

Chen, L., Zonneveld, K. A., & Versteegh, G. J. (2011). Short term climate variability during "Roman classical period" in the eastern Mediterranean. *Quaternary Science Reviews, 30*(27–28), 3880–3891.

Christian, C., & Elbourne, L. (2018). Shocks to military support and subsequent assassinations in ancient Rome. *Economics Letters, 171*, 79–82.

Cornell, T. (2012). *The beginnings of Rome: Italy and Rome from the Bronze Age to the Punic Wars (c. 1000–264 BC)*. London: Routledge.

Cosme, P. (2015). Civil wars: Principate. In Y. Le Bohec, G. Brizzi, E. Deschler-Erb, G. Greatrex, B. N. Rankov, & M. Reddé (Eds.), *The encyclopedia of the Roman Army* (pp. 205–211). Oxford, UK: Wiley Blackwell.

Crawford, M. (1969). Coin hoards and the pattern of violence in the late republic. *Papers of the British School at Rome, 37*, 76–81.

Crawford, M. (1974). *Roman Republican coinage*. London: Cambridge University Press.

Daly, M. (2017). *Killing the competition: Economic inequality and homicide*. New York: Routledge.

Daly, M., Wilson, M., & Vasdev, S. (2001). Income inequality and homicide rates in Canada and the United States. *Canadian Journal of Criminology, 43*, 219–236.

De Nardis, M. (2015a). Economy and the army: Late republic. In Y. Le Bohec, G. Brizzi, E. Deschler-Erb, G. Greatrex, B. N. Rankov, & M. Reddé (Eds.), *The encyclopedia of the Roman Army* (pp. 351–388). Oxford, UK: Wiley Blackwell.

De Nardis, M. (2015b). Economy and the army: Principate. In Y. Le Bohec, G. Brizzi, E. Deschler-Erb, G. Greatrex, B. N. Rankov, & M. Reddé (Eds.), *The encyclopedia of the Roman Army* (pp. 358–362). Oxford, UK: Wiley Blackwell.

Drinkwater, J., & Lee, A. D. (2015). Civil wars: Late empire. In Y. Le Bohec, G. Brizzi, E. Deschler-Erb, G. Greatrex, B. N. Rankov, & M. Reddé (Eds.), *The encyclopedia of the Roman Army* (pp. 211–219). Oxford, UK: Wiley Blackwell.

Duncan, M. (2017). *The storm before the storm: The beginning of the end of the Roman Republic*. New York: Public Affairs.

Duplá, A. (2011). Consules populares. In H. Beck, A. Duplá, M. Jehne, & F. P. Polo (Eds.), *Consuls and res publica: Holding high office in the Roman Republic* (pp. 279–298). Cambridge, UK: Cambridge University Press.

Eisner, M. (2011). Killing kings: Patterns of regicide in Europe, AD 600–1800. *The British Journal of Criminology, 51*(3), 556–577.

Farrokh, K. (2007). *Shadows in the desert: Ancient Persia at war*. New York: Osprey.

Figueredo, A. J., Cabeza de Baca, T., Fernandes, H. B. F., Black, C. J., Peñaherrera, M., Hertler, S. C., et al. (2017). A sequential canonical cascade model of social biogeography: Plants, parasites, and people. *Evolutionary Psychological Science, 3*, 40–61.

Golden, G. K. (2013). *Crisis management during the Roman Republic: The role of political institutions in emergencies*. Cambridge, UK: Cambridge University Press.

Harper, K. (2017). *The fate of Rome: Climate, disease, and the end of an empire*. Princeton, NJ: Princeton University Press.

Harper, K., & McCormick, M. (2018). Reconstructing the Roman climate. In W. Scheidel (Ed.), *The science of Roman history: Biology, climate, and the future of the past* (pp. 11–52). Princeton, NJ: Princeton University Press.

Hekster, O. (2015). *Emperors and ancestors: Roman rulers and the constraints of tradition*. New York: Oxford University Press.

Keppie, L. (2002). *The making of the Roman army: From republic to empire*. London: Routledge.

Koutsoyiannis, D., Montanari, A., Lins, H., & Cohn, T. (2009). Climate, hydrology and freshwater: Towards an interactive incorporation of hydrological experience into climate research. *Hydrological Sciences Journal, 54*(2), 394–405.

Levick, B. (2002). *The government of the Roman Empire: A sourcebook*. New York: Routledge.

Liebeschuetz, W. (2007). Warlords and landlords. In P. Erdkamp (Ed.), *A companion to the Roman Army* (pp. 477–494). Oxford, UK: John Wiley & Sons.

Livy. (1982). *Rome and Italy: Books VI–X of the history of Rome from its foundation*. London: Penguin Classics.

Livy. (2006). *The history of Rome, books 1–5* (With introduction and notes, by V. Warrior, Trans.). Indianapolis, IN: Hackett Publishing.

Lockyear, K. (2018). *Coin hoards of the Roman Republic online*, version X. New York: American Numismatic Society. Data retrieved from http://numismatics.org/chrr/analyze

Martin, M. (1995). Treasure: Interpreting Roman hoards. *Theoretical Roman Archaeology Journal, 0*, 99–106.

Matyszak, P. (2008). *Chronicle of the Roman Republic: The rulers of ancient Rome from Romulus to Augustus*. New York: Thames & Hudson.

McElreath, R., Boyd, R., & Richerson, P. (2003). Shared norms and the evolution of ethnic markers. *Current Anthropology, 44*(1), 122–130.

Oakley, S. P. (2004). The early republic. In H. Flower (Ed.), *The Cambridge companion to the Roman Republic* (pp. 15–30). New York: Cambridge University Press.

Peñaherrera-Aguirre, M., Hertler, S. C., Figueredo, A. J., Fernandes, H. B. F., Cabeza de Baca, T., & Matheson, J. (2019). A social biogeography of homicide: Multilevel and sequential canonical examinations of intragroup unlawful killings. *Evolutionary Behavioral Sciences, 13*(2), 158–181.

Potter, D. (2014). *Emperors of Rome: The story of Imperial Rome from Julius Caesar to the last emperor*. London: Quercus Publishing.

Rankov, B. (2015). Death in battle: Principate. In Y. Le Bohec, G. Brizzi, E. Deschler-Erb, G. Greatrex, B. Rankov, & M. Reddé (Eds.), *The encyclopedia of the Roman Army: A-EAS* (pp. 292–293). Oxford, UK: Wiley Blackwell.

Rawlings, L. (2007). Army and battle during the conquest of Italy (350–264 BC). In P. Erdkamp (Ed.), *A companion to the Roman Army* (pp. 45–62). Oxford, UK: John Wiley & Sons.

Reece, R. (1988). Interpreting Roman hoards. *World Archaeology, 20*(2), 261–269.

Scarre, C. (1995). *Chronicle of the Roman emperors: The reign-by-reign record of the rulers of Imperial Rome*. New York: Thames and Hudson.

Scheidel, W. (2017). *The great leveler: Violence and the history of inequality from the stone age to the twenty-first century* (Vol. 74). Princeton, NJ: Princeton University Press.

Steinhilber, F., Beer, J., & Fröhlich, C. (2009). Total solar irradiance during the Holocene. *Geophysical Research Letters, 36*, L19704. https://doi.org/10.1029/2009GL040142

Stickler, T. (2007). The foederati. In P. Erdkamp (Ed.), *A companion to the Roman Army* (pp. 495–514). Oxford, UK: John Wiley & Sons.

Taylor, D. (2016). *Roman Empire at war: A compendium of Roman battles from 31 BC to AD 565*. Barnsley, Yorkshire: Pen and Sword.

Taylor, D. (2017). *Roman Republic at war: A compendium of Roman battles from 502 to 31 BC*. Barnsley, Yorkshire: Pen and Sword.

Thorne, J. (2015a). Ambush: Principate. In Y. Le Bohec, G. Brizzi, E. Deschler-Erb, G. Greatrex, B. N. Rankov, & M. Reddé (Eds.), *The encyclopedia of the Roman Army* (pp. 37–38). Oxford, UK: Wiley Blackwell.

Thorne, J. (2015b). Guerilla warfare: Principate. In Y. Le Bohec, G. Brizzi, E. Deschler-Erb, G. Greatrex, B. N. Rankov, & M. Reddé (Eds.), *The Encyclopedia of the Roman Army* (pp. 453–454). Oxford, UK: Wiley Blackwell.

Thyssen-Bornemisza, L. (2018). *Coin hoards of the Roman Empire online*. Oxford University. Data retrieved from http://chre.ashmus.ox.ac.uk/hoard/

Turchin, P. (2003). *Historical dynamics: Why states rise and fall* (Vol. 26). Princeton, NJ: Princeton University Press.

Turchin, P. (2007). *War and peace and war: The rise and fall of empires*. New York: Penguin.

Turchin, P., & Nefedov, S. A. (2009). *Secular cycles*. Princeton, NJ: Princeton University Press.

Whatley, C. (2015). Death in battle: Late empire. In Y. Le Bohec, G. Brizzi, E. Deschler-Erb, G. Greatrex, B. N. Rankov, & M. Reddé (Eds.), *The encyclopedia of the Roman Army* (pp. 293–294). Oxford, UK: Wiley Blackwell.

11

Dear Enemies: French and English Power Ratios

Aurelio José Figueredo, Steven C. Hertler, and Mateo Peñaherrera-Aguirre

1 Introduction

A recent technical article of ours on which this chapter is based is entitled *War and Peace: A Diachronic Social Biogeography of Life History Strategy and Between-Group Relations in Two Western European Populations* (Figueredo et al., 2019a). From an excess of ambition and length, this article spawned a daughter paper on which the present chapter also relies, entitled *The Ecology of Empire: The Dynamics of Strategic Differentiation-Integration in Two Competing Western European Biocultural Groups* (Figueredo et al., 2019b). Together, the two articles amass hundreds of years' worth of quantified data, the analysis of which is presented across fifteen statistical tables. These complexities render inaccessible to a general audience a cache of interesting findings relevant to the power dynamics subsisting between two long-standing rivals. Rendered more accessible and freed from the confining format of a journal article, this chapter dilates on Gallo-Britannic relations as they competed for the role of European *hegemon*. Extending back to the days of the Roman Empire, neither Gallic nor Britannic biocultural groups were major power players, and so did not significantly clash with one another. Thereafter, in

S. C. Hertler et al., *Multilevel Selection*, https://doi.org/10.1007/978-3-030-49520-6_11

what might be called the post-World War II *Pax Americana*, imposed by the most powerful Britannic successor state, these nations are once again at peace. However, during much of the time between these periods, one or the other nation was the premiere military might. Thence, the two nations shaped one another, evolutionarily as well as culturally, as they vied for hegemonic status, first gained by preindustrial France and then by postindustrial Britain. The Battle of Waterloo marked France's final bid for dominance, after which that nation's waning demographic, economic, and martial fortunes marked it as the whetstone upon which Britain's edge would be honed.

Both papers, *War and Peace* and *The Ecology of Empire*, analyze Gallo-Britannic populations as manifest within their territorial homelands, as well as those conquered colonies or settled territories peopled by these nations. As both papers study the same populations, both implement a diachronic perspective allowing for the detection of biocultural and demographic change through time. Accordingly, we provide some instructive historical circumstances in the following sections that assist in conceptually understanding Gallo-Britannic competition from a multilevel selectionist standpoint and in contextualizing the analyses and their results presented in the latter portions of this chapter. Not extending back in time before AD 1800 due to insufficient quantitative data, the biohistorical statistical analyses that we present may well truncate a larger process, and so we begin our narrative some centuries before to properly contextualize the analysis of the later developments described in Chap. 12.

2 Historical Review

Within certain species of fiddler crab (*Uca annulipes*), resident in coastal Indochina, individuals stake territorial claims to burrows. Neighboring crabs compete among one another, though conflict is restrained at times via territorial coalitions temporarily established to defend against intruders. The enemy of my enemy is my friend, so says the fiddler crab. Allowing a small local rival to be displaced with a larger interloper leaves territory exposed, inducing the collaboration of erstwhile enemies. Thus, with nations as it is with crabs, multilevel selection theory provides a framework for understanding conflict and coalition formation as seen in Gallo-Britannic relations throughout the early and late modern eras.

The complicated relations between Britain and France are captured in historical titles, such as *That Sweet Enemy: Britain and France: The History of a Love-Hate Relationship* (Tombs & Tombs, 2006) and *Best of Enemies: Anglo-French Relations Since the Norman Conquest* (Gibson, 2004). Nonetheless, as seen in Tombs' and Tombs' separate sections on *struggle* and *coexistence*, there is a clear pattern wherein the two nations closed ranks against distant powers early on, as in the Third Crusade against Sultan *Salah ad-Din*. This coalition degenerated into war over territory in the thirteenth century and war over succession in the fourteenth century. The Hundred Years' War, involving the historic battles of Crécy, Poitiers, and Agincourt, featured the Plantagenet Kings of England pressing dynastic claims against a France whose King, Louis X, had failed to produce a male heir. France annexed the city of Bordeaux following England's crushing defeat in the Battle of Castillon in AD 1453, which ended the Hundred Years' War and secured the former's continental dominion. Then came the many successive decades of French hegemony wherein Britain was subordinate.

The annals of *Merovingian* and *Carolingian* Kings mark the early stability of the French polity, which was preserved by a secure alliance between church and state, and an enduring monarchy, allowing France to come to the fore ahead of England. Building on these national traditions, France was a continental power at once internally cohesive and formidable to rivals. Under the reign of the *Sun King*, Louis *XIV* (AD 1643–1715), pre-industrial France radiated power, whether judged in terms of military might, agricultural productivity, or political eminence. With respect to the projection of power, King Louis XIV alone led several armed conflicts between AD 1661 and 1715, including the *War of Devolution* (AD 1667–1668) with Spain over the Spanish Netherlands; the *Dutch War* (AD 1672–1678), an attempt to conquer the United Provinces of the Netherlands; the *War of the Grand Alliance* (AD 1688–1697) pitting almost all the European powers against incessant French expansionism; and the *War of the Spanish Succession* (AD 1707–1714) waged during the reign of Charles II with the Austrians, the Dutch, and the British, over various territories formerly belonging to the Spanish Empire. Gallic populations were thus under much greater pressure from between-group competition than Britannic ones, and this permitted the growing internal disunity of the latter.

In contrast to the British, the French were subduing internal factionalism while projecting power across the continent during this broad historical period. For example, Roman Catholic France persecuted so-called heretics (e.g., Huguenots), whereas Protestant England did not; and the latter therefore suffered from internal disunity as a result of the proliferation of radical Protestant sects (e.g., Puritans), who eventually rebelled against the Crown and permanently undermined its authority (Sharpe, 1992). This internal strife within Britannic populations included conflicts such as the *Irish Rebellion* (AD 1641), the *Great Rebellion* subsuming the first and second *English Civil Wars* (AD 1642–1651), the *Cromwellian Reconquest of Ireland* (AD 1649–1653), and the *Glorious Revolution* (AD 1688), which altered the previously legitimate line of royal succession by actually facilitating a foreign invasion from the Netherlands by William and Mary of Orange (see Turchin, 2016), who usurped the English throne. After the events of AD 1688, the King of England and Scotland was further crippled by parliament and became something of a pensioner to the King of France. As seen through the lens of multilevel selection theory, without a higher-order threat coming from a rival nation, the British at this time were not under the degree of danger presented by external groups as were the French. The channel seems to have buffered England from the many continental wars (Macfarlane, 2003) that were experienced early on by the French. Over the course of the seventeenth century, Britain was involved in very few conflicts outside the British Isles. Prior to the aforementioned reign of Louis XIV, Britannic forces were indirectly involved in the *Thirty Years' War* against the Catholic Holy Roman Empire from AD 1619 to 1622 and then again from AD 1628 to 1630. Britannic troop deployments, however, were usually made as small and perhaps token parts of larger multinational coalitions. For example, one English-Dutch regiment was deployed early in the conflict to the Palatinate, and one Scottish-Dutch and one English-Scottish regiment were deployed in support of Calvinist Bohemia in their revolt against the Holy Roman Empire. Although it is difficult to demonstrate a negative, it appears that Britannic forces were only directly involved in fighting for the so-called Protestant cause from AD 1634 to 1638, once again in small and possibly token numbers, such as the three-regiment Scottish Brigade that was deployed to France in support of the Dutch forces fighting there. Other than such minor interventions, there are no indications of significant British involvement in the *Thirty Years' War* (Bonney, 2014; Parker, 2006; see also Murdoch, 2001).

During the reign of the Sun King, France and England were not at war, very possibly because England had acquiesced to France's clear dominance. Near the end of his reign, Britannic involvement in overseas military adventures was gradually escalating. For example, in the *War of the Spanish Succession*, Britain was now aligned with the Holy Roman Empire against France and Spain and made more substantial military commitments (Tincey, 2004). In the Battle of Blenheim alone, for instance, a joint military force commanded by the Duke of Marlborough and the Prince of Savoy was no less than two-thirds British and included 51 infantry battalions and 92 cavalry squadrons, totaling 56,000 men. This is a lot more than just the previously committed regiment or two, presumably provided for mostly moral support. Wars over royal succession and national borders ceased, only to be replaced with wars over colonial territories. Indeed, the same century that brought a close to the Hundred Years' War inaugurated the Age of Exploration; thus, not long after France and England stopped fighting over the Old World, they began fighting for the New World.

The latter part of the eighteenth century saw a waxing England and a waning France crossing each other's paths, leaving them in something of an equal position for a spate of decades. Prior power dynamics were eventually inverted, but not before a sanguinary century of nearly equal power. War commenced only as France's clear hegemonic status eroded. These times were marked by the several aforementioned wars over colonial possessions, as well as the French Revolutionary Wars and the subsequent Napoleonic Wars. Relations stabilized again only once one biocultural group had established hegemony over the other—this time it was an industrialized Britain that radiated its power through invention, commerce, and finance. The balance of power has remained in Britain's favor ever since. The AD 1815 tipping point is described eloquently in Tombs and Tombs (2006):

> Meanwhile, the Congress of Vienna, where sovereigns and statesmen met to decide the future of Europe, continued in session. How much had changed since 1688, when the Three Kingdoms had been hustled into European affairs as a minor auxiliary against Louis XIV! Now the United Kingdom was predominant in Europe, it was the sole global power, and it had become the prototype of economic transformation. France, still formidable, was no longer menacing. Though it took nearly another century for it to become entirely clear, the Franco-British war was over, and with it, the series of world wars it had spawned. (p. 288)

At the beginning of the eighteenth century, France was widely reputed to be the greatest European power, a superiority manifested in its territorial extent, colonial possessions, stable regime, expansive population, and bounteous fields. Waning Habsburg supremacy had created a power vacuum that France first filled, only to cede the position of hegemon to England following the Seven Years' War, the American Revolutionary War, the French Revolutionary Wars, and the Napoleonic Wars. The French Revolution in particular proved devastating for France's economy, the precipitous decline of which partially enabled Britain's rise to European dominance (Crouzet, 1990).

Once uncontested national borders were drawn, and colonial holdings affixed, and with the Bourbon Restoration stifling the remainder of revolutionary excess, Britannic-Gallic relations gave way to spats over colonial possessions (Brailey, 2002; Goldman, 1972), banking (Boyce, 2002), and trade (Marsh, 2002) through the remainder of the nineteenth century. In the Post-Napoleonic War period spanning AD 1815–1999, there appear to have been no more than three incidents that brought Britain and France to the precipice of armed conflict (Tombs & Tombs, 2006): (1) the *Fashoda Incident* (1898), which was a dispute over colonial possessions in East Africa; (2) the *Dreyfus Affair* (1894–1906), a political scandal that triggered international outrage and prominently included pointed criticism of the court martial proceedings by the Lord Chief Justice of England; and (3) the *Second Boer War* (1899), in which some French citizens fought as foreign volunteers on the Boer side against Britain. None of these incidents seems to have resulted in anything more than diplomatic strain and mutual resentment. In these exchanges, and more generally in the decades of the nineteenth century preceding them, Britain remained a more significant force than France in international politics and economic power.

The second part of this 200-year period—the twentieth century—witnessed increasing cooperation wherein relations between these nations transitioned from that of *best of enemies* to *rival companions* (Chassaigne & Dockrill, 2002; Mallaby, 2002). This shift came through the settlement of conflicts over national and colonial territories, after the collapse of nationalism and empire-building following World War II, which was replaced by an era of irenic economic cooperation and egalitarianism in the West (Westbrook, 2004; Woodley of Menie et al., 2017). The industrialization of Germany, and the unification of the *Triple Alliance*, formed between Germany, Austria, and Italy in 1882, once again, and more than ever before, induced France and England to close ranks in opposition to a common threat during World War I. But it was the aftermath of the next great conflict that did the most to ensure long-term cooperative interactions between France and Britain and among Western countries more generally. Following the end of World War II, the Allied Powers imposed a variety of policies in Europe with the explicit purpose of fostering economic interdependence, with the hope that this would put an end to nationalism and its global wars (Westbrook, 2004). Their effort apparently has been highly successful or has at least coincided with other factors that have promoted between-group peace (Gat, 2008; Woodley of Menie et al., 2017). Thus, the past millennium opened with competition and conflict but closed with pacific cooperation. Even as in AD 1815, the two were locked in bloody combat in the Battle of Waterloo, the empires of France and Britain were united during the twentieth century in opposing common foes, such as the National Socialist Germany. Within the twentieth century, cooperation between Britain and France followed from their place among other nations, with alliance formation first being facilitated in opposition to a stronger rival, and later from being jointly superseded by a hegemonic global power.

3 Biohistorical Analyses

As cautioned above, our biohistorical statistical analyses could not be extended back in time before AD 1800 due to insufficient quantitative data having been collected on the requisite biodemographic information hitherto. Nevertheless, this 200-year historical period covers most of the late modern era from the climax of the age-old conflict, the Napoleonic Wars (AD 1803–1815), to the period of closest alliance spanning World War I (AD 1914–1918), World War II (AD 1939–1945), and the Cold War (AD 1947–1991).

These empirical tests rely quite strongly on an understanding of evolutionary life history (LH) theory, thereby necessitating some description of the concept before going further. As a syndrome is a collection of symptoms, so life histories are collections of coadapted traits. Many LH traits exist along a continuum of evolved developmental speed, such that organisms with a slow LH mature slowly, expend more energies in long-term parental care than in early and exhaustive reproduction, and invest more in somatic maintenance to stave off senescence, disease, and death, while quite the opposite is found among organisms with fast LHs. Your average scurrying rodent exemplifies a fast LH strategy, while great apes, for example, have comparatively slower LH strategies, accomplishing the necessities of the life cycle over much greater periods of time. While life histories range most widely between species, they range modestly within species. Accordingly, the relevance of LH theory at present relates to the importance of this much smaller but nontrivial degree of within-species variance in life history strategies among human individuals and human groups. While all humans have slower LHs than most mammals, some are somewhat slower than others and, as small as this within-species differential may be, it amounts to a substantial explanatory factor within social scientific inquiry, being relevant to social deviance, deferral of gratification, time orientation, emotional regulation, alliance formation, sensation seeking, and conscientiousness, among other fitness-relevant traits. The LH speed of a population, taken as an aggregate, will bear upon its competitive capacity, making it difficult to speak of intergroup competition without speaking of the general advantage accruing to individual slow LH strategists competing within relatively stable environments.

Further research is needed to explore the dynamic relationship between these two concepts, though at present it suffices to say they are sufficiently independent of each other to appear uncoupled in many historical populations. At the same time, we observe that the two variables overlap in that fast LH strategists tend to express less altruistic behavior, engage in more exploitative relationships, which are less stable and enduring, and thus are most often considering what their country can do for them, not what they can do for their country. Beyond this basic association, in the course of studying the cycling of nations and empires, it seems that group-selected traits are more prominent early on, during the stages of growth and aggregation, with the stability of selective regimes evoked by the success of such societies precipitating the slowing of LH strategies. As might be imagined, some of the most formidable and stable societies like those of ancient Athens and Victorian Britain, merged relatively high levels of group-selected traits with slow LH speeds. Relying on other sources to provide readers with a thoroughgoing explanation of LH theory, we here lastly mention LH theory's specific relevance to the following analyses derives from its complex relationship with group selection, as discussed above.

Data from AD 1800 to 1999 were collected for the following Britannic populations: the United Kingdom, the United States, Canada, New Zealand, and Australia; corresponding data from Gallic populations were also collected, including Continental France as well as its several overseas departments in French Algeria (North Africa) and French Guiana (South America), all counted as part of the French Republic in the national census (Figueredo et al., 2019a). Population sizes were obtained for both biocultural groups from the *Maddison Project* database (Bolt, Inklaar, de Jong, & van Zanden, 2018), a repository curated by the *Groningen Growth and Development Center* (GGDC). Warfare mortality estimates were gathered from the *Correlates of War* database (Sarkees & Wayman, 2010); although this database contains both *inter*state and *intra*state (civil) wars, we excluded all intrastate conflicts and included only conflict between states for present purposes. Wars containing at least one Britannic polity were kept in the database; similar procedures were employed with the Gallic sample. Standardized rates (per 100,000) were computed after accounting for population size, as population size confounds the intensity of warfare due to the fact that societies with a larger population experience a greater absolute number of deaths. The proportion of the world population was estimated based on *Roser's* demographic database (Roser & Ortiz-Ospina, 2017). Total fertility rates (Ajus, Lindgren, & Rosling, 2015), infant mortality rates (Johansson, Lindgren, & Rosling, 2015), and life expectancy information (Lindgren, 2015) were obtained from the *Gapminder* database repository.

For all lexicographic measures, the diachronic utilization of specific classes of words by each biocultural group was evaluated via their relative frequencies of usage through Google Ngram Viewer (Michel et al., 2011), an interactive textual corpus encompassing over 5.9 million texts and 500 billion written words from AD 1500 to 2008. The data were obtained in the form of frequency counts of each word within its respective language across the 200 years spanning AD 1800–1999. *The Descent of Man* Altruism Words was a list of words employed by Darwin (1871) to describe within-group altruism and between-group competition in humans, harvested from the original text by Woodley of Menie, Figueredo, Sarraf, Hertler, Fernandes, and Peñaherrera-Aguirre (2017). The slow LH words and fast LH words were lists of words harvested from the collections of words observed by Sherman, Figueredo, and Funder (2013) to be employed disproportionally in conversation by either slower or faster life history strategists, respectively, which had been identified by non-lexicographic methods. Unit-weighted factor scales (Gorsuch, 1983) were estimated as the means of the standardized scores for the lexicographic items on each scale (Figueredo, McKnight, McKnight, & Sidani, 2000). The words then used as items in each of these lexicographic scales were psychometrically selected on the basis of obtaining adequate part-whole correlations for each word to the corresponding aggregate scale score for each lexicographic scale. All lexicographic scales used in this study were originally created in English and then translated into French for the cross-cultural comparison. As we suspected that using simple literal translations might miss important cultural differences in the contexts of their usage, we instead generated lists of plausible synonyms in French for all the original words in English and then selected the French synonym that had the highest part-whole correlation to its respective lexicographic scale in the French language. By this psychometric process of selection, we obtained the optimal French-language equivalent to each English-language word based on its consistency with all the other synonyms selected for each lexicographic scale.

Part-whole correlations of the Darwin *Descent of Man* Altruism Words ranged from 0.29 to 0.92 ($p < 0.05$) for the English-language version, with the overall factor scale explaining 55% of the variance, and ranged from 0.40 to 0.94 ($p < 0.05$) for the French-language version, with 62% of the variance explained by the factor scale. Part-whole correlations of the fast LH words ranged from 0.30 to 0.96 ($p < 0.05$) for the English-language version, with the overall factor scale explaining 62% of the variance, and ranged from −0.29 to 0.92 (all positive and significant except for *songes*) for the French-language version, with 54% of the variance explained by the factor scale. Part-whole correlations of the slow LH words ranged from 0.82 to 0.97 ($p < 0.05$) for the English-language version, with the overall factor scale explaining 86% of the variance, and ranged from −0.11 to 0.94 (all positive and significant except for *victoire*) for the French-language version, with 65% of the variance explained by the factor scale. The two negatively loaded items in the French-language versions were not eliminated to maintain the integrity of the selection procedures. Nevertheless, the convergent validities of the two scales were generally quite acceptable.

The five life history (LH) strategy indicators were aggregated into two lower-order "method" factors: (1) biodemographic; and (2) lexicographic. The biodemographic factor comprised three scales: (1) infant mortality, reversed; (2) total fertility, reversed; and (3) life expectancy. The lexicographic LH factor comprised two scales: (1) fast LH words, reversed, and (2) slow LH words. Figure 11.1 displays the latent hierarchical structure of the LH strategy nexus.

As with the *asabiyyah* analyses presented previously in Chap. 6, three nested MLMs were estimated to test the need for increasing parameterization as alternative hypotheses: (1) *MLM1* estimated a single intercept and a single logarithmic slope (unconditional LH "nexus") for all LH methods and indicators over time, as well as the same intercepts and logarithmic slopes for all LH indicators nested within each LH method; (2) *MLM2* estimated a separate intercept and a separate logarithmic slope for each LH method over time but the same intercept and logarithmic slopes

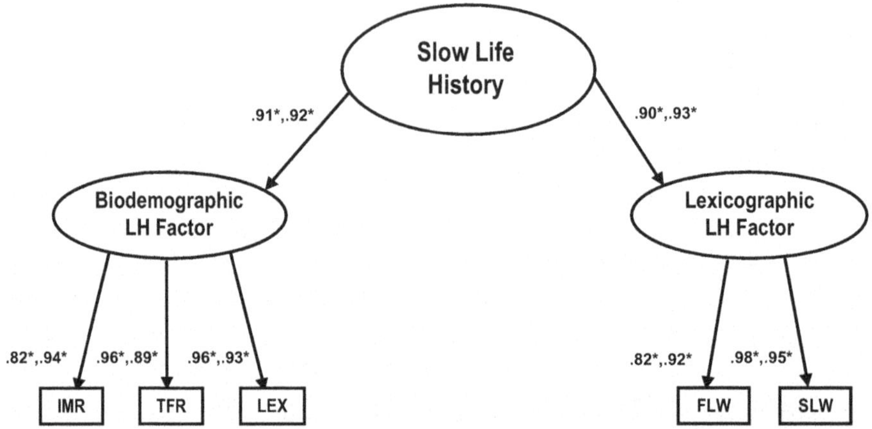

Fig. 11.1 The latent hierarchical structure of the life history strategy nexus. (IMR = infant mortality, reversed; TFR = total fertility, reversed; LEX = life expectancy; FLW = fast LH words, reversed; SLW = slow LH words. Factor loading coefficients = Britannic, Gallic. * $p < 0.05$)

for all LH indicators nested within each LH method; and (3) *MLM3* a separate intercept and a separate logarithmic slope for each LH method over time as well as a separate intercept and a separate logarithmic slope for each LH indicator over time within each LH method.

Table 11.1 displays the pertinent nested model comparisons. The systematic AIC and -2RLL comparisons performed among the nested models representing the specific variance components of the LH methods and LH indicators revealed the following: (1) the specific variance components for LH methods were not statistically significant for either Britannic or Gallic populations, and (2) the specific variance components for LH indicators were nonetheless statistically significant for the Britannic but not the Gallic population ($p < 0.05$). Comparisons of squared multiple correlations among the three nested MLMs yielded essentially the same results. The magnitudes of the specific variances (ΔR^2) of the LH methods and LH indicators were found to be negligibly small (<<1%) in contrast with the common factor variance of the "unconditional" LH nexus, representing the general slow LH construct, which was found to be quite large for both the Britannic (82%) and Gallic populations (86%). This implied that there was no systematic difference between the biodemographic method and lexicographic method LH indicators.

Table 11.1 Nested MLM comparisons for level 1 and level 2 with level 3 latent chronometric life history constructs as natural logarithmic functions of time with Britannic and Gallic populations across the 200 years spanning AD 1800–1999

	Common factor variance (MLM1)	Common method variance (MLM2)	Specific indicator variance (MLM3)
Britannic population			
AIC	1102.1	1105.9	1088.2
-2RLL	1094.1	1093.9	1064.2
	$\Delta\chi^2$ =	**0.2**	**29.7***
R^2	0.824*	0.824*	0.829*
	ΔR²=	**0.000**	**0.005***
	ΔModel df =	**2**	**7**
Gallic population			
AIC	894.4	898.3	908.4
-2RLL	886.4	886.3	884.4
	$\Delta\chi^2$ =	**0.1**	**1.9**
R^2	0.857*	0.857*	0.858*
	ΔR²=	**0.000**	**0.001**
	ΔModel df =	**2**	**7**

*$p < 0.05$

The logarithmic intercepts (*a*) and slopes (*b*) of this unitary higher-order slow LH construct over time was statistically significant ($p < 0.05$): $a = -236^*$, $b = 31^*$ for Britannic populations and $a = -240^*$, $b = 32^*$ for Gallic populations. These model parameters were surprisingly similar, to the point of being nearly identical, and indicated progressively slowing LH speed for both populations. No significant serially autoregressive effects were found (*ARH1* = 0) for either biocultural group. This can all be taken to mean that the measurement model for our slow LH construct is virtually identical for the Britannic and Gallic biocultural groups, combining both lexicographic and biodemographic indicators; furthermore, the level of this slow LH factor is increasing at virtually identical rates for both populations over the specified historical period.

As shown in Fig. 11.2, the measurement models for the between-group competition (BGC) factor were also quite similar. The logarithmic intercepts (*a*) and slopes (*b*) of this BGC construct over time were statistically significant ($p < 0.05$) for Gallic but not Britannic populations: $a = 13$, $b = -2$ for Britannic populations and $a = 173^*$, $b = -23^*$ for Gallic populations. These model parameters were surprisingly quite different and

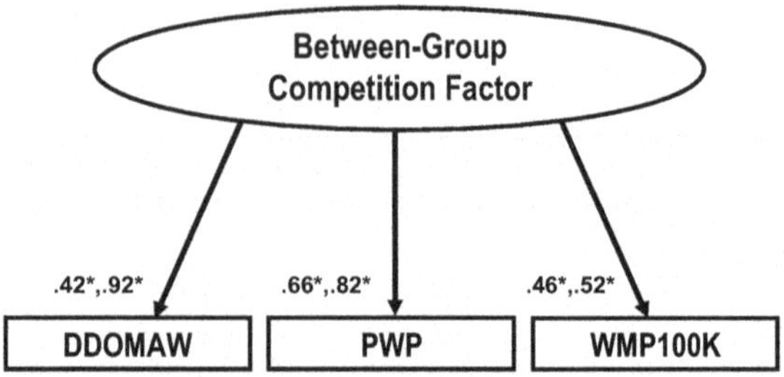

Fig. 11.2 The latent structure of the between-group competition factor. (DDOMW = Darwin's *Descent of Man* Altruism Words; PWP = proportion of the world's population; WMP100K = war mortality per 100,000. Factor loading coefficients = Britannic, Gallic. * $p < 0.05$)

indicated decreasing levels of BGC for Gallic but not Britannic populations. No significant serially autoregressive effects were found ($ARH1 = 0$) for either biocultural group. As with the measurement model for slow LH, the single lexicographic indicator (the Darwin *Descent of Man* Altruism Words) used in the BGC factor model converged very well with the two biodemographic ones (proportion of the world's population and war mortality) for both the Britannic and the Gallic populations. The difference in the slopes of BGC over time might be taken to reflect the slowing of Gallic imperial expansion after their defeat in the Napoleonic Wars (AD 1815), decreasing their levels of BGC, in contrast to the unabated Britannic imperial expansion, continuing especially throughout the Victorian era (AD 1837–1901), maintaining their levels of BGC.

As with the *asabiyyah* analyses presented previously in Chap. 6, MLM residuals were then exported for both slow LH and BGC and used for subsequent general linear modeling. MLM residuals were thus statistically adjusted for the logarithmic effect of time as well as of any single-lagged heterogeneous autoregressive serial dependencies among successive data prior to regression modeling, thus circumventing this potential problem as a threat to the validity of correlational analysis. It was especially important to statistically control for the effects of time to ascertain that any association was not a simply coincidental one of slow LH increasing and BGC simultaneously but independently decreasing over the same period of time, at least for the Gallic sample. As seen in Fig. 11.3, the

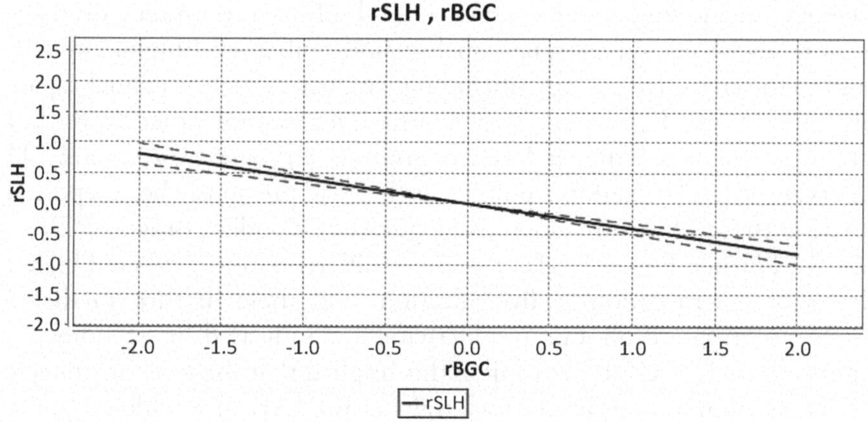

Note: Dashes are upper and lower confidence intervals.

Fig. 11.3 Time-adjusted MLM residuals of between-group competition predicting MLM residuals of slow life history (SLH) (AD 1800–1999)

semipartial correlation of the time-adjusted MLM residuals of slow LH with those of BGC was $r = -0.41$ (90% *CI*: −0.49, −0.33), $F(1,396) = 80.37$, $p < 0.0001$, with no statistically significant differences between the Gallic and Britannic populations, empirically supporting the hypothesis that declining BGC is historically associated with slowing LH speed, independently of the effects of time. The possible causal directionality of this effect, however, remains uncertain.

4 The Role of Limiting Similarity Theory

In *A Sequential Canonical Cascade Model of Social Biogeography: Plants, Parasites, and People*, Figueredo et al. (2017) performed a cross-sectional or *synchronic* analysis of sixty-six national polities to document the evolutionary cascade of consequences stemming from the physical ecology (including parameters such as average temperature, annual precipitation, altitude, and latitude) to the community ecology (including parameters such as dominant forms of vegetation, total parasite burden, parasite diversity, population density, and life history strategies), to the social ecology (including parameters such as levels of macroeconomic diversification, social equality, sexual equality, and embodied human capital), and to the cognitive ecology (including parameters such as cranial capacity and national IQ scores). They reported, for example, that slower LH strategies predicted higher levels of *strategic differentiation* among LH traits at the level of national polities, as they had previously been reported to do at the level of individuals (see Figueredo, Woodley, Brown, & Ross, 2013). Further, higher levels of strategic differentiation predicted higher levels of macroeconomic diversification, and these in turn predicted higher aggregate economic productivities, as indicated by econometric measures such as GDP per capita. This implied that the level of strategic differentiation among life history traits at the level of a national polity could be used to gauge the *niche breadth* of a population.

Hutchinson's (1957, 1959) *Theory of Limiting Similarity* described the maximum allowable overlap between two ecologically similar species. A longitudinal or *diachronic* comparison between the relative sizes of the Britannic and the Gallic populations over the biohistorical study period supported the application of this ecological cross-species principle to the results of competition between human biocultural groups: The Britannic-Gallic population ratio started at barely over 0.5:1 in AD 1800 and rose to nearly 4:1 by AD 1999. We therefore predicted that over this same period of time, the niche breadth of the Gallic biocultural group should have contracted relative to that of the increasingly victorious Britannic biocultural group, as indicated by the relative degrees of strategic differentiation among LH traits evidenced by the two competing populations over time. Increasing or decreasing niche breadths are to be expected as population adaptations to territorial expansions or contractions, especially across diverse geographical habitats, as were experienced by the Britannic and Gallic biocultural groups, respectively, during this period of time.

To assess strategic differentiation of LH strategy over time within each of the two biocultural groups, cross-trait sample variances ("mean-squares across traits" or MSTRT values) were computed in parallel across the standardized (z) scores of each of the five convergent LH indicators for each cross-sectional time point spanning the years from AD 1800 to 1999 (Figueredo et al., 2019b). The bivariate linear slopes of this MSTRT construct over time were found to be opposite in direction for the two biocultural groups: $r = -0.58$ (90% *CI*: −0.67, −0.48), $F(1,198) = 100.46$, $p < 0.0001$ for Gallic populations, as shown in Fig. 11.4, and $r = 0.12$ (90% *CI*: −0.02, 0.26), $F(1,198) = 2.97$, $p = 0.09$ for Britannic populations, as shown in Fig. 11.5. The parameters of these two growth curves were formally tested against each other and were found to be significantly different from each other: $F(1,396) = 340.20$, $p < 0.0001$ for the intercepts and $F(1,396) = 90.08$, $p < 0.0001$ for the slopes.

Once again using the time-adjusted MLM residuals of BGC as a predictor, the bivariate linear slopes of the MSTRT construct as a function of rBGC were likewise found to be opposite in direction for the two biocultural groups: $r = -0.19$ (90% *CI*: −0.32, −0.05), $F(1,198) = 7.08$, $p = 0.008$ for Gallic populations, as shown in Fig. 11.6, and $r = 0.14$ (90% *CI*: 0.00, 0.28), $F(1,198) = 4.15$, $p = 0.04$ for Britannic populations, as shown in Fig. 11.7. The parameters of these two growth curves were formally tested against each other and were found to be significantly different from each other: $F(1,396) = 254.94$, $p < 0.0001$ for the intercepts and $F(1,396) = 10.80$, $p = 0.001$ for the slopes.

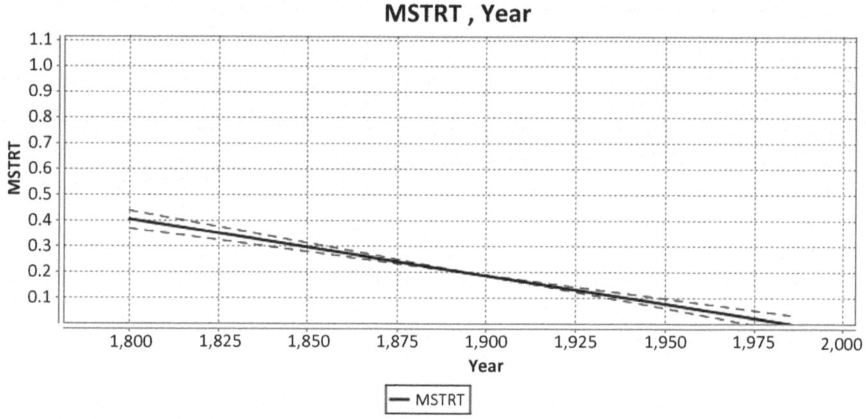

Note: Dashes are upper and lower confidence intervals.

Fig. 11.4 Cross-sectional, cross-trait variances among convergent indicators of slow life history as a function of time for Gallic populations across the 200 years spanning AD 1800–1999

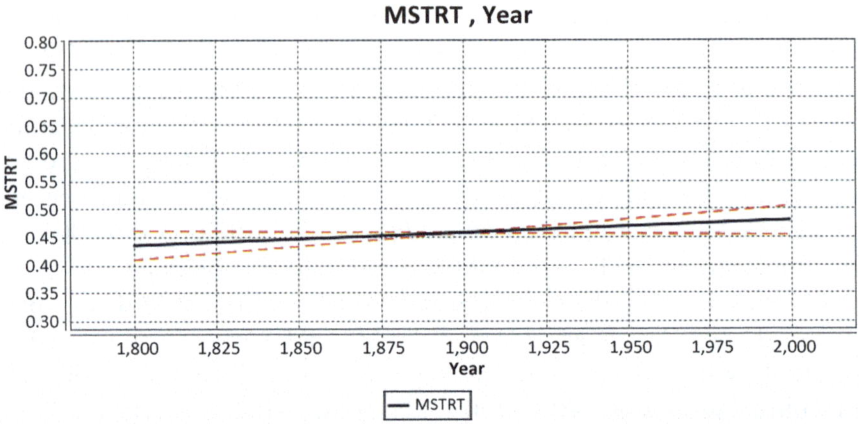

Note: Dashes are upper and lower confidence intervals.

Fig. 11.5 Cross-sectional, cross-trait variances among convergent indicators of slow life history as a function of time for Britannic populations across the 200 years spanning AD 1800–1999

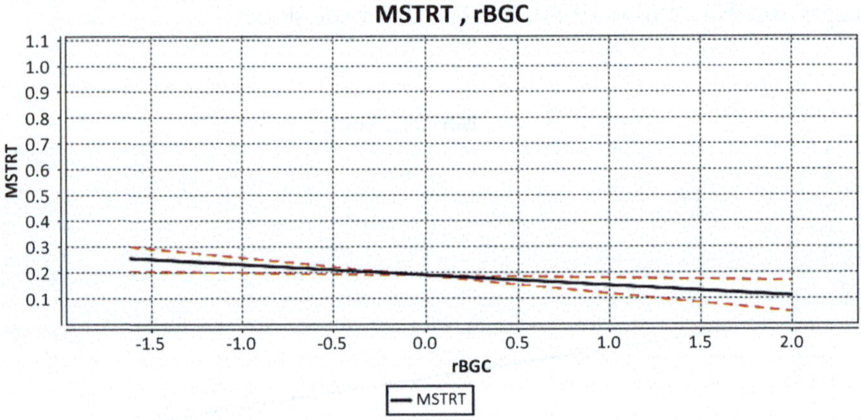

Note: Dashes are upper and lower confidence intervals.

Fig. 11.6 Cross-sectional, cross-trait variances among convergent indicators of slow life history as a function of between-group competition, residualized by MLM for any logarithmic effects of time, for Gallic populations across the 200 years spanning AD 1800–1999

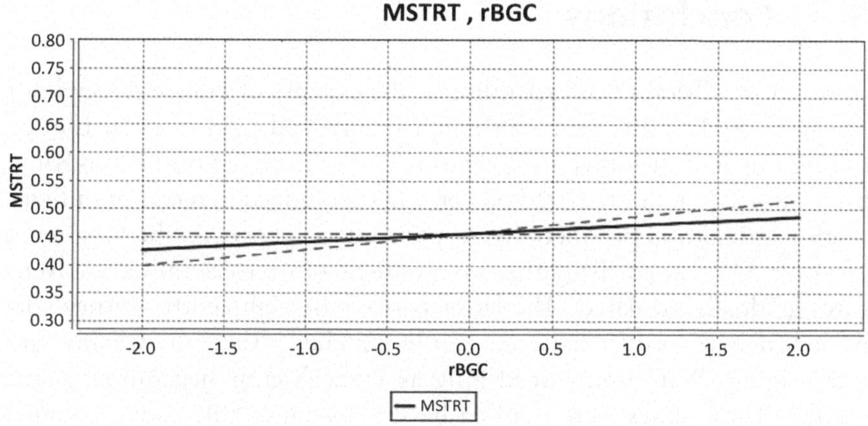

Note: Dashes are upper and lower confidence intervals.

Fig. 11.7 Cross-sectional, cross-trait variances among convergent indicators of slow life history as a function of between-group competition, residualized by MLM for any logarithmic effects of time, for Britannic populations across the 200 years spanning AD 1800–1999

What all this can be taken to mean is that intense between-group competition can have entirely different effects upon the strategic diversification of LH strategy depending on whether any given group *wins* or *loses* (independent of any secular temporal trends that may exist). Losing the between-group competition dramatically reduced the strategic diversification among the Gallic biocultural group's LH parameters over the historical period examined, both in relation to the Britannic biocultural group and in absolute terms as well. We therefore interpret these results to support the prediction that the aggregate population niche breadth of the Gallic biocultural group did in fact contract, as expected by theory, relative to that of the increasingly victorious Britannic biocultural group across the 200 years spanning AD 1800–1999.

5 Conclusions

In summary, Britain was expanding at the expense of France throughout much of the last 200 years, although France had held sway in Europe during much of the previous 200 years. Under the reign of the Sun King, pre-industrial France radiated power, whether judged in terms of military might, agricultural productivity, or political eminence. As the long reign of Louis XIV ended, Britannic involvement in overseas military adventures gradually escalated. The latter part of the eighteenth century witnessed intense conflict between evenly matched rivals, one waxing and one waning. War commenced only as France's clear hegemonic status eroded. These times were marked by the several conflicts over colonial possessions, as well as the French Revolutionary Wars and the subsequent Napoleonic Wars. With the banishment of Emperor Napoleon to Saint Helena, the great struggles between France and England ended, inaugurating an age of relative peace between these two great rival nation-states. Relations stabilized again only once one biocultural group had established hegemony over the other—this time it was an industrialized Britain that radiated its power through invention, commerce, and finance. The balance of power has remained in Britain's favor ever since, even as both nations transitioned to allies against Axis powers and thereafter became partners in the European Union amid the backdrop of the *Pax Americana*.

Based on these historical considerations, we have provided results from some diachronic statistical analyses testing evolutionary hypotheses jointly derived from multilevel selection theory and more general principles of quantitative theoretical ecology, mostly drawn from two recent publications (Figueredo et al., 2019a, 2019b). The results described herein are not exhaustive of all those reported in the corresponding academic papers, but instead summarize their main findings, graphically where possible. Although any definitive proof is elusive in science, these results are generally supportive of our application of multilevel selection theory to the historical competition between such rival biocultural groups.

References

Ajus, F., Lindgren, M., & Rosling, O. (2015). *Total fertility rate*. Retrieved March, 2018, from https://www.gapminder.org/data/

Bolt, J., Inklaar, R., de Jong, H., & van Zanden, J. L. (2018). Rebasing 'Maddison': New income comparisons and the shape of long-run economic development. *GGDC Research Memorandum, 174*, Maddison Project Database, version 2018.

Bonney, R. (2014). *The thirty years' war 1618–1648*. London: Bloomsbury Publishing.

Boyce, R. (2002). Breaking the banque: The great crisis in Franco-British central bank relations between the wars. In P. Chassaigne & M. Dockrill (Eds.), *Anglo-French relations 1898–1998* (pp. 80–93). London: Palgrave Macmillan.

Brailey, N. J. (2002). Anglo-French rivalry over Siam and the treaties of April 1904. In P. Chassaigne & M. Dockrill (Eds.), *Anglo-French relations 1898–1998* (pp. 56–68). London: Palgrave Macmillan.

Chassaigne, P., & Dockrill, M. (2002). *Anglo-French relations 1898–1998*. London: Palgrave Macmillan.

Crouzet, F. (1990). *Britain ascendant: Comparative studies in Franco-British economic history*. Cambridge, UK: Cambridge University Press.

Darwin, C. (1871). *The descent of man, and selection in relation to sex*. London: John Murray.

Figueredo, A. J., Cabeza de Baca, T., Fernandes, H. B. F., Black, C. J., Peñaherrera, M., Hertler, S. C., et al. (2017). A sequential canonical cascade model of social biogeography: Plants, parasites, and people. *Evolutionary Psychological Science, 3*, 40–61.

Figueredo, A. J., McKnight, P. E., McKnight, K. M., & Sidani, S. (2000). Multivariate modeling of missing data within and across assessment waves. *Addiction, 95*(11s3), 361–380.

Figueredo, A. J., Peñaherrera-Aguirre, M., Fernandes, H. B. F., Lomayesva, S. L., Hertler, S. C., & Sarraf, M. A. (2019a). War and peace: A diachronic social biogeography of life history strategy and between-group relations in two Western European populations. *Journal of Methods and Measurement in the Social Sciences, 10*(1), 36–75.

Figueredo, A. J., Peñaherrera-Aguirre, M., Fernandes, H. B. F., Lomayesva, S. L., Hertler, S. C., & Sarraf, M. A. (2019b). The ecology of empire: The dynamics of strategic differentiation-integration in two competing Western European biocultural groups. *Politics and the Life Sciences, 38*(2), 210–225.

Figueredo, A. J., Woodley, M. A., Brown, S. D., & Ross, K. C. (2013). Multiple successful tests of the Strategic Differentiation-Integration Effort (SD-IE) hypothesis. *Journal of Social, Evolutionary, and Cultural Psychology, 7*(4), 361.

Gat, A. (2008). *War in human civilization*. Oxford, UK: Oxford University Press.

Gibson, R. (2004). *Best of enemies: Anglo-French relations since the Norman conquest*. Exeter, UK: Impress Books.

Goldman, M. F. (1972). Franco-British rivalry over Siam, 1896–1904. *Journal of Southeast Asian Studies, 3*(2), 210–228.

Gorsuch, R. L. (1983). *Factor analysis* (2nd ed.). Hillsdale, NJ: Lawrence Erlbaum Associates.

Hutchinson, G. E. (1957). *A treatise on limnology*. Hoboken, NJ: Wiley.

Hutchinson, G. E. (1959). Homage to Santa Rosalia or why are there so many kinds of animals? *The American Naturalist, 93*, 145–159.

Johansson, K., Lindgren, M., & Rosling, O. (2015). *Infant mortality*. Retrieved March, 2018, from https://www.gapminder.org/data/

Lindgren, M. (2015). *Life expectancy*. Retrieved March, 2018, from https://www.gapminder.org/data/

Macfarlane, A. (2003). *The savage wars of peace: England, Japan and the Malthusian trap*. New York: Palgrave Macmillan.

Mallaby, C. (2002). Britain and France: Some comments on a complex relationship. In P. Chassaigne & M. Dockrill (Eds.), *Anglo-French relations 1898–1998* (pp. 7–10). London: Palgrave Macmillan.

Marsh, P. T. (2002). The end of the Anglo-French commercial alliance, 1860–1894. In P. Chassaigne & M. Dockrill (Eds.), *Anglo-French relations 1898–1998* (pp. 34–43). London: Palgrave Macmillan.

Michel, J. B., et al. (2011). Quantitative analysis of culture using millions of digitized books. *Science, 331*(6014), 176–182.

Murdoch, S. (Ed.). (2001). *Scotland and the thirty years' war: 1618–1648* (Vol. 6). Leiden, The Netherlands: Brill.

Parker, G. (2006). *The thirty years' war*. London: Routledge.

Roser, M., & Ortiz-Ospina, E. (2017). *World population growth*. Retrieved February, 2017, from https://ourworldindata.org/world-population-growth

Sarkees, M. R., & Wayman, F. W. (2010). *Resort to war: 1816–2007*. Washington, DC: CQ Press.

Sharpe, K. (1992). *The personal rule of Charles I*. New Haven, CT: Yale University Press.

Sherman, R. A., Figueredo, A. J., & Funder, D. C. (2013). The behavioral correlates of overall and distinctive life history strategy. *Journal of Personality and Social Psychology, 105*(5), 873.

Tincey, J. (2004). *Blenheim 1704: The Duke of Marlborough's masterpiece* (Vol. 141). Oxford, UK: Osprey Publishing.

Tombs, R., & Tombs, I. (2006). *That sweet enemy: Britain and France: The history of a love hate relationship*. New York: Vintage.

Turchin, P. (2016). *Ages of Discord: A Structural-Demographic Analysis of American History*. Chaplin, CT: Beresta Books.

Westbrook, D. A. (2004). *City of gold: An apology for global capitalism in a time of discontent*. New York: Routledge.

Woodley of Menie, M. A., Figueredo, A. J., Sarraf, M. A., Hertler, S. C., Fernandes, H. B. F., & Aguirre, M. P. (2017). The rhythm of the west: A biohistory of the modern era, AD 1600 to present. *Journal of Social, Political and Economic Studies Monograph Series, 37*.

12

Expansion, Fission, and Decline: England and Anglo America

Aurelio José Figueredo, Steven C. Hertler, and Mateo Peñaherrera-Aguirre

1 Introduction

In 2017, our research team produced a technical, statistically-driven, monograph entitled *The Rhythm of the West: A Biohistory of the Modern Era, AD 1600 to the Present*. Therein, general intelligence, life history, and other topics were treated alongside multilevel selection theory. Here, after providing a general overview for the sake of context, we extract findings and discussion points from *The Rhythm of the West* directly relevant to demonstrating the reality of group selection within the history of the Britannic peoples. In a colloquial and qualitative manner, displaying essential analyses and separated from ancillary topics, we explain the dynamics of multilevel selection among the Britannic peoples as they have transitioned through stages of expansion, fission, and decline. Wealth, cognitive capacity, subjective well-being, poverty, and longevity are among the oblique markers of civilizational pulse. These correlate with, and are corroborated by, demographic decline. Declining evolutionary pressures for group-selected behaviors within mild industrial and postindustrial environments, operating for generations, have had a causal role in population decline, both in absolute and relative terms. To describe

S. C. Hertler et al., *Multilevel Selection*, https://doi.org/10.1007/978-3-030-49520-6_12

and explain the aforementioned decline, we review a variety of changes to Europe's *Early Modern Era* selective regime, including climatic changes during the *Little Ice Age*, niche expansion and modification during the Age of Exploration, nutritional advances gained during the *British Agricultural Revolution*, and technological advances gained during the *Industrial Revolution*.

2 Malthusianism: A Temporary Respite

Writing at the very end of the eighteenth century, Malthus correctly describes historical demographic trends, thereafter documented by economic historian David Hackett Fischer (1996), who found population growth to be the primary driver of *price revolutions*, characterized by decreasing returns to labor, rising inequality, increasing costs of necessary commodities, augmented crime, as well as strain to family units and the general social order. Fischer's price revolutions, in turn, are reminiscent of Turchin's (2016) applications of *Structural Demographic Theory*. Though a cleric, Malthus wrote against the scriptural injunction to be *fruitful and multiply*, for he saw, as did Fischer and Turchin after him, how increasing population can quickly tax available resources. He is famous for comparing the *arithmetical* rate of resource growth with the *geometric* rate of population growth, doing the world the service of highlighting the consequences of the contrast. However, classic Malthusian constraints characterized societies existing prior to the mid-eighteenth century, and times further past, but not those of late modernity, as we noted in Woodley of Menie et al. (2017):

> With respect to Malthusian fears, at least in the West, agricultural yields outstripped population growth, not the other way around (Mayhew, 2014)—or as Winch (1992, p. xxxi) put it, "the tortoise of food production overtook the hare of population growth," resulting in a rising tide of prosperity. The world did not become Easter Island writ large. In consequence, Malthusian predictions were [putatively] denuded of credibility, like an unfulfilled biblical prophecy. (p. 12)

As in the last sentence in the quote above, Malthus is often unjustly depicted as being proven wrong as a prognosticator, and even as a promulgator of general sociodemographic principles, in light of the nearly sustained global growth taking place within the last 200 years.[1] In spite of criticism, it may well be that he was never controverted, but simply unfortunate in the timing of his publication, as just then selective pressures were temporarily altered, and populations rapidly expanded without the catastrophic consequences previously foreseen. The temporal divide between the Malthusian past and the Modern present closely coincides with the 1798 publication of *An Essay on the Principle of Population.* In turn, the broad significance of this study is traceable to some few decades just before and after, as from thence came spectacular changes in the realized carrying capacity, the global population, and the prevailing selective regime. In subsequent sections leading up to our analyses, we review climatic warming, the impact of colonial expansion, agricultural advances, industrialization, and related developments allowing the carrying capacity to be raised rapidly in the nineteenth and twentieth centuries, before explaining how these features collectively altered prevailing selective regimes, which in turn, not only enabled sustained numeric growth but changed the composition of societies.

3 Climate Change

Following the generous climatic conditions of the *Medieval Warm Period*, marked by high gothic devotional cathedrals, the *Little Ice Age*, reigning from the fourteenth to nineteenth centuries[2] (Fagan, 2000), witnessed a drop in global mean annual temperatures (Mann et al., 2009) along with glacial advances (Grove, 2019). As per some estimates based upon varying criteria, the Little Ice Age lasted until the *middle* to *late* nineteenth century (e.g., Fagan, 2000). However, the *beginning* of the nineteenth century was the inflection point at which the current trend of global warming was initiated. This inflection point occurred roughly after 1816, the *year without a summer*, which killed 65,000 Europeans. This final major cold blast of the Little Ice Age was attributable to volcanism, which reduced insolation, making an already cold climate colder. It took some decades after that for temperatures to recover from the previous period of cold that started in the early fourteenth century, despite the upward trend, and this accounts for some of the differences in chronology. Prior to that, there was a high degree of variability around an otherwise nearly "flat" (or only minimally sloped) line extending between the middle of the fourteenth and the beginning of the nineteenth century.

During the Little Ice Age, traditional agriculture in Europe was severely and repeatedly disrupted (Fagan, 2000). Just as tree ring growth was slowed (White, 2013), there was a corresponding effect on agricultural output due to this period's reduced growing seasons and restricted crop yields, both of which are reflected in agricultural prices in England (Cressy, 2006). Across Europe at large, there were no less than 30 major famines[3] before the aforementioned 1816 inflection point: 1315–1317, the *Great Famine* throughout Europe; 1504, Spain; 1518, Venice, Italy; 1528, Languedoc, France; 1586, England; 1601–1603, Russia; 1601–1603, Estonia; 1618–1648, throughout Europe as a consequence of the Thirty Years' War; 1648–1660, Poland; 1649, Northern England; 1650–1652, Eastern France; 1651–1653, Ireland during Cromwell's conquest; 1670–1680, Spain; 1680, Sardinia, Italy; 1690, Scotland; 1693–1694, France; 1695–1697, Estonia and Livonia; 1695–1697, Sweden; 1696–1697, Finland; 1708–1711, East Prussia; 1709–1710, France; 1727–1728, England; 1740–1741, *Great Irish Famine*; 1764, Naples, Italy; 1770–1771, Czechia; 1771–1772, Saxony and Southern Germany; 1773, Sweden; 1783, Iceland; 1788, France just two years prior to the French Revolution; and 1811–1812, Madrid, Spain. Each of these famines, along with their associated plagues (as the malnourished become vulnerable to disease), took the lives of anywhere between tens of thousands to multiple millions.

The period of global warming spanning most of what historians call the Late Modern Era saw a gradual increase in crop yields, partially as a result of the more favorable climatic conditions for the growth of vegetation. This natural increase in biological productivity, however, was enhanced by two anthropogenic factors: (1) the British Agricultural Revolution and (2) the Industrial Revolution. Prior to reviewing the effects of either of these important revolutions, we first review the consequences of colonial expansion.

4 Expanding Ecological Niche Space in the Age of Exploration

Roughly contemporaneous with the Little Ice Age, Europe's *Early Modern Era* is bracketed by the Medieval Era on one side and the Industrial Revolution on the other. In ranging from the fifteenth century to the late eighteenth century, the Early Modern Era's beginning essentially coincides with the Fall of Constantinople to the Ottoman Turks in 1453. As the overland trade route between Europe and East Asia was thus closed due to domination by a hostile and expansionist imperial power, all European exploration, discovery, and colonization subsequent to that traumatic event were necessarily maritime. With traditional trade routes controlled by the Ottoman Turks, and pressed by the Little Ice Age's harsh cold to garner necessary resources through trade or conquest, Early Modern Europeans inaugurated the *Age of Exploration*, including both the explorations of Africa and Asia and the discovery and colonization of the Americas.

As we reviewed in our third chapter of *Life History Evolution: A Biological Meta-Theory for the Social Sciences*, this period saw a catastrophic change in the evolution of biogeographically regional *Symbiotic Portmanteau Assemblages* (SPAs),[4] which had hitherto proceeded mostly in mutual isolation. Ecologically, the European maritime expansions of the Early Modern Era saw the construction of an increasingly global network of "sea bridges," connecting all major biogeographical regions and their local SPAs. Through both intentional and unintentional transportation and exchange of human symbiota, this development put many regionally coevolved human-constructed SPAs of the world into direct contact and competition for the first time since the breakup of Pangaea. With the so-called Columbian exchange (Crosby, 1972), European explorers and colonizers transported *ecomorphs* originating in different SPAs, with cattle and horses grazing New World grasses, just as maize and potatoes grew in Old World soils. The term "ecomorph" (Williams, 1972; p. 72) denotes "species with the same structural habitat/niche, similar in morphology and behavior, but not necessarily close phyletically." Thus, ecomorphs, or species occupying the same or similar ecological niches, hitherto isolated by the Atlantic Ocean, came into direct competition during the Early Modern Era. Columbus' 1492 voyage began the invasion of Native American SPAs by Eurasian ecomorphs, but it also began the counter-invasion of Eurasian SPAs by Native American ecomorphs. Alternative SPAs contained functionally equivalent constructed niches for such *ecomorphs*. These mutual invasions of ecomorphs from different SPAs sometimes took the form of *competitive exclusion*, but other times took the form of *niche-splitting*, eventually evolving into *mutualisms*. Gause's (1932, 1934) *Law of Competitive Exclusion* states that two species competing for the same resources cannot coexist *sympatrically*,[5] as one will inevitably drive the other to local extinction. Grinnell (1904) had formulated the principle of competitive exclusion as follows: "Two species of approximately the same food habits are not likely to remain long evenly balanced in numbers in the same region. One will crowd out the other" (p. 377). MacArthur and Levins (1967) predicted that the ecological overlaps cannot exceed a certain *limiting similarity*, in which roughly ecomorphic species are too ecologically similar, selection will lead to *character displacement* and thereby exploitation of different resources. Character displacement is produced by selection against individuals occupying the zones of maximal overlap. *Niche-splitting* is synonymous with the terms *niche differentiation*, *niche segregation*, *niche separation*, and *niche partitioning* and refers to the process by which competing species use the environment differently in ways that permit them to coexist. Two species that differentiate their niches tend to compete less strongly.

The Columbian exchange (Crosby, 1972) therefore saw the creation of selectively "blended" Native American-Eurasian SPAs, with niche-splitting ecomorphs derived from both Eurasian SPAs and Native American SPAs. These associations evolved into mutualisms only after the *British Agricultural Revolution* of the late eighteenth century, with the development of highly bioproductive crop rotation methods, frequently involving mixed Old and New World cultivars. This later British Agricultural Revolution would not have been possible without the availability of these hybridized or "blended" SPAs. The early niche-splitting among ecomorphs derived from both Eurasian and Native American SPAs was made possible by the filling of vacant niches within each SPA by species from the other assemblage. For example, there were vacant microhabitats in Southern China and the Indonesian Archipelago that were not suitable for rice farming, but were suitable for the cultivation of New World sweet potatoes. In addition, the global cooling of the Northern Hemisphere during the Early Modern Era made it difficult, and in some cases impossible, to grow many traditional Old World cultivars in Northern Europe, freeing those niches for invasion by New World white potatoes.

5 The British Agricultural Revolution

According to Clark (2007), the intraspecific selection pressures on European populations during the Early Modern Era (or Little Ice Age) were quite severe and consequential. Until the early nineteenth century, the European upper classes produced over twice the number of surviving offspring than the European lower classes. These selective pressures altered the cognitive and conative characteristics of European populations in systematic ways by the mechanisms of gene-culture coevolution. For Clark, as famine and disease thinned the ranks of the impoverished classes, they were replaced in the population by the offspring of the wealthy. What Clark calls "middle-class values," such as nonviolence, literacy, and hard work, were thus spread throughout the population both culturally and genetically. Clark (2008) further argues that "the rich in pre-industrial England had to be different in personality and culture from the poor" (p. 16) and that consequently "the rich in modern industrial society are genetically different from the poor" (p. 19). Such differences must have been relevant to economic success and could have been passed on by culture, genetics, or a combination of the two.

Clark's (2007) theory is evolutionarily plausible, as every subsistence economy selects phenotypes (and, indirectly, genotypes) that are better suited to survival and reproduction under its defining "material conditions of existence" (Marx & Engels, 1848, p. 496). These selective pressures affect the evolution of regional populations of human and nonhuman animals alike, as well as their associated plants. These selective pressures also necessarily affect the coevolution of symbiotic human and nonhuman animals and plants, as these inevitably constitute part of each other's *adaptively relevant environments* (Irons, 1998).

According to Toffler's (1980) *Three Wave Theory*: (1) the First Wave is Agricultural Age Society, (2) the Second Wave is Industrial Age society, and (3) the Third Wave is Postindustrial Age Society. If Clark (2007) is correct, the new "Second Wave" mentality and vision of the world was thus favored by social selection, which Toffler (1980) has characterized as *indust-reality* (p. 97):

> The Second Wave Society is industrial and based on mass production, mass distribution, mass consumption, mass education, mass media, mass recreation, mass entertainment, and weapons of mass destruction. You combine those things with standardization, centralization, concentration, and synchronization, and you wind up with a style of organization we call bureaucracy.[6]

This industrialized worldview involved the *commodification* of the natural as well as social world. This commodification included all nonhuman animals and plants and also included other human animals (e.g., enslaved Africans). The commodification of the biological world naturally led to the direct selective breeding of nonhuman animals for more specific purposes than ever hitherto envisioned, as well as some expressed intentions and unsuccessful attempts (e.g., Eugenics movements) to apply similar principles to the "improvement" of human populations by means of artificial selection, based on traits presumably conferring social utility to the group (Woodley & Figueredo, 2013). Thus, the intensified, intentional, and directed artificial selection of nonhuman animals did not occur until the beginning of the Late Modern Era in Western civilization and was part and parcel of the industrialization process of European societies. There appears to be little evidence that it occurred at any time before then, at least on anything even remotely approaching the modern scale.

The Industrial Revolution actually began with the industrialization of agriculture, sometimes called the *British Agricultural Revolution* (AD ~1700–1850). This intensification of agricultural technologies was a direct response to the existential threats to European agriculture (and, hence, the food supply) posed by adverse climate change during the Little Ice Age. The necessary adaptations to European agricultural production practices created new selective pressures for different cognitive and conative phenotypes in the affected/afflicted populations.

The transformation of modern farming practices promoted by the British Agricultural Revolution was based on four pragmatic principles: (1) enclosure; (2) mechanization; (3) four-field crop rotation; and (4) selective breeding. Agriculture across Europe had previously been characterized by the feudal open field system, within which farmers worked on strips of land in fields they held in common. This was later viewed as "inefficient" and as reducing the individual incentive to improve productivity. British yeomen thus began to *enclose* and then optimize the use of *their* land. This process of land reform accelerated in the eighteenth century with special acts of the British Parliament expediting the consolidation of larger and *privately owned* holdings, encouraging experiments in increased productivity by more entrepreneurial *landowners*.

The second Viscount Charles Townshend (Frey & Frey, 2019) introduced the four-field crop rotation in the eighteenth century, and these new patterns of land use resulted in substantial expansions to the available area of arable land, producing both fodder crops and grazing crops that enabled livestock to be bred year-round. The use of nitrogen-rich manure and nitrogen-fixing crops, such as clover, increased yields of cereal crops by enhancing the amount of available nitrogen in the soil. This removed the major limiting factor on cereal production existing up until the early nineteenth century. For example, the productivity of wheat in England increased from approximately nineteen bushels per acre in 1720 to approximately thirty bushels per acre by 1840. The changes in agriculture implemented in Great Britain during this period subsequently affected agricultural practices around the world. These new agricultural technologies and cultivars multiplied yields per land unit to many times those produced in the Medieval Era.

Selective breeding of animals was also first established as a scientific practice during this historical period. For example, the second Viscount Robert Bakewell (see Wood, 1973) improved the Lincoln Longwool by the selective breeding of native sheep stock and later used it to develop the hornless and meatier Dishley Leicester. Bakewell was the first to breed cattle primarily for *beef*, as British cattle were previously kept mostly as *oxen* for pulling ploughs. To accomplish this, he crossed long-horned heifers with Westmoreland bulls. From this hybrid stock, he eventually developed the Dishley Longhorn, which he afterwards replaced with shorter-horned versions. Such innovations led to dramatic increases in the size and quality of farm animals. For example, the average weight of a bull sold for slaughter was 168 kg in 1700, but had more than doubled to 381 kg by 1786.

The British Agricultural Revolution brought about large excesses of calories that fueled dense settlements and allowed all manner of specializations to flourish. More could now become mechanics, inventors, scientists, naturalists, and chemists, being sated by agricultural surplus and liberated from directly working the land. In this way, the British Agricultural Revolution fostered the Industrial Revolution that further changed the selective regime and thus the carrying capacity of the environment.

6 The Industrial Revolution

The Field and the Forge: Population, Production, and Power in the Pre-Industrial West, written by John Maxwell Landers, draws on a distinction between *organic* and *mineral* economies. As reviewed in our seventh chapter of *Life History Evolution: A Biological Meta-Theory for the Social Sciences*, organic economies are defined by seasonal and agricultural rhythms and subject to subsistence living under Malthusian constraints. They impose temporal torpors, periods of inactivity imposed by seasonal cold and darkness. These organic societies were also limited in travel, trade, and war because all of these activities ultimately require fuel and energy, both of which are in short supply in organic societies. "Ultimately," Landers (2003; p. 17) writes, "everything depended on the efficiency of plant photosynthesis and the energy conversion of biological 'engines', and both are low by mechanical standards." From the scarcity of energy came restricted productivity, rendering scarce the provisions necessary to sustain and reproduce life. Laboring to keep dry, warm, and fed, the mass of the peasantry could not then contribute to economic diversification, arts, letters, and research, or otherwise stimulate the economy with demands for luxury goods. Undifferentiated societies, lacking in basic resources and existing amidst climatic variability and harshness, then phased into their opposites with the transition to mineral economies. On the cusp of industrialization, England had high labor rates and rich coal deposits, conditions fortuitously following the Enlightenment and its scientific revolution (Allen, 2009). Harnessing this power in blast furnaces and steam engines, Britons inaugurated an era of sustained growth, manifest in its expanding economy, territory, and population. The unprecedented excesses of energy unlocked from fossil fuels allowed work to be accomplished; such work allowed infrastructure, foodstuffs, and houses to be created that would not have otherwise existed.[7]

Relevant to this discussion, and that pursued in the subsequent section, Clark makes the point that, in 1800, the average European was saddled with material conditions that were in many ways worse than their hunter-gatherer counterparts, as measured by, for instance, longevity and material consumption. Yet, Clark insists that since then, industrialized countries have become ten or twenty times as wealthy as industrialization spread in earnest, with the main beneficiaries being the poor.[8] Fertility rates in pre-industrial Europe were low only because there was a late age for females marrying (between 24% and 26%), because some (between 10% and 25%) never married, and because there was a low (between 3% and 4%) illegitimacy rate, which implied that sex was largely confined to marriage. These reproductive restraints mark the continued operation of more or less Malthusian conditions, which became unrecognizable only when all foregoing factors combined with advanced industrialization. To this point, consider the following passage from Gat (2017):

> An exponential increase in wealth has been central to the rise of industrial-technological society. It has been fueled by a steep and continuous growth in per capita production and marked a sharp break from the Malthusian trap that characterized human history until then. Premodern increases in productivity were largely absorbed by population growth, leaving the vast majority of people in dire poverty, precariously close to subsistence level. With the outbreak of the industrial-technological revolution, however, that

> changed dramatically. Average growth in the industrial world has become about ten times faster than in pre-industrial times, with production per capita for the first time registering substantial and sustained real growth at an average annual rate of 1.5–2.0 percent. (p. 154)

It has been said that humans convert resources into offspring. So it was that after 1850, with the combined effects of climatic warming, adaptive introgressions from New World SPAs, and the British Agricultural and Industrial Revolutions, the ceiling on the carrying capacity rose dramatically, and the European population increased accordingly. Beyond simple increases in population, however, the diversification of social roles and economic specializations very importantly followed from population growth and energy inputs. With economic diversification, Ricardo's (1817) *Law of Comparative Advantage* operated at unprecedented levels, fostering the expansion of trading networks. Thus, the selective regime was changed, with more people participating in reproduction, flourishing of social supports, opportunities to accrue personal fortunes, more levers of power to grab, and less emphasis on martial valor and ingroup loyalty *vis-à-vis* other groups.

7 Compositional Changes Following from Demographic Growth

Malthusian theory is most accurate when discussed in the context of scarcity, rather than in the relative abundance found in many world regions within the last two centuries. Consequently, we can review a variety of factors making it appear that Malthusian predictions were falsified: warming trends following the Little Ice Age, the abundance of the New World absorbing excess migration, New World crops supplementing and sustaining Old World populations, advances in crop and animal selection, systematization, collectivization, and industrialization of agriculture, as well as the scientific revolution and related industrial revolution unlocking the power of machinery and fossil fuels. As profound as these changes were, especially in their collective and dynamic impact, none of them invalidated Malthusian logic, so much as raised the carrying capacity. Understanding this is valuable in and of itself, as it instills a proper understanding of history, demographics, and tamps down ebullient optimism about a future without constraints. More importantly for present purposes, we must connect the carrying capacity to the selective regime. Doing so connects gross growth with compositional changes. Resultant populations were not just larger, they were different. More precisely, changes in the carrying capacity did not simply create a larger population with the same attributes, metrics, and configuration across biopsychological variables, it instead promoted population growth with proportional changes in the composition of societies among these biopsychological variables. The complexities of the dynamic interplay between carrying capacity, technological innovations, and population growth are beyond the scope of this chapter. Generally speaking, and most true of Britain and eventually other Western European countries, we find selective pressures favoring group-selected traits due to heightened levels of resource competition between groups prior to the middle decades of the nineteenth century. Thereafter, we instead find selective pressures favoring individually selected traits. Severe group-selective pressures waned, allowing individually selected traits to become more prominent, beginning in the nineteenth century, and accelerating into the twentieth century. An attempt was made to analytically detect the effects of this shift in selective pressures.

An extension on a prior publication focusing on waning heritable general intelligence since the 1850s (Woodley & Figueredo, 2013), *The Rhythm of the West*, was organized around two diachronic analyses, each labeled for the span of historical time under review. There was the *Nexus 200*, reviewing AD 1810 to 2010, and the *Nexus 400*, reviewing AD 1600 to 1999. The word *nexus* as used above is an abbreviation of the *co-occurrence nexus*, which refers to an overarching, coherent higher-order factor comprising *heritable general intelligence* (gh), *specialized intelligences* (se), and *somatic modifications* (sm). In turn, these three lower-order factors rest upon, and are marked by, their relationship to fifteen convergent indicators. Again, indicators and lower-order variables alike cohere and covary within the aforementioned co-occurrence nexus.

As will be explained in further detail subsequently, the *Nexus 200*'s longitudinal analysis, controlling statistically for serial autocorrelations, found this suite of convergent factors causally linked to directional selective pressures, wherein extant ratios of group to individual selection were inverted. Whereas in prior centuries, group selection prevailed over individual selection, by the latter portion of the nineteenth century, one sees individual selection prevailing over group selection. The inversion of selective pressures itself was partially related to the aforementioned climatic warming and stabilization, both of which reduced the motive force of intergroup competition driving group selection. To these climatic changes were added the anthropogenic alteration of the selective pressures to which humans were historically exposed, courtesy of the harnessing of fossil fuels by industrialization. Taken together, we see a historical divide, on one side of which were organic economies with restricted reproductive capacities engaging in frequent war, and on the other side mineral economies with increased energetic and economic inputs being converted into demographic surges, now less restrained by the culling and limiting factors of war and cold.

The *Nexus 400* extended the period of analysis by approximately 200 years, reproducing and expanding those findings yielded by the Nexus *200*. The Nexus *400* analysis was partially based on a latent common factor representing group-selective pressure via between-group competition, combining two biodemographic indicators (national war mortality and national proportion of world population) with a single lexicographic indicator (frequency of altruistic word usage). War mortality was thought to be a gross marker of martial virtue and thereby linked to group selection. Population levels, relative to competing societies, mark demographic expansion, which is an indirect indicator of group selection (Okasha, 2006). Lastly, language communicates ideas and thus can be a marker for realities that are not simply semantic. This is the essence of the lexicographic endeavor within psychology, which has successfully been applied to other substantive problems such as those of personality measurement. Here *altruistically valenced* words were assumed to be associated with altruistic impulses expressed among members of a society. These three indicators then produced the between-group competition factor, indexing the generative pressures of group selection, the rationale for which rested on the correlations among these three indicators. As stated, these three indicators were aggregated into a latent common group selection factor, which was then found to negatively correlate with mean global temperatures, such that lower temperatures were associated with higher degrees of intergroup competition for resources. The group selection factor was also associated with higher levels of heritable intelligence at the population level. Some of the methodology and underlying analytics behind these findings will be elucidated subsequently; suffice it to say in this section that, from a general decline in group-selective pressures in that 400-year period, results suggested an inversion among the different forms of intelligence, with general intelligence declining in favor of more specialized forms of intelligence. However, unlike *The Rhythm of the West*, which focuses on intelligence as an outcome of shifting selective regimes, we are herein concerned principally with the selective regime itself. In other words, we here focus on waning group selection, rather than intelligence as a product of that process.

8 Biohistorical Analyses

We performed two separate biohistorical statistical analyses: (1) the *Nexus 200*, testing the co-occurrence model across the 200 years spanning AD 1810–2010, and (2) the *Nexus 400*, testing our climate-driven multilevel selection model of the evolution of intelligence across the 400 years spanning AD 1600–1999. The second period of time spans approximately 200 years of the Early Modern Period, comprising the end of the Little Ice Age, and 200 years of the Late Modern Period, comprising the period of global warming that followed and continues to this day. As in the biohistorical analyses presented in Chap. 11, data were collected for the following Britannic nations: the United Kingdom, the United States, Canada, New Zealand, and Australia.

For the *Nexus 200* analyses, fifteen hypothesized indicators of the co-occurrence nexus were collected, as detailed in Woodley of Menie, Figueredo, Sarraf, Hertler, Fernandes, and Peñaherrera-Aguirre (2017): (1) male fluctuating asymmetry; (2) sinistrality rate; (3) body mass index; (4) average height; (5) brain weight; (6) GDP per capita (micro-innovation rate); (7) concretization in language; (8) forward digit span; (9) psycholinguistic word use; (10) WORDSUM easy word use; (11)

Descent of Man altruism words use; (12) male reaction time; (13) backward digit span; (14) WORDSUM hard word use; and (15) macro-innovation rate per capita. These fifteen indicators of the co-occurrence nexus were aggregated into three lower-order factors: (1) somatic modifications; (2) specialized abilities; and (3) general intelligence. By reverse-scoring the somatic modifications and specialized abilities factors, we further aggregated these three lower-order factors into a single higher-order factor (*the co-occurrence nexus*). Figure 12.1, adapted from Woodley of Menie, Figueredo, Sarraf, Hertler, Fernandes, and Peñaherrera-Aguirre (2017), displays these relationships in graphical form and also provides quantitative information for the convergent validity coefficients among the indicators at each level of the hierarchy.

As with the *asabiyyah* analyses presented previously in Chap. 6, three nested MLMs were estimated to test the need for increasing

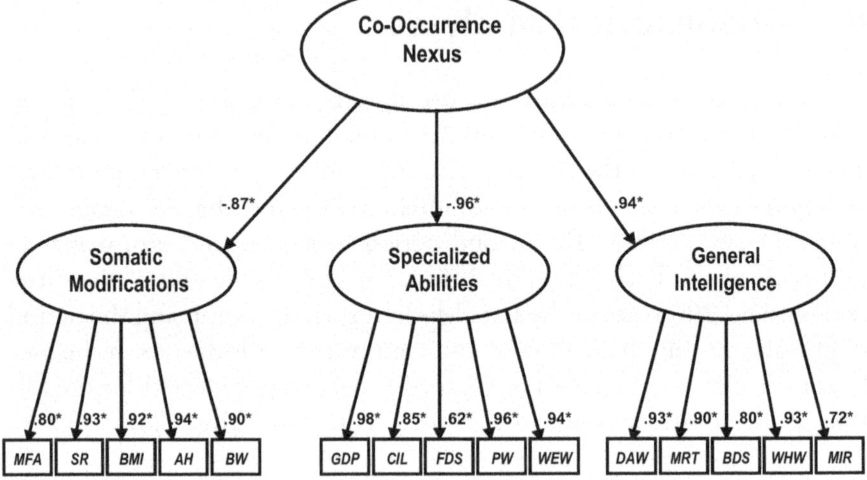

Fig. 12.1 The latent hierarchical structure of the co-occurrence nexus from AD 1810 to 2010: (1) somatic modifications: *MFA* male fluctuating asymmetry, *SR* sinistrality rate, *BMI* body mass index, *AH* average height, *BW* brain weight; (2) specialized abilities: *GDP*, GDP per capita (micro-innovation rate), *CIL* concretization in language, *FDS* forward digit span, *PW* psycholinguistic word use, *WEW* WORDSUM easy word use; and (3) general intelligence: *DAW* descent of man altruism words use, *MRT* male reaction time, *BDS* backward digit span, *WHW* WORDSUM hard word use, *MIR* macro-innovation rate per capita. $^*p<0.05$

parameterization as alternative hypotheses: (1) *MLM1* estimated a single intercept and a single logarithmic slope (the "unconditional" co-occurrence nexus) for all nexus factors and indicators over time, as well as the same intercepts and logarithmic slopes for all nexus indicators nested within each nexus factor; (2) *MLM2* estimated a separate intercept and a separate logarithmic slope for each nexus factor over time, but the same intercept and logarithmic slopes for all nexus indicators nested within each nexus factor; and (3) *MLM3* a separate intercept and a separate logarithmic slope for each nexus factor over time as well as a separate intercept and a separate logarithmic slope for each nexus indicator over time within each nexus factor.

Table 12.1 displays the pertinent nested model comparisons. The systematic AIC and -2RLL comparisons performed among the nested models representing the specific variance components of the nexus factors and nexus indicators indicated the following: (1) the specific variance components for nexus factors were statistically significant; and (2) the specific variance components for nexus indicators within nexus factors were also incrementally statistically significant ($p<0.05$). Comparisons of squared multiple correlations among the three nested MLMs yielded essentially the same results. The incremental magnitudes of the specific variances (ΔR^2) were found to be non-negligible but relatively small for both the nexus factors (~2%) and the nexus indicators (~6%), in contrast with the common factor variance of the "unconditional" co-occurrence nexus, representing the higher-order construct, which was found to be quite

Table 12.1 Fit indices for nested multilevel models of co-occurrence nexus, lower-order factors, and specific indicators as natural logarithmic functions of time with Britannic populations across the 200 years spanning AD 1810–2010

The co-occurrence nexus	MLM1: Year	MLM2: + Factor + Factor*Year	MLM2: + Indicator + Indicator*Year
AIC	2012.0	1957.9	1690.0
-2RLL	2008.0	1949.9	1682.0
	$\Delta\chi^2$=	**58.1***	**267.9***
R^2	0.645*	0.663*	0.731*
	ΔR^2=	**0.018***	**0.059***
	ΔNDF=	**4**	**24**

*$p<0.05$

large (~65%). These results indicated that the unitary co-occurrence nexus did a reasonably good job of accounting for the temporal covariation among both factors and indicators over time.

Given the strength of these findings, we chose to report and interpret the model parameters for only the unconditional co-occurrence nexus (*MLM1*), as the extra model parameters added by the lower levels of aggregation (*MLM2* and *MLM3*) did not add very much explanatory power to our account of the diachronic variances in the co-occurrence nexus factors and indicators. The logarithmic intercepts (*a*) and slopes (*b*) of this unitary higher-order unconditional co-occurrence nexus construct over time were statistically significant: $a = 250^*$, $b = -33^*$ ($p<0.05$).

For the *Nexus 400* analysis, as with the biohistorical analyses presented previously in Chap. 6, MLM residuals were then exported for the between-group competition (BGC) factor, which was constructed identically as in Chap. 11 but estimated separately for the present sampling frame (see Fig. 12.2).

This MLM residualization is also done for a single indicator of the co-occurrence nexus, the lexicographic WORDSUM hard word use, which was the only nexus indicator that we could obtain for as far back as AD 1600. This served as an indicator of high verbal ability and indirectly of general intelligence. MLM residuals were thus statistically adjusted for the logarithmic effect of time as well as of any unstructured autoregressive serial dependencies among successive data prior to regression

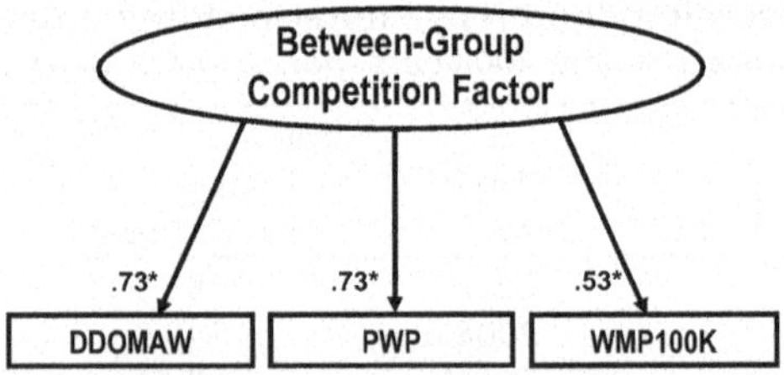

Fig. 12.2 The latent structure of the between-group competition (*BGC*) factor from AD 1600 to 1999 (*DDOMW* Darwin's *Descent of Man* altruism words, *PWP* proportion of the world's population, and *WMP100K* war mortality per 100,000). Factor loading coefficients = Britannic, Gallic. *$p<0.05$

modeling, thus circumventing this potential problem as a threat to the validity of correlational analysis.

Both of these were modeled as a function of the twenty-five-year floating averages of mean global temperature (TempMN25), aggregated from three sources, as detailed in Woodley of Menie, Figueredo, Sarraf, Hertler, Fernandes, and Peñaherrera Aguirre (2017): (CRUTEM3) global annual land-surface air temperature anomalies (Brohan, Kennedy, Harris, Tett & Jones, 2006); (ERSST v3) global land and sea surface temperature anomalies from the GISS Surface Temperature Analysis project (Hansen, Ruedy, Sato & Reynolds, 1996; Smith, Reynolds, Peterson & Lawrimore, 2008); and (HadCET) surface temperature for Central England, measured in a roughly triangular area enclosed by Lancashire, London, and Bristol (Parker, Legg & Folland, 1992).

The semipartial correlation of the time-adjusted MLM residuals of BGC with TempMN25 was $r = -0.13$ (90% *CI*: −0.22, −0.03), $F(1{,}398) = 6.56$, $p<0.01$, empirically supporting the hypothesis that declining BGC is historically associated with rising mean annual temperatures, independently of the effects of time, as depicted in Fig. 12.3.

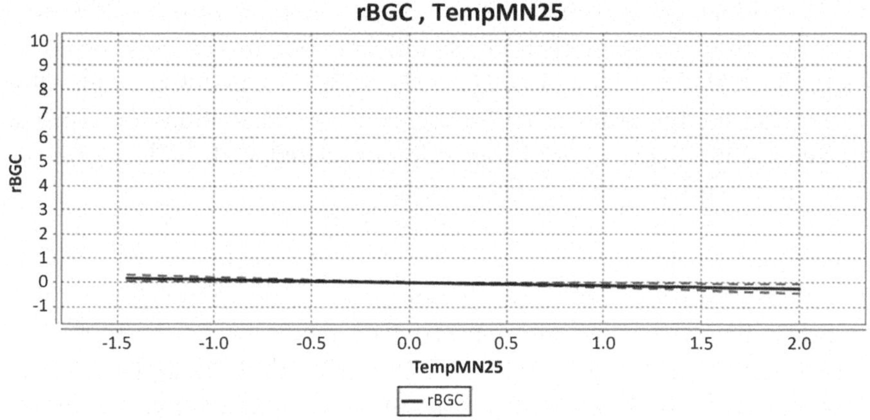

Note: Dashes are upper and lower confidence intervals.

Fig. 12.3 Time-adjusted MLM residuals of between-group competition (*rBGC*) predicted by the twenty-five-year floating averages of mean global temperature (*TempMN25*) from AD 1600 to 1999. (Note: dashes are upper and lower confidence intervals)

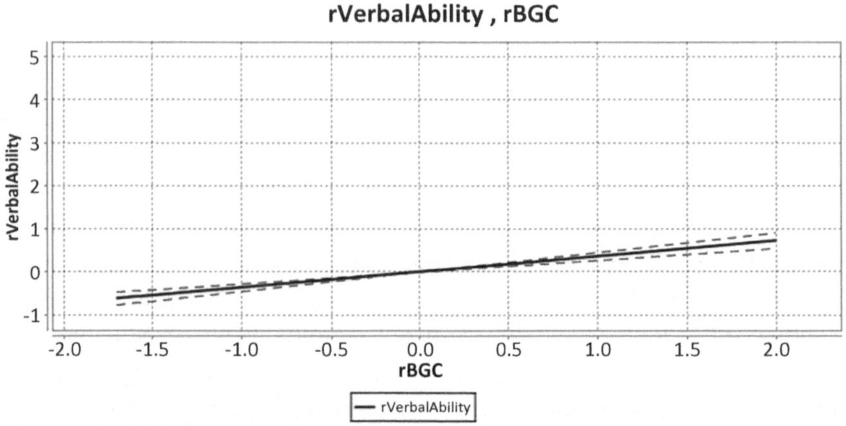

Note: Dashes are upper and lower confidence intervals.

Fig. 12.4 Time-adjusted MLM residuals of verbal ability (*rVerbalAbility*) predicted by those of between-group competition (*rBGC*) from AD 1600 to 1999. (Note: dashes are upper and lower confidence intervals)

As seen in Fig. 12.4, now using the time-adjusted MLM residuals of BGC as a predictor, the semipartial correlation of the time-adjusted MLM residuals of the verbal ability indicator as a function of rBGC was likewise found to be statistically significant, but in this case positive, $r = 0.37$ (90% *CI*: 0.28, 0.45), $F(1,397) = 67.05$, $p<0.0001$, empirically supporting the hypothesis that higher levels of verbal ability (as a proxy for general intelligence) are historically associated with higher levels of BGC, independently of the effects of time.

Statistically controlling for the effect of rBGC on rVerbalAbility, we then estimated the semipartial correlation of the residual direct effect of TempMN25 on rVerbalAbility. This effect, as presented in Fig. 12.5, was found to be statistically significant and negative, $r = -0.24$ (90% *CI*: −0.33, −0.15), $F(1,397) = 28.61$, $p<0.0001$, empirically supporting the hypothesis that lower levels of verbal ability (as a proxy for general intelligence) are historically associated with higher levels of TempMN25, independently of the effects of time as well as of the selective pressure of between-group competition. Thus, global warming (TempMN25) has two negative effects on verbal ability: (1) one indirect and negative effect

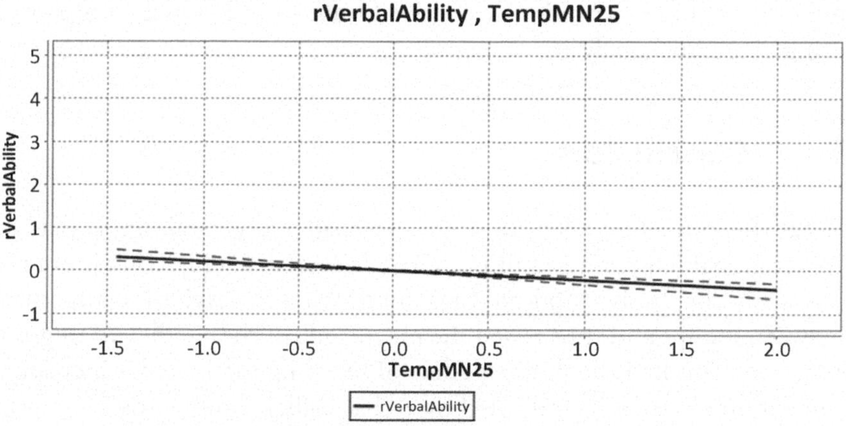

Notes: Dashes are upper and lower confidence intervals. Prior partialled variable(s): rBGC

Fig. 12.5 Time-adjusted MLM residuals of verbal ability (*rVerbalAbility*), statistically controlled for those of between-group competition (*rBGC*), predicted by the twenty-five-year floating averages of mean global temperature (*TempMN25*) from AD 1600 to 1999. (Notes: dashes are upper and lower confidence intervals. Prior partialled variable(s): *rBGC*)

through BGC, representing relaxation of group selection pressure; and (2) one direct and negative effect, representing relaxation of individual selection pressure.

From these results, we drew the theoretical conclusion that higher levels of general intelligence (as indicated by rVerbalAbility) were partially under group selection for the historical period in question. In our predictive models, about 14% of the variance in intelligence is accounted for by group selection (as indicated by rBGC), whereas only about 6% of the residual variance is accounted for by individual selection (as indicated by the residual direct effect of TempMN25), once statistically controlling the effects of temperature for any indirect effects through group selection (rBGC). This analysis is analogous to the one that we presented for chimpanzee intercommunity conflict, in that it breaks down the proportions of variance in different outcome variables attributable to group and individual selection, respectively. For example, we had found that individual selection accounts for 22% of the variance and group selection accounts for 11% of the variance in the relative fitness of individual chimpanzees using contextual analysis.

9 Conclusions

What this can all be taken to mean, biohistorically speaking, is that the colder global temperatures of the Little Ice Age (roughly contemporaneous with the Early Modern Era) increased the level of competition between groups and thus raised the magnitude of the coefficient of group selection. This had the indirect effect of selecting for a secular increase in general intelligence, as indicated by verbal ability. Afterwards, the warming temperatures of the Late Modern Era relaxed the group-selective pressure exerted by between-group competition, leading to a secular decrease in general intelligence and an increase in specialized intelligences that continues to this day. Direct climatic changes, as we have seen, were amplified by an approximate coincidence with New World trade surpluses, agricultural advances, and industrialization, the cumulative effect of which was to increase the carrying capacity, thus altering the proportions of group to individual selective pressure. These findings are consistent with those presented in Chap. 6, wherein a general reduction in *asabiyyah* was associated with rising GDPs. Moreover, this general reduction in competition between groups was illustrated in Chap. 11 through the waning rivalry between the Gallic and Britannic biocultural groups during the last two centuries.

Even as we are here most concerned with the changing selective regime itself, before closing, we turn again to intelligence, for it represents one of many potential consequences of waning group-selective pressures. In *A Farewell to Alms*, Clark (2007) makes the point that the most famous innovators of the Industrial Revolution contributed greatly to the welfare of their social group but "typically benefited little from their endeavors" (p. 235) as individuals. Clark (2007) lists how many of them instead died in poverty in spite of the riches they produced for others by their efforts. In the monograph *Historical Variability in Heritable General Intelligence*, Woodley and Figueredo (2013), after reviewing evidence from Clark and other sources in the literature, found the following two propositions empirically well-supported. First:

> the "genius fraction" of individuals disproportionately making intellectual contributions to society are either not benefiting personally or are actually sacrificing personal success, and thus putting themselves at a competitive disadvantage in within-group competition between individuals. (p. 69)

Second:

> that the societies in which these intellectual products are being generated benefit in comparison with other societies, and thus gain a significant competitive advantage in between-group competition. (p. 69)

What this adds up to is the theoretical prediction that higher levels of heritable general intelligence should be *disfavored* by the pressures of individual selection, but *favored* by the pressures of group selection. These predictions were followed up on in *The Rhythm of the West*, and it is evident that the results of the quantitative biohistorical analyses reviewed in the present chapter strongly support those conclusions, thus demonstrating the utility of multilevel selection theory in the study of human cognitive evolution.

Notes

1. https://www.scientificamerican.com/article/why-malthus-is-still-wrong/
2. These are approximate dates. Fagan actually allows this period to extend to 1850, while other authors prescribe slightly different dates, even as there is broad overlap across sources.
3. https://en.wikipedia.org/wiki/List_of_famines
4. As explained in Life History Evolution: A Biological Meta-Theory for the Social Sciences, SPAs are an expansion of Alfred Crosby's *portmanteau assemblages* (1986), which are "co-adapted ecological associations between humans and domesticated flora and fauna, vermin, weeds, and pathogens, which act together as a unit in competition with rival assemblages upon contact." Our modified term simply recognizes the extent to which human

agency and humans themselves are not always central to the ways in which rival species assemblages interact, at least after being initially transported by humans.
5. Within the same territory.
6. Swann, N. (1998). Interview with Alvin Toffler, Australian Broadcasting Corporation Radio National, "Life Matters," 5 March.
7. Of course, the British Agricultural Revolution and Industrial Revolution overlapped in time and are in some senses therefore confounded. However, there was a fair amount of time wherein agricultural advances were present prior to appreciable industrialization, especially in the United States.
8. Within societies, industrialization has brought great wealth and was a rising tide that raised all boats. This is even true of non-industrialized societies, even as industrialization has increased income gaps between industrialized and non-industrialized countries.

References

Allen, R. C. (2009). *The British industrial revolution in global perspective.* New York: Cambridge University Press.

Brohan, P., Kennedy, J. J., Harris, I., Tett, S. F. B., & Jones, P. D. (2006). Uncertainty estimates in regional and global observed temperature changes: a new data set from 1850. *Journal of Geophysical Research, 111*, D12106.

Clark, G. (2007). *A farewell to alms: A brief economic history of the world.* Princeton: Princeton University Press.

Clark, G. (2008). In defense of the Malthusian interpretation of history. *European Review of Economic History, 12*(2), 175–199.

Cressy, D. (2006). *England on edge: Crisis and revolution 1640–1642.* New York: Oxford University Press.

Crosby, A. W. (1972). *The Columbian exchange: Biological and cultural consequences of 1492.* Westport: Greenwood Publishing Group.

Crosby, A. W. (1986). *Ecological imperialism: The biological expansion of Europe, 900–1900.* New York: Cambridge University Press.

Fagan, B. (2000). *The little ice age.* New York: Basic Books.

Fischer, D. H. (1996). *The great wave: Price revolutions and the rhythms of history.* New York: Oxford University Press.

Frey, L., & Frey, M. (2019). *Townshend, Charles, second Viscount Townshend (1674–1738)*. Oxford Dictionary of National Biography (online ed.). Oxford University Press.

Gat, A. (2017). *The causes of war and the spread of peace*. New York: Oxford University Press.

Gause, G. F. (1932). Experimental studies on the struggle for existence: 1. Mixed population of two species of yeast. *Journal of Experimental Biology, 9*, 389–402.

Gause, G. F. (1934). *The struggle for existence*. Baltimore: Williams & Wilkins.

Grinnell, J. (1904). The origin and distribution of the Chestnut-Backed Chickadee. *The Auk, 21*(3), 364–382.

Grove, J. M. (2019). *The little ice age*. New York: Routledge.

Hansen, J., Ruedy, R., Sato, M., & Reynolds, R. (1996). Global surface air temperature in 1995: Return to pre-Pinatubo level. *Geophysical Research Letters, 23*, 1665–1668.

Irons, W. (1998). Adaptively relevant environments versus the environment of evolutionary adaptedness. *Evolutionary Anthropology, 6*(6), 194–204.

Landers, J. (2003). *The field and the forge: Population, production, and power in the pre-industrial West*. New York: Oxford University Press.

MacArthur, R., & Levins, R. (1967). The limiting similarity, convergence, and divergence of coexisting species. *The American Naturalist, 101*(921), 377–385.

Mann, M. E., Zhang, Z., Rutherford, S., Bradley, R. S., Hughes, M. K., Shindell, D., et al. (2009). Global signatures and dynamical origins of the little ice age and medieval climate anomaly. *Science, 326*(5957), 1256–1260.

Marx, K., & Engels, F. (1848). *Manifesto of the Communist Party*. Chicago: Charles H. Kerr & Company.

Okasha, S. (2006). *Evolution and the levels of selection*. Oxford: Oxford University Press.

Parker, D. E., Legg, T. P., & Folland, C. K. (1992). A new daily Central England temperature series, 1772–1991. *International Journal of Climatology, 12*, 317–342.

Ricardo, D. (1817). *On the principles of political economy and taxation* (1st ed.). London: John Murray.

Smith, T. M., Reynolds, R. W., Peterson, T. C., & Lawrimore, J. (2008). Improvements to NOAA's historical merged land-ocean surface temperature analysis (1880–2006). *Journal of Climate, 21*, 2283–2296.

Toffler, A. (1980). *The third wave: The classic study of tomorrow*. New York: Bantam Books.

Turchin, P. (2016). *Ages of discord: A structural-demographic analysis of American History*. Chaplin: Beresta Books.

White, S. (2013). The real little ice age. *Journal of Interdisciplinary History, 44*(3), 327–352.

Williams, E. E. (1972). The origin of faunas. Evolution of lizard congeners in a complex island fauna: A trial analysis. In T. Dobzhansky, M. K. Hecht, & W. C. Steere (Eds.), *Evolutionary biology* (pp. 47–89). New York: Springer.

Wood, R. J. (1973). Robert Bakewell, pioneer animal breeder and his influence on Charles Darwin. *Folia Mendeliana, 58*, 231–242.

Woodley, M. A., & Figueredo, A. J. (2013). *Historical variability in heritable general intelligence: Its evolutionary origins and socio-cultural consequences.* Buckingham: University of Buckingham Press.

Woodley of Menie, M.A, Figueredo, A. J., Sarraf, M. A., Hertler, S. C., Fernandes, H. B. F., & Aguirre, M. P. (2017). The rhythm of the West: A biohistory of the modern era, AD 1600 to present. *Journal of Social, Political and Economic Studies Monograph Series, 37*.

Epilogue: The Case for Multilevel Selection

Aurelio José Figueredo, Steven C. Hertler,
and Mateo Peñaherrera-Aguirre

To attempt to summarize the considerable array of theory and evidence reviewed in the entirety of this volume would be an unrealistic expectation for the present epilogue. Instead, we will restate our original objectives and come to some kind of self-assessment as to whether or not we believe we have accomplished them. We will therefore not try to characterize the field of multilevel selection as a whole, this being beyond our present abilities, but concentrate on our specific contributions to it as encapsulated in the present volume.

Our main goal in writing Part I was to create a coherent narrative of the history of ideas underlying the theory of multilevel selection without getting bogged down in the various polemics that have plagued this area of research. We have instead focused on the series of intellectual contributions, both theoretical and empirical, that have been made by various scientists to contemporary thinking in this field of study. While not glossing over the controversies, we have tried to provide our own best synthesis of these ideas and facts in a manner that best highlights the cumulative nature of the positive and lasting contributions while discarding any that have been refuted by subsequent reasoning or disconfirmed by subsequent findings.

S. C. Hertler et al., *Multilevel Selection*, https://doi.org/10.1007/978-3-030-49520-6

Our main goal in the preparation of Part II was to revisit the historical, sociological, and anthropological literature on so-called civilizational cycles, these being the observed successions of sociopolitical aggregation and disaggregation that chart the course of hierarchically stratified societies over long periods of time. We have tried to reinterpret much of the conventional literature on this topic through the evolutionary lens of multilevel selection theory, assessing the adaptive costs and benefits of various purported group-level adaptations rather than simply bemoan the so-called decline of civilization as if sociopolitical complexity were the *summum bonum* of human existence.

Our main goal in compiling Part III was to marshal a corpus of recently reported empirical evidence, most of it from our own research collaborations, subjecting hypotheses derived from theories reviewed in Parts I and II to various potentially disconfirmatory tests. We did this by examining relevant data from anthropoid apes, prestate human societies, empires of classical antiquity, and rival nation-states transitioning from competition to cooperation over the course of the early to late modern eras. We finished this progression of empirical tests by assessing the evidence for the cognitive and conative beginnings of decline in the victorious faction among those two latter biocultural groups.

This work was never designed to be either complete or comprehensive in its treatment of this weighty and voluminous topic. We have not either reviewed all of the theoretical literature or presented all of the empirical evidence that currently exists either for or against multilevel selection theory within the confines of this volume. Furthermore, there is always room for disagreement in the scientific process, and it is this healthy process of continuing discourse that powers intellectual progress. Thus, we did not unequivocally *sort this thing out once and for all*, as Dobzhansky had hoped, but did endeavor to advance our collective understanding such that students of multilevel selection might more confidently begin writing and working in the *post-resolution phase* that D. S. Wilson has invoked.

We offer the present work as our own contribution to this emerging body of literature, providing our own overarching perspectives based on our synthetic theoretical constructions of the current state of knowledge in this area of research. We believe that we have accomplished our three major objectives. This volume presents our best compendium of the theoretical and empirical rationales behind the way that we three researchers see it, no more and no less.

Index[1]

[1] Note: Page numbers followed by 'n' refer to notes.

S. C. Hertler et al., *Multilevel Selection*, https://doi.org/10.1007/978-3-030-49520-6

The manufacturer's authorised representative in the EU is Springer Nature Customer Service Centre GmbH, Europaplatz 3, 69115 Heidelberg, Germany. If you have any concerns regarding our products, please contact ProductSafety@springernature.com

Printed and bound by CPI Group (UK) Ltd, Croydon, CR0 4YY
05/10/2026
02228123-0007